Compendium of Plant Genomes

Series Editor

Chittaranjan Kole, Raja Ramanna Fellow, Government of India, ICAR-National Research Center on Plant Biotechnology, Pusa, New Delhi, India

Whole-genome sequencing is at the cutting edge of life sciences in the new millennium. Since the first genome sequencing of the model plant Arabidopsis thaliana in 2000, whole genomes of about 100 plant species have been sequenced and genome sequences of several other plants are in the pipeline. Research publications on these genome initiatives are scattered on dedicated web sites and in journals with all too brief descriptions. The individual volumes elucidate the background history of the national and international genome initiatives; public and private partners involved; strategies and genomic resources and tools utilized; enumeration on the sequences and their assembly; repetitive sequences; gene annotation and genome duplication. In addition, synteny with other sequences, comparison of gene families and most importantly potential of the genome sequence information for gene pool characterization and genetic improvement of crop plants are described.

Interested in editing a volume on a crop or model plant?

Please contact Dr. Kole, Series Editor, at ckoleorg@gmail.com

More information about this series at http://www.springer.com/series/11805

Shanfa Lu
Editor

The *Salvia miltiorrhiza* Genome

Springer

Editor
Shanfa Lu
Institute of Medicinal Plant Development
Chinese Academy of Medical Sciences
& Peking Union Medical College
Beijing, China

ISSN 2199-4781 ISSN 2199-479X (electronic)
Compendium of Plant Genomes
ISBN 978-3-030-24715-7 ISBN 978-3-030-24716-4 (eBook)
https://doi.org/10.1007/978-3-030-24716-4

This Springer imprint is published by the registered company Springer Nature Switzerland AG
The registered company address is: Gewerbestrasse 11, 6330 Cham, Switzerland

This book series is dedicated to my wife Phullara, and our children Sourav, and Devleena
Chittaranjan Kole

Preface to the Series

Genome sequencing has emerged as the leading discipline in the plant sciences coinciding with the start of the new century. For much of the twentieth century, plant geneticists were only successful in delineating putative chromosomal location, function, and changes in genes indirectly through the use of a number of "markers" physically linked to them. These included visible or morphological, cytological, protein, and molecular or DNA markers. Among them, the first DNA marker, the RFLPs, introduced a revolutionary change in plant genetics and breeding in the mid-1980s, mainly because of their infinite number and thus potential to cover maximum chromosomal regions, phenotypic neutrality, absence of epistasis, and codominant nature. An array of other hybridization-based markers, PCR-based markers, and markers based on both facilitated construction of genetic linkage maps, mapping of genes controlling simply inherited traits, and even gene clusters (QTLs) controlling polygenic traits in a large number of model and crop plants. During this period, a number of new mapping populations beyond F2 were utilized, and a number of computer programs were developed for map construction, mapping of genes, and for mapping of polygenic clusters or QTLs. Molecular markers were also used in the studies of evolution and phylogenetic relationship, genetic diversity, DNA fingerprinting, and map-based cloning. Markers tightly linked to the genes were used in crop improvement employing the so-called marker-assisted selection. These strategies of molecular genetic mapping and molecular breeding made a spectacular impact during the last one and a half decades of the twentieth century. But still, they remained "indirect" approaches for elucidation and utilization of plant genomes since much of the chromosomes remained unknown and the complete chemical depiction of them was yet to be unraveled.

Physical mapping of genomes was the obvious consequence that facilitated the development of the "genomic resources" including BAC and YAC libraries to develop physical maps in some plant genomes. Subsequently, integrated genetic–physical maps were also developed in many plants. This led to the concept of structural genomics. Later on, the emphasis was laid on EST and transcriptome analysis to decipher the function of the active gene sequences leading to another concept defined as functional genomics. The advent of techniques of bacteriophage gene and DNA sequencing in the 1970s was extended to facilitate sequencing of these genomic resources in the last decade of the twentieth century.

As expected, the sequencing of chromosomal regions would have led to too much data to store, characterize, and utilize with the-then available computer software could handle. But the development of information technology made the life of biologists easier by leading to a swift and sweet marriage of biology and informatics, and a new subject was born—bioinformatics.

Thus, the evolution of the concepts, strategies, and tools of sequencing and bioinformatics reinforced the subject of genomics—structural and functional. Today, genome sequencing has traveled much beyond biology and involves biophysics, biochemistry, and bioinformatics!

Thanks to the efforts of both public and private agencies, genome sequencing strategies are evolving very fast, leading to cheaper, quicker, and automated techniques right from clone-by-clone and whole-genome shotgun approaches to a succession of second-generation sequencing methods. The development of software of different generations facilitated this genome sequencing. At the same time, newer concepts and strategies were emerging to handle sequencing of the complex genomes, particularly the polyploids.

It became a reality to chemically—and so directly—define plant genomes, popularly called whole-genome sequencing or simply genome sequencing.

The history of plant genome sequencing will always cite the sequencing of the genome of the model plant Arabidopsis thaliana in 2000 that was followed by sequencing the genome of the crop and model plant rice in 2002. Since then, the number of sequenced genomes of higher plants has been increasing exponentially, mainly due to the development of cheaper and quicker genomic techniques and, most importantly, the development of collaborative platforms such as national and international consortia involving partners from public and/or private agencies.

As I write this preface for the first volume of the new series "Compendium of Plant Genomes," a net search tells me that complete or nearly complete whole-genome sequencing of 45 crop plants, eight crop and model plants, eight model plants, 15 crop progenitors and relatives, and three basal plants is accomplished, the majority of which are in the public domain. This means that we nowadays know many of our model and crop plants chemically, i.e., directly, and we may depict them and utilize them precisely better than ever. Genome sequencing has covered all groups of crop plants. Hence, information on the precise depiction of plant genomes and the scope of their utilization are growing rapidly every day. However, the information is scattered in research articles and review papers in journals and dedicated Web pages of the consortia and databases. There is no compilation of plant genomes and the opportunity of using the information in sequence-assisted breeding or further genomic studies. This is the underlying rationale for starting this book series, with each volume dedicated to a particular plant.

Plant genome science has emerged as an important subject in academia, and the present compendium of plant genomes will be highly useful both to students and teaching faculties. Most importantly, research scientists involved in genomics research will have access to systematic deliberations on the plant genomes of their interest. Elucidation of plant genomes is of interest not only for the geneticists and breeders, but also for practitioners of an array of plant science disciplines, such as taxonomy, evolution, cytology, physiology, pathology, entomology, nematology, crop production, biochemistry, and obviously bioinformatics. It must be mentioned that information regarding each plant genome is ever-growing. The contents of the volumes of this compendium are, therefore, focusing on the basic aspects of the genomes and their utility. They include information on the academic and/or economic importance of the plants, description of their genomes from a molecular genetic and cytogenetic point of view, and the genomic resources developed. Detailed deliberations focus on the background history of the national and international genome initiatives, public and private partners involved, strategies and genomic resources and tools utilized, enumeration on the sequences and their assembly, repetitive sequences, gene annotation, and genome duplication. In addition, synteny with other sequences, comparison of gene families, and, most importantly, the potential of the genome sequence information for gene pool characterization through genotyping by sequencing (GBS) and genetic improvement of crop plants have been described. As expected, there is a lot of variation of these topics in the volumes based on the information available on the crop, model, or reference plants.

I must confess that as the series editor, it has been a daunting task for me to work on such a huge and broad knowledge base that spans so many diverse plant species. However, pioneering scientists with a lifetime experience and expertise on the particular crops did excellent jobs editing the respective volumes. I myself have been a small science worker on plant genomes since the mid-1980s and that provided me the opportunity to personally know several stalwarts of plant genomics from all over the globe. Most, if not all, of the volume editors are my longtime friends and colleagues. It has been highly comfortable and enriching for me to work with them on this book series. To be honest, while working on this series I have been and will remain a student first, a science worker second, and a series editor last. And I must express my gratitude to the volume editors and the chapter authors for providing me the opportunity to work with them on this compendium.

I also wish to mention here my thanks and gratitude to Springer staff particularly, Dr. Christina Eckey and Dr. Jutta Lindenborn, for the earlier set of volumes and presently Ing. Zuzana Bernhart for all their timely help and support.

I always had to set aside additional hours to edit books beside my professional and personal commitments—hours I could and should have given to my wife, Phullara, and our kids, Sourav, and Devleena. I must mention that they not only allowed me the freedom to take away those hours from them but also offered their support in the editing job itself. I am really not sure whether my dedication of this compendium to them will suffice to do justice to their sacrifices for the interest of science and the science community.

New Delhi, India Chittaranjan Kole

Preface

Salvia miltiorrhiza is a member of Lamiaceae, the sixth largest family of flowering plants. It is a deciduous perennial plant widely distributed in China. It also grows in other countries, such as Korea, Japan, Vietnam, and Australia. *S. miltiorrhiza* is a well-known traditional Chinese medicine (TCM) material and an emerging model medicinal plant. It was recorded in *Shen Nong's Classic of the Materia Medica*, the oldest materia medica book in China composed in about the second century BC. It was also recorded in many other ancient Chinese books. Nowadays, over 10% of the traditional Chinese patent medicines and simple preparations recorded in the Chinese Pharmacopoeia 2015 edition contain *S. miltiorrhiza*. A huge market has been developed for *S. miltiorrhiza* and its products. Annual consumption of *S. miltiorrhiza* materials has over 16 million kilograms in China.

Although the knowledge on medicinal plants is relatively poor compared with model plants and crops, significant progresses have been achieved in cultivation, bioactive compounds, pharmacological activities, and genes and genomes of *S. miltiorrhiza* after the studies of more than 1000 years, particularly the extensive studies of recent 30 years. With the decoding of *S. miltiorrhiza* genomes, studies on this plant are being accelerated. The information obtained from *S. miltiorrhiza* will be useful for understanding the genomes and bioactive components of other medicinal plants.

This is the first book on the genome of *S. miltiorrhiza*. It dedicates to recent research progresses on molecular maps, whole-genome sequencing, chloroplast and mitochondria genomes, epigenetics, transcriptomics, and functional genomics of *S. miltiorrhiza*. It also describes the taxonomy, distribution, morphology, and growth requirements of this significant medicinal plant and provides useful information on its resources, cultivation, and breeding. The current knowledge of biochemistry and biosynthesis of tanshinones and phenolic acids, two main classes of bioactive components produced in this plant species, is summarized. The technology of hairy root induction, tissue culture, and genetic transformation of *S. miltiorrhiza* is reviewed and discussed. We hope the book will be useful for students, teachers, and scientists in academia and industry interested in medicinal plants and pharmacy. We welcome any criticisms and comments.

We thank the support from the CAMS Innovation Fund for Medical Sciences (CIFMS) (2016-I2M-3-016), the National Natural Science Foundation of China, and other sources of funding. We highly appreciate Series Editor Prof. Chittaranjan Kole for inviting us to prepare this book. We are very grateful to the Springer team for their great assistance.

Beijing, China Shanfa Lu

Contents

Contributors

Haimei Chen Key Laboratory of Bioactive Substances and Resources Utilization of Chinese Herbal Medicine, Ministry of Education, Institute of Medicinal Plant Development, Chinese Academy of Medical Sciences and Peking Union Medical College, Beijing, People's Republic of China

Juan Guo State Key Laboratory Breeding Base of Dao-Di Herbs, National Resource Center for Chinese Materia Medica, China Academy of Chinese Medical Sciences, Beijing, China

Bing Han Laboratory of Medicinal Plant Biotechnology, College of Pharmacy, Zhejiang Chinese Medical University, Hangzhou, Zhejiang, People's Republic of China

Xiaolong Hao Laboratory of Medicinal Plant Biotechnology, College of Pharmacy, Zhejiang Chinese Medical University, Hangzhou, Zhejiang, People's Republic of China

Guoyin Kai Laboratory of Medicinal Plant Biotechnology, College of Pharmacy, Zhejiang Chinese Medical University, Hangzhou, Zhejiang, People's Republic of China

Xingfeng Li State Key Laboratory of Crop Biology, Agronomy College, Shandong Agricultural University, Taian, China

Caili Li Institute of Medicinal Plant Development, Chinese Academy of Medical Sciences and Peking Union Medical College, Beijing, China

Shucan Liu Laboratory of Medicinal Plant Biotechnology, College of Pharmacy, Zhejiang Chinese Medical University, Hangzhou, Zhejiang, People's Republic of China

Zhixiang Liu Laboratory of Medicinal Plant Biotechnology, College of Pharmacy, Zhejiang Chinese Medical University, Hangzhou, Zhejiang, People's Republic of China

Chang Liu Key Laboratory of Bioactive Substances and Resources Utilization of Chinese Herbal Medicine, Ministry of Education, Institute of Medicinal Plant Development, Chinese Academy of Medical Sciences and Peking Union Medical College, Beijing, People's Republic of China

Shanfa Lu Institute of Medicinal Plant Development, Chinese Academy of Medical Sciences and Peking Union Medical College, Haidian District, Beijing, China

Hongmei Luo Institute of Medicinal Plant Development, Chinese Academy of Medical Sciences and Peking Union Medical College, Beijing, China

Ying Ma State Key Laboratory Breeding Base of Dao-Di Herbs, National Resource Center for Chinese Materia Medica, China Academy of Chinese Medical Sciences, Beijing, China

Deyou Qiu State Key Laboratory of Tree Genetics and Breeding, Key Laboratory of Tree Breeding and Cultivation of State Forestry Administration, Research Institute of Forestry, Chinese Academy of Forestry, Beijing, China

Xiaoxiao Qiu Institute of Medicinal Plant Development, Chinese Academy of Medical Sciences and Peking Union Medical College, Beijing, China

Fenjuan Shao State Key Laboratory of Tree Genetics and Breeding, Key Laboratory of Tree Breeding and Cultivation of State Forestry Administration, Research Institute of Forestry, Chinese Academy of Forestry, Beijing, China

Min Shi Laboratory of Medicinal Plant Biotechnology, College of Pharmacy, Zhejiang Chinese Medical University, Hangzhou, Zhejiang, People's Republic of China

Zhenqiao Song State Key Laboratory of Crop Biology, Agronomy College, Shandong Agricultural University, Taian, China

Chun Sui Key Laboratory of Bioactive Substances and Resources Utilization of Chinese Herbal Medicine, Ministry of Education and National Engineering Laboratory for Breeding of Endangered Medicinal Materials, Institute of Medicinal Plant Development (IMPLAD), Chinese Academy of Medical Sciences and Peking Union Medical College, Beijing, China

Meizhen Wang Institute of Medicinal Plant Development Chinese Academy of Medical Sciences and Peking Union Medical College, Haidian District, Beijing, China

Jianhua Wang State Key Laboratory of Crop Biology, Agronomy College, Shandong Agricultural University, Taian, China

Liqiang Wang Key Laboratory of Bioactive Substances and Resources Utilization of Chinese Herbal Medicine, Ministry of Education, Institute of Medicinal Plant Development, Chinese Academy of Medical Sciences and Peking Union Medical College, Beijing, People's Republic of China

Zhichao Xu Institute of Medicinal Plant Development, Chinese Academy of Medical Sciences, Beijing, China

Hong Zhou Institute of Medicinal Plant Development, Chinese Academy of Medical Sciences and Peking Union Medical College, Beijing, China

Salvia miltiorrhiza: An Economically and Academically Important Medicinal Plant

1

Shanfa Lu

Abstract

Salvia miltiorrhiza Bunge is a perennial plant in the genus *Salvia* of the family Lamiaceae. It has a long medicinal history with the first recorded in about the second century BC. *S. miltiorrhiza* has great economic and medicinal value. Over 10% of the traditional Chinese patent medicines and simple preparations contain *S. miltiorrhiza*. Annual consumption of *S. miltiorrhiza* has exceeded 16 million kilograms in China. So far, more than two hundred chemical compounds have been isolated from this plant. *S. miltiorrhiza* and its ingredients have been clinically used for managing vascular diseases and academically shown potential in treating various other diseases. Moreover, *S. miltiorrhiza* has great academic value and is emerging as a model system in medicinal plant biology. This plant has a relatively short life cycle and a relatively small genome size. It is easy to propagate and cultivate. There are whole genome and transcriptome sequences, in vitro tissue culture systems, hairy root and crown gall induction systems and genetic transformation systems available for the research community of *S. miltiorrhiza*. This chapter provides an overview of the taxonomy, distribution, morphology, propagation, growth requirements, chemical constituents, and medicinal and academic significance of *S. miltiorrhiza*.

1.1 Introduction

Salvia miltiorrhiza Bunge, also known as Danshen, Zi Danshen, Tan Shen, red sage and Chinese sage, is a perennial plant in the genus *Salvia* with great economic and medicinal value and a long medicinal history. *S. miltiorrhiza* was first recorded in the oldest materia medica book in China, *Shen Nong's Classic of the Materia Medica* (Shen Nong Ben Cao Jing), composed in about the second century BC. In this book, it was classified as a top-tier medicine. Afterward, it was recorded in many other ancient Chinese books, such as *Collection of Commentaries on the Classic of the Materia Medica* (Ben Cao Jing Ji Zhu) written over 1500 years ago, *Illustrated Classic of the Materia Medica* (Tu Jing Ben Cao) published in Song Dynasty, *Essentials of Materia Medica Distinctions* (Ben Cao Pin Hui Jing Yao) of Ming Dynasty, and *Compendium of Materia Medica* (Ben Cao Gang Mu) written by Shizhen Li in Ming Dynasty. Nowadays, *S. miltiorrhiza* has become one of the most popular and extensively used traditional Chinese

S. Lu (✉)
Institute of Medicinal Plant Development, Chinese Academy of Medical Sciences & Peking Union Medical College, No. 151 Malianwa North Road, Haidian District, Beijing 100193, China
e-mail: sflu@implad.ac.cn

S. Lu (ed.), *The Salvia miltiorrhiza Genome*, Compendium of Plant Genomes,
https://doi.org/10.1007/978-3-030-24716-4_1

medicine (TCM) materials. It is used alone or often mixed with other herbs for treating dysmenorrhoea, amenorrhoea, and cardiovascular diseases in China and other countries, such as Japan, the USA, and European countries (Zhou et al. 2005; Cheng 2007; Xu et al. 2007).

1.2 Taxonomy

S. miltiorrhiza is a member of Lamiaceae, the sixth largest family of flowering plants. Lamiaceae contains more than 7000 species across about 236 genera. In an earlier study, 226 of the 236 genera was classified into seven subfamilies, including Viticoideae, Symphorematoideae, Ajugoideae, Prostantheroideae, Nepetoideae, Scutellarioideae, and Lamioideae, whereas the other ten could not be placed in a subfamily (Harley et al. 2004). Recently, using large-scale chloroplast sequences, Li et al. (2016a) reconstructed the phylogenetic relationship of Lamiaceae species and described three new subfamilies, including Cymarioideae, Peronematoideae, and Premnoideae. Based on this analysis, the Lamiaceae family was divided into ten subfamilies and two unassigned genera (Li et al. 2016a). Among these subfamilies, Nepetoideae, which is characterized by hexacolpate, three-nucleate pollen, an investing embryo and the presence of rosmarinic acid, is one of the most clearly defined subfamilies (Harley et al. 2004; Bräuchler et al. 2010). This subfamily contains about 3400 species across 118 genera and is recovered as monophyletic (Harley et al. 2004; Bräuchler et al. 2010; Li et al. 2016a). Nepetoideae is divided into three tribes, including Elsholtzieae, Ocimeae, and Mentheae, of which tribe Mentheae is the largest and includes many well-known genera, such as *Rosmarinus* (rosemary), *Salvia* (sage), *Mentha* (peppermint), and *Nepeta* (catnip).

Salvia is the largest genus of the family Lamiaceae. It is characterized by the well-known stamina lever mechanism of the flower. Two monothecic stamens are modified to levers with a thin ligament between the connective and the filament to form a joint, which enables a lever-like reversible movement causing pollen transfer (Wester and Claßen-Bockhoff 2007). Filaments are short, horizontal, or erect. Connectives are prolonged, linear, and T-shaped. Upper arms have fertile elliptic or linear anther cells. Lower arms are robust or slender, with fertile or sterile anther cells, separated or connected to each other. Two staminodes are small or absent (Flora of China Editorial Committee 1994).

Salvia includes approximately 1000 species and is well-known for its medicinal, ornamental, esculent, or hallucinogenic plants (Will and Claßen-Bockhoff 2017). It has radiated extensively in three regions of the world: Central and South America (500 spp.), western Asia (200 spp.), and eastern Asia (100 spp.). The most recent common ancestor (MRCA) of *Salvia* existed in the later Eocene (34 million years ago) to the middle Oligocene (26 million years ago) (Will and Claßen-Bockhoff 2017). *Salvia* is polyphyletic. Walker et al. (2004) and Walker and Sytsma (2007) revealed three distinct evolutionary lineages (Clade I–III). Later, Will and Claßen-Bockhoff (2014) recognized the Clade III described by Walker et al. (2004) and Walker and Sytsma (2007) as two independent lineages (Clades III and IV). The four clades described strongly reflect the geographical distribution of *Salvia* species, with Clade I mainly in Europe, Southwest and Central Asia, and Southern Africa, Clade II mainly in Central and South America, Clade III mainly in Southwest Asia and Northern Africa, and Clade IV mainly in East Asia (Will and Claßen-Bockhoff 2017).

S. miltiorrhiza is a Clade IV species native to China (Will and Claßen-Bockhoff 2017). China is the center of diversity in East Asia, with 84 *Salvia* species (Flora of China Editorial Committee 1994). Among them, 70 species are endemic to China, whereas the other 14 are cosmopolitan species (Wei et al. 2015). The majority of *Salvia* species in China, such as *S. miltiorrhiza*, *Salvia yunnanensis*, *Salvia plebeia*, *Salvia bowleyana,* and *Salvia trijuga*, are members of Clade IV (Will and Claßen-Bockhoff 2017). Few of the others, such as *Salvia deserta* and *Salvia officinalis*, belong to Clade I.

In addition, there are several species introduced from America and Europe, such as *Salvia coccinea*, *Salvia splendens*, and *Salvia sclarea*.

1.3 Distribution

S. miltiorrhiza is widely distributed in China. Wild and domestic *S. miltiorrhiza* can be found in at least nineteen provinces and cities of China. It includes Liaoning, Hebei, Beijing, Tianjin, Jiangsu, Shandong, Shanghai, Anhui, Zhejiang, Jiangxi, Hunan, Hubei, Henan, Shanxi, Shaanxi, Gansu, Sichuan, Guizhou, and Yunnan, covering a total of 192 counties of China (Wei et al. 2015). *S. miltiorrhiza* also grows in Mongolia, Korea, Japan, Vietnam, and Australia (Xu et al. 2007; Sheng et al. 2009; Li et al. 2011; Tung et al. 2017). It grows at about 100–1300 m elevation and prefers sunny and wet places in forests, hillsides, and streamsides (Flora of China Editorial Committee 1994).

1.4 Morphology

S. miltiorrhiza is a deciduous perennial plant with quadrangular, erect, and branched stems (Fig. 1.1). The stems are about 40–80 cm tall and covered with hairs and sticky glands. The cotyledons and the true leaf are simple leaf. Compound leaves developed after the true leaf are odd-pinnate compound with 3–5(–7) leaflets. Each of the leaflets has 1.5–8 × 1–4 cm in size. Leaves are ovate or broadly lanceolate and pilose. Leaf margin is crenate. Leaf apex is acute to acuminate. Petiole is 1.3–7.5 cm in length. Petiolule is 2–14 mm in length. *S. miltiorrhiza* has a verticillaster inflorescence with densely villous or glandular villous (Flora of China Editorial Committee 1994).

The flowers of *S. miltiorrhiza* grow in whorls. Each whorl has 6–many flowers with light purple to lavender blue or white corollas. The corolla is about 2.5 cm in length, glandular pubescent, and two-lipped (Fig. 1.2). The upper lip of corolla is vertical and falcate. The lower lip is shorter, and its terminal is lobulated and trifid. The central lobe is longer and larger than the two side lobes. Corolla tube is exserted with imperfectly pilose annulate inside. The tube is shorter than limb. The calyxes are dark purple and slightly bell-shaped and divided into two parts or lips. The upper lip of calyxes is entire and triangular with three-mucronate apex. The lower lip is almost as long as upper and is two-toothed (Flora of China Editorial Committee 1994).

As the other species of the genus *Salvia*, *S. miltiorrhiza* is also characterized by the well-known stamina lever mechanism of the flower (Fig. 1.3) (Wester and Claßen-Bockhoff 2007). The filaments of *S. miltiorrhiza* are 3.5–4 mm, and the connectives are 1.7–2 cm. Both filaments and connectives are much exserted (Flora of China Editorial Committee 1994).

S. miltiorrhiza plants bloom from April to August and fruit from September to October. The nutlets of *S. miltiorrhiza* are ellipsoid with about 3.2 × 1.5 mm in size (Flora of China Editorial Committee 1994). *S. miltiorrhiza* roots are the medicinal parts of the plant and are well-known TCM materials. It consists of taproot and lateral roots. The taproot is thickened, succulent, and usually scarlet outside (Flora of China Editorial Committee 1994).

1.5 Propagation

S. miltiorrhiza can be propagated from seeds, cuttings, and root segments. It can also be propagated through tissue culture.

S. miltiorrhiza seeds are small, dark brown nutlets. It cannot be stored for a long time at room temperature. Seed germination rate will decrease significantly after three-month storage at room temperature (Li et al. 2016b). Under normal circumstances, the optimal storage temperature is 4 °C. For long-term storage, it requires −20 °C (Li et al. 2016b). Except for temperature, water content is the other key factor affecting the vitality of stored seeds. The optimal water content is about 5% for storage of *S. miltiorrhiza* seeds (Li et al. 2016b). In practice, the seeds are usually sown from late March to early April in the spring or late September in the

Fig. 1.1 *S. miltiorrhiza* plants and its Chinese name

Fig. 1.2 Flowers of *S. miltiorrhiza*. The entire flower (**a**), style (**b**), calyx (**c**), upper lip of corolla (**d**), lower lip of corolla (**e**), corolla tube, and lever-like stamens (**f**) are shown

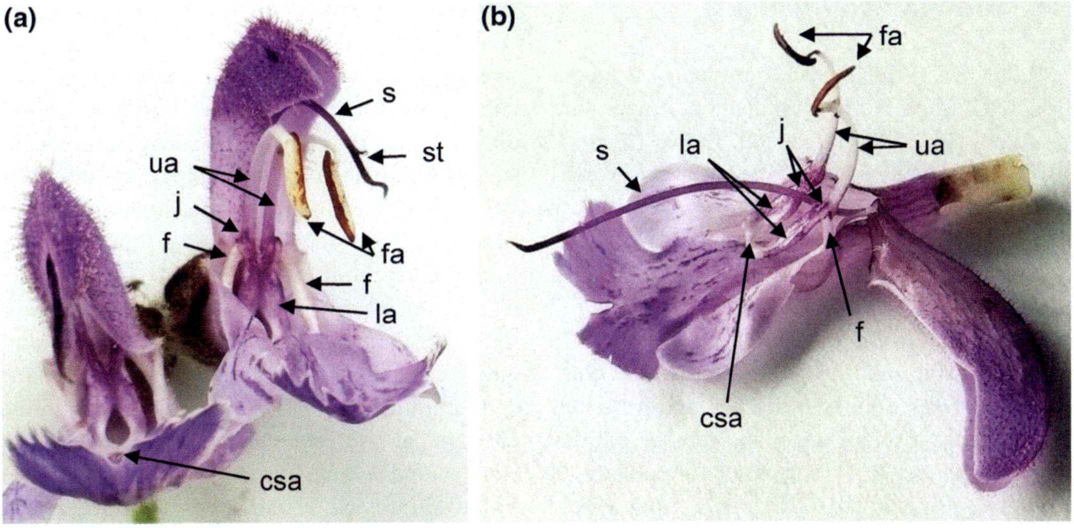

Fig. 1.3 Lever-like stamens of *S. miltiorrhiza*. **a** The entire flowers; **b** A flower with the upper lip of corolla removed. f, filament; fa, fertile anther; j, joint between filament and connective; la, lower lever arm of stamen; s, style; csa, connected sterile anthers; st, stigma; ua, upper lever arm of stamen

autumn and can be sown into the soil immediately after harvest (He et al. 2014).

To propagate *S. miltiorrhiza* from cuttings, select healthy and strong stems of vigorous plants during growing season, cut them into about 10 cm segments, excise the entire leaves on the lower part and half of the leaves on the upper part, and slantingly insert the cuttings about one-third to half of their length into moist and free-draining soil (Xu 2002; Zhang 2018; Wang 2018). Arrange the cuttings with space between each one to avoid overcrowded. Provide some shade and water regularly to keep the soil consistently moist. Avoid letting the soil dry out or waterlogging the cuttings. Once the roots form, transfer them to the field.

Propagation from root segments is one of the most popular propagation methods for *S. miltiorrhiza*. To propagate using this method, select healthy and strong roots with about 1.0 cm in diameter, cut them into about 5 cm segments, put the segments into pre-prepared soil holes, keep the segments slantingly with the upper part upward, and then cover the segments with about 2 cm of soil (Xu 2002; Zhang 2018; Wang 2018). Root segment propagation can be done in February or March. The roots used for propagation can be harvested in November and stored in damp sand until use in February or March of the next year (Zhang 2018). Alternatively, it can be kept in the soil until the next February or March (Xu 2002; Zhang 2018; Wang 2018). Middle and upper parts of the roots and one-year-old roots will be greater for shoot emergence (Xu 2002). In addition, rootstalks can also be used for propagation (Zhang 2018).

Tissue culture is the other useful method for *S. miltiorrhiza* propagation (Jia et al. 2016). It includes suspension cell culture, hairy root culture, crown gall tissue culture, in vitro shoot culture, and callus culture. Tissue culture technology can be widely used for rapid propagation and elite breeding of *S. miltiorrhiza*, industrial production of active compounds, and academic purposes, such as functional characterization of genes and mutant library construction (Zhang et al. 1995; Lee et al. 2008; Cui et al. 2011; Li et al. 2017a).

1.6 Growth Requirements

S. miltiorrhiza seeds can germinate at about 10–30 °C. The optimal germination temperature is around 20–25 °C (He et al. 2014; Li et al. 2016b). The seeds usually take about two weeks to germinate. *S. miltiorrhiza* begins to grow when soil temperature reaches to 10 °C. Adventitious buds start to develop on root division when soil temperature reaches to 15–17 °C in spring (Xu 2002). *S. miltiorrhiza* plants prefer sunny and wet soil and relatively tolerate cold, but they are susceptible to drought and waterlog. The optimal air temperature and relative humidity for *S. miltiorrhiza* plant growth are 20–26 °C and 80%, respectively. The plants grow vigorously during growing season and take about 200 days from transplanting to harvesting. The aerial parts of *S. miltiorrhiza* plants start to wither when the air temperature falls below 10 °C in autumn. The underground parts can overwinter safely even if the temperature drops to 15 °C below zero (Xu 2002).

1.7 Chemical Constituents

S. miltiorrhiza mainly contains lipophilic diterpenoid tanshinones and hydrophilic phenolic acids, which are two major classes of bioactive compounds. According to the Chinese Pharmacopoeia 2015 edition (Pharmacopoeia Commission of People's Republic of China 2015), total content of tanshinone IIA, cryptotanshinone, and tanshinone I should not be less than 0.25% and salvianolic acid B should not be less than 3% in dry *S. miltiorrhiza* medicinal materials. In addition to tanshinones and phenolic acids, *S. miltiorrhiza* also contains monoterpenes, sesquiterpenes, triterpenes, alkaloids, flavonoids, anthocyanidins, sterols, saccharides, quinones, and so on. So far, more than two hundred chemical compounds have been isolated from this plant (Wang et al. 2017a; Mei et al. 2019).

Diterpenoid tanshinones are the representatives of lipophilic components in *S. miltiorrhiza*. They can be divided into three types, including diterpenoid tanshinones, tricyclic diterpenoid tanshinones, and royleanone tanshinones (Luo2015; Wang et al. 2017a). The basic structure of diterpenoid tanshinones contains a 1,2-*o*-naphthoquinone mother nucleus with a furan or dihydrofuran ring. Many bioactive tanshinones, such as tanshinone I, tanshinone IIA, tanshinone IIB, cryptotanshinone, and dihydrotanshinone (Zhang and Lu 2017), belong to this type. Tricyclic diterpenoid tanshinones are characterized by ligation of an isopropyl group, instead of a furan or dihydrofuran ring, to the naphthoquinone mother nucleus. Miltirone, miltirone I, dehydromiltirone and methylene miltirone are typical tricyclic diterpenoid tanshinones. Royleanone tanshinones are characterized by 1,4-*p*-quinone. Compounds included in this type are isotanshinone I, isotanshinone IIA, isotanshinone IIB, isocryptotanshinone, 2-hydroxyisodihydrotanshinone, danshenxinkun A, danshenxinkun B, and danshenxinkun C. Besides these diterpenoid tanshinones, there are other diterpenes in *S. miltiorrhiza*, such as ferruginol, sugiol, neotanshinlactone, danshenol A and danshenol B (Mei et al. 2019). Recently, new diterpenoid compounds, including salmiltiorins A–F, normiltirone, and isosalviamides F–H, have been identified from *S. miltiorrhiza* (Wei et al. 2017; Ngo et al. 2019). In total, about 90 diterpenoid compounds have been isolated from *S. miltiorrhiza* (Wei et al. 2017; Mei et al. 2019; Ngo et al. 2019).

Water-soluble phenolic acids are the other major class of bioactive compounds in *S. miltiorrhiza*. It includes more than forty compounds, which can be classified into single phenolic acids and polyphenolic acids (Mei et al. 2019). Single phenolic acids include danshensu, caffeic acid, ferulic acid, isoferulic acid, protocatechuic aldehyde, protocatechuic acid, and β-(3,4-dihydroxyphenyl) lactic acid. Many of them contain a core skeleton of C6–C3. Polyphenolic acids includes salvianolic acids A–G, salvianolic acids I–L, salvianolic acid N, salvianolic acid C1, lithospermic acid B, 8″-*epi*-lithospermic acid C, and rosmarinic acid, most of which are depsides of danshensu and derivatives or dimer of caffeic acid (Zhang et al. 2017; Mei et al. 2019). Rosmarinic acid is a simple depside formed by the condensation of a molecule of danshensu and a molecule of caffeic acid. Salvianolic acid A and salvianolic acid C are depsides composed of a molecule of rosmarinic acid and a molecule of danshensu. Salvianolic acid C contains a 2-arylbenzofuran, which may be converted from salvianolic acid A through cyclization. Both salvianolic acid B and salvianolic acid E are composed of two molecules of rosmarinic acid. Salvianolic acid B, the major phenolic acid component in *S. miltiorrhiza*, contains a 2-aryldihydrobenzofuran with 2*R*, 3*R* configuration (Luo 2015). It may be converted from salvianolic acid E through cyclization. Lithospermic acid B has the same planar structure as salvianolic acid B, whereas its 2-aryldihydrobenzofuran skeleton possesses a 2*S*, 3*S* configuration (Luo 2015).

Among the diterpenoids and phenolic compounds isolated from *S. miltiorrhiza*, some are nitrogen-containing. These nitrogen-containing chemical compounds are also known as alkaloids. So far, various alkaloids have been isolated from *S. miltiorrhiza*. They include salvianan, neosalvianen, salvianen, salviadione, isosalviamines F, salviamiltamide, 3-(3′,4′-dihydroxyphenyl)lactamide, and 5-(methoxymethyl)-1*H*-pyrrole-2-carbaldehyde (Choi et al. 2001; Don et al. 2005; Pan et al. 2018).

Through the analysis of volatile terpenes in root, stem, leaf, and inflorescence of *S. miltiorrhiza* by extraction of fresh tissues with hexane and subjected to GC-MS, Fang et al. (2017) detected eight monoterpenes and eleven sesquiterpenes, including α-pinene, camphene, sabinene, β-pinene, β-myrcene, β-thujene, *trans*-β-ocimene, *cis*-β-ocimene, α-copaene, β-elemene, β-caryophyllene, β-copaene, cadina-3.5-diene, α-humulene, *trans*-cadina-1(6),4-diene, germacrene D, β-cadinene, germacrene D-4-ol, and germacrene B. Among the four organs analyzed, leaf has the most diversified terpenes, with eight monoterpenes and ten sesquiterpenes. Analysis of the composition of essential oil of *S. miltiorrhiza* flower showed that the main components are monoterpenes, sesquiterpenes, fatty acids, and alkanes. Among them, the sesquiterpenes, such as β-caryophyllene, β-caryophyllene oxide, α-caryophyllene, and cadinadiene, are the most abundant (Qian et al. 2009).

Triterpenes are the other class of bioactive secondary metabolisms produced in *S. miltiorrhiza* (Steinkamp-Fenske et al. 2007; Zhang et al. 2008; Leng et al. 2017; Tung et al. 2017). They are produced not only in the aerial parts of *S. miltiorrhiza* (Zhang et al. 2008; Zeng et al. 2017), but also in the root part of the plant (Tung et al. 2017). The main triterpenes produced in *S. miltiorrhiza* are oleanolic acid and ursolic acid (Zeng et al. 2017). Other triterpenes, such as urs-12-ene-2α,3β,7β,16α-tetraol, 2β-hydroxypomolic acid, maslinic acid, and asiatic acid, can also be detected in *S. miltiorrhiza* (Tung et al. 2017).

In addition to terpenoids and phenolic acids, *S. miltiorrhiza* also contains abundant flavonoid, anthocyanidin, saccharide, and quinone compounds (Ravipati et al. 2012; Zeng et al. 2017; Deng et al. 2018; Li et al. 2019; Liu et al. 2019a). Flavonoids and anthocyanidins, such as rutin, isoquercitrin, astragalin, cyanidin-3,5-di-O-glucoside, delphinidin chloride, are mainly produced in the aerial parts of *S. miltiorrhiza* (Zeng et al. 2017; Li et al. 2019). In *S. miltiorrhiza* roots, most of the saccharides are stachyose, whereas in the aerial parts, the saccharides were mainly monosaccharides (Zeng et al. 2017). Plastoquinone (PQ) and ubiquinone (UQ, Coenzyme Q, CoQ) are two major lipid-soluble prenylquinones in plants. The structural feature of PQ and UQ is a medium/long trans-polyprenyl side chain attached to the benzoquinone skeleton. HPLC-UV analysis showed that *S. miltiorrhiza* contains PQ and UQ. The side chain of PQ produced in *S. miltiorrhiza* has nine isoprenoid units (PQ-9), whereas the side chain of *S. miltiorrhiza* UQ can be nine (UQ-9) or ten (UQ-10) isoprenoid units (Liu et al. 2019a).

1.8 Medicinal Significance

S. miltiorrhiza medicinal materials can activate blood circulation to remove blood stasis, induce menstruation to relieve meralgia, clear the heart to remove troubles, and cool blood to eliminate carbuncle (Pharmacopoeia Commission of People's Republic of China 2015). They can be used to treat chest stuffiness and pain, abdominal hypochondriac pain, gynecologic abdominal lump, arthralgia caused by heat pathogen, vexation and sleeplessness, menstrual disorder, dysmenorrhea, and sore ulcer in traditional Chinese medicine (Pharmacopoeia Commission of People's Republic of China 2015). In the Chinese Pharmacopoeia 2015 edition (Pharmacopoeia Commission of People's Republic of China 2015), a total of 1493 traditional Chinese patent medicines and simple preparations were recorded, of which about 157, accounting for over 10%, contain *S. miltiorrhiza*. It indicates the huge market for *S. miltiorrhiza* and its products (Cheng 2007). Indeed, annual consumption of *S. miltiorrhiza* has exceeded 16 million kilograms in China. In addition, *S. miltiorrhiza* is the first Chinese medicinal material entering the international market. The Compound Danshen Dripping Pill, a *S. miltiorrhiza*-containing traditional Chinese patent medicine officially listed in the Chinese Pharmacopoeia 2015 edition (Pharmacopoeia Commission of People's Republic of China 2015), had been approved for phase II and III clinical trials in the USA by the Food and Drug Administration (FDA).

In China, various *S. miltiorrhiza*-containing traditional Chinese patent medicines and simple preparations, such as Compound Danshen dripping pills, Fufang Danshen tablets, Tongxinluo capsule, and Danhong injection, have been clinically used for management of vascular diseases, such as coronary heart disease, ischemic cardiovascular disease, congestive heart failure, chronic heart failure, hyperlipidemia, cardiovascular neurosis (Wang et al. 2017a). *S. miltiorrhiza* and its ingredients can lower hypertension and hyperlipidemic and improve atherosclerosis and myocardial ischemia preperfusion (Wang et al. 2017a; Su et al. 2015). Tanshinone IIA, tanshinone VI, cryptotanshinone, danshensu, salvianolic acid B, protocatechuic acid, protocatechualdehyde, catechin, and polysaccharides are responsible for its cardioprotective effects through various mechanisms, such as anti-inflammation, anti-oxidation, anti-thrombosis, anti-proliferation of vascular smooth muscle cells, improvement of acute myocardial ischemia, and inhibition of adhesion molecule expression in vascular endothelium and leukocytes (Wang et al. 2005, 2017a; Fang et al. 2008; Han et al. 2008; Hur et al. 2008; Zhao et al. 2008; Ho and Hong 2011; Xia et al. 2014; Zhang et al. 2015c; Lu et al. 2015).

In addition to clinical use for vascular disease treatment, *S. miltiorrhiza* and its ingredients also showed activities in anti-Alzheimer's disease, anti-Parkinson's disease, anti-neuropathic pain, anticancer, anti-hepatocyte injury, and so on (Su et al. 2015). The pharmacological activities of *S. miltiorrhiza* and its ingredients have been reviewed in details in many papers published recently (Su et al. 2015; Wang et al. 2017a; Chong et al. 2019; Jia et al. 2019; Jiang et al. 2019; Mei et al. 2019; Wang et al. 2019), suggesting the significance of *S. miltiorrhiza* in medicinal research.

1.9 Academic Significance

S. miltiorrhiza is one of the medicinal plants extensively studied using modern techniques. In addition to its medicinal value, *S. miltiorrhiza* has great academic value and is emerging as a model system in medicinal plant biology due to its relatively short life cycle, easy to propagate and cultivate, relatively small genome size, and the availability of in vitro tissue culture systems, hairy root and crown gall induction systems and genetic transformation systems for biotechnology and functional genomics studies (Song et al. 1998, 2000, 2013; Wang et al. 1998, 2009, 2017b; Qiu et al. 2004; Yan and Wang 2007; Lee et al. 2008; Song and Wang 2011; Zhao et al. 2011; Liu et al. 2015).

So far, the whole genomes of two *S. miltiorrhiza* lines have been sequenced and are available for the research community (Zhang et al. 2015a; Xu et al. 2016b). One of the *S. miltiorrhiza* lines is known as 99-3, whereas the name of the other one is unknown (Zhang et al. 2015a; Xu et al. 2016b). The genome size of 99-3 estimated by flow cytometry analysis is about

615 Mb. Its genome was sequenced by Illumina Hiseq 2000 platform, PacBio RS platform, and Roche/454. The final size of genome assembly is about 538 Mb, with contig and scaffold N50 of 12.38 kb and 51.02 kb, respectively (Xu et al. 2016b). A total of 30,478 protein-coding genes were predicted from the genome assembly and validated by RNA-seq data (Xu et al. 2015). The genome size of the other line was estimated by *k*-mers (Zhang et al. 2015a). The calculated genome size is about 645.78 Mb. Genome sequencing with Illumina and PacBio RS platforms and subsequent de novo assembly resulted in a final draft genome of 641 Mb, with a contig N50 of 82.8 kb and scaffold N50 of 1.2 Mb (Zhang et al. 2015a). A total of 34,598 protein-coding genes were predicted from this genome assembly (Zhang et al. 2015a). Although the two genome assemblies are draft and their quality needs to be improved, the sequence data provide basic information for gene identification and comparative analysis.

In addition to the whole genome sequences, a huge amount of transcriptome sequences of *S. miltiorrhiza* have been sequenced using high-throughput sequencing platforms, such as Illumina HiSeq 2500, Illumina HiSeq 2000, Illumina Genome Analyzer, 454 GS FLX Titanium, and PacBio RS, and deposited in the National Center of Biotechnology Information (NCBI) databases (https://www.ncbi.nlm.nih.gov/sra). These sequences were obtained from a wide variety of organs and tissues, such as root, stem, leaf, flower, root periderm, root phloem, and root xylem, and from plants or hairy roots treated with salicylic acid, yeast extract and Ag^{+}, methyl jasmonate and drought stress (Li et al. 2010; Wenping et al. 2011; Yang et al. 2013; Gao et al. 2014; Luo et al. 2014; Xu et al. 2015, 2016a; Zhang et al. 2016b; Song et al. 2017; Zhou et al. 2017; Wang et al. 2018; Wei et al. 2018).

There are more than two hundred chemical compounds isolated from *S. miltiorrhiza* (Wang et al. 2017a; Mei et al. 2019). These compounds can be divided into various classes, such as terpenoids, phenolic acids, flavonoids, anthocyanidins, quinones, alkaloids, and saccharides, which are the main chemical classes of bioactive compounds. Therefore, the knowledge obtained from *S. miltiorrhiza* may allow us to design experiments of other medicinal plants for addressing important questions about their biology, particularly the biosynthesis and regulatory mechanisms of bioactive compounds. So far, the biosynthetic pathways and biosynthesis-related genes of diterpenoid tanshinones, phenolic acids and flavonoids have been intensively studied and much knowledge has been gained (Fig. 1.4) (Ma et al. 2012; Guo et al. 2013; Hou et al. 2013; Gao et al. 2014; Cui et al. 2015; Wang et al. 2015; Deng and Lu 2017; Zhang and Lu 2017; Deng et al. 2018; Shi et al. 2018).

The biosynthetic pathway of tanshinones may be generally divided into three stages (Ma et al. 2012; Zhang and Lu 2017). At the first stage, the universal isoprene precursor isopentenyl diphosphate (IPP) and its isomer dimethylallyl diphosphate (DMAPP) are synthesized from pyruvate and glyceraldehyde-3-phosphate through the 2-C-methyl-D-erythritol 4-phosphate (MEP) pathway and/or from acetyl-CoA through the mevalonate (MVA) pathway under the catalysis of fourteen enzymes. At the second stage, the intermediate diphosphate precursor geranylgeranyl diphosphate (GGPP) is synthesized under the catalysis of GGPP synthase. At the last stage, GGPP is sequentially converted to copalyl diphosphate (CPP), miltiradiene and ferruginol under the catalysis of CPP synthase (CPS), kaurene synthase-like (KSL), and CYP76AH1, respectively (Gao et al. 2009; Cui et al. 2015; Guo et al. 2013). Then, ferruginol is converted into 11,20-dihydroxy ferruginol and 11,20-dihydroxy sugiol under the catalysis of SmCYP76AH3 and SmCYP76AK1 (Guo et al. 2016). The downstream pathway of tanshinone biosynthesis remains to be elucidated.

The biosynthesis of bioactive phenolic acids requires two independent upstream pathways, including the phenylpropanoid pathway and the tyrosine-derived pathway. The phenylpropanoid pathway leads to the production of 4-coumaroyl-CoA, caffeic acid, and caffeoyl-CoA, whereas the tyrosine-derived pathway is devoted to the biosynthesis of 4-hydroxyphenyllactic acid and 3,4-dihydroxyphenyllactic acid (danshensu).

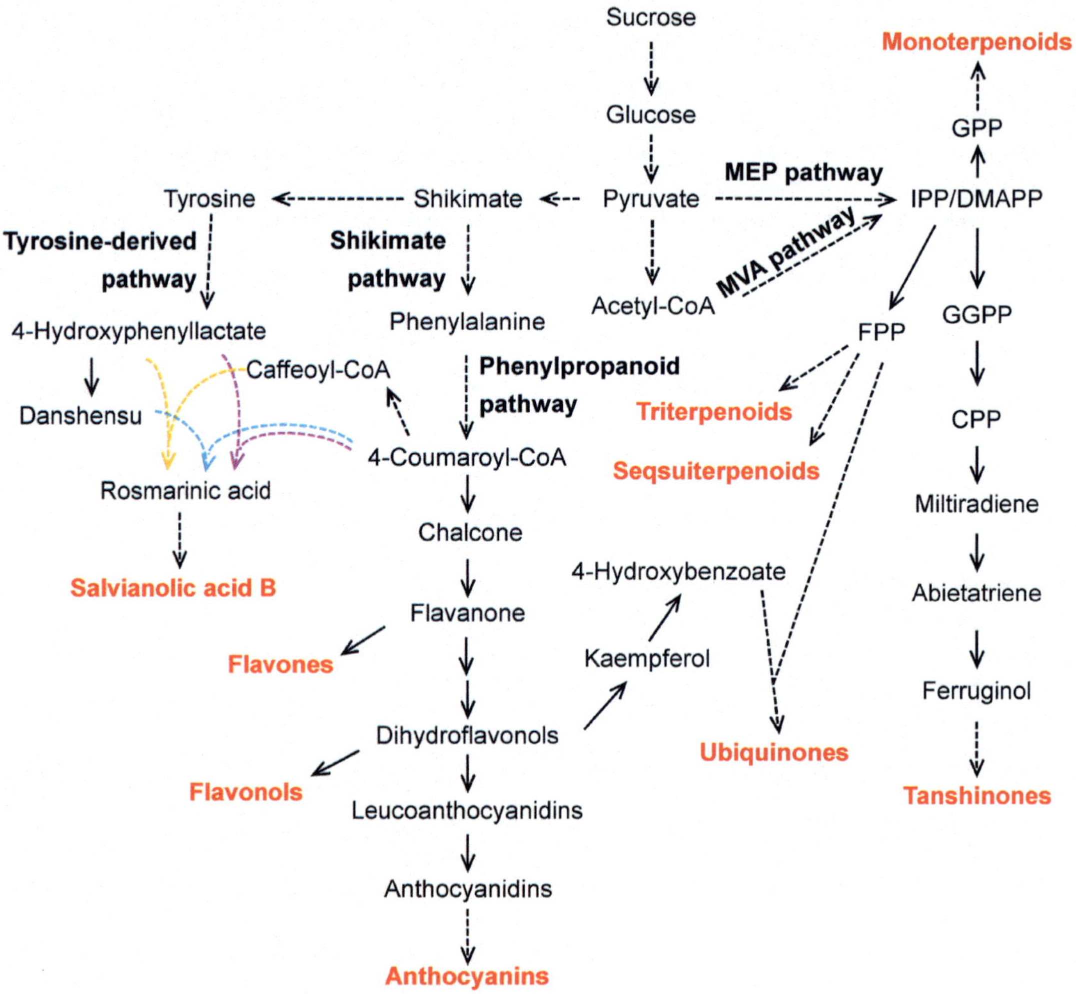

Fig. 1.4 Biosynthetic pathways of bioactive compounds in *S. miltiorrhiza*. Solid arrows represent single biosynthetic steps. Dashed arrows denote multiple steps. Three proposed pathways for rosmarinic acid biosynthesis are shown with pink, blue and yellow arrows, respectively. Bioactive compounds are shown in red. CPP, copalyl diphosphate; DMAPP, dimethylallyl diphosphate; FPP, farnesyl diphosphate; GGPP, geranylgeranyl diphosphate; GPP, geranyl diphosphate; IPP, isopentenyl diphosphate; MEP, 2-C-methyl-D-erythritol 4-phosphate; MVA, mevalonate

These compounds were used to synthesize the important intermediate, rosmarinic acid. There are three proposed pathways for the biosynthesis of rosmarinic acid: (1) the intermediary precursors, 4-coumaroyl-CoA and 4-hydroxyphenyllactate, are coupled to 4-coumaroyl-4′-hydroxyphenyllactic acid under the catalysis of rosmarinic acid synthase, and then the 3- and 3′-hydroxyl groups are introduced to 4-coumaroyl-4′-hydroxyphenyllactic acid to form rosmarinic acid under the catalysis of cytochrome P450-dependent monooxygenases (Petersen et al. 2009); (2) rosmarinic acid is synthesized from 4-coumaroyl-CoA and 3,4-dihydroxyphenyllactate (danshensu) under the catalysis of rosmarinic acid synthase and CYP98A14 (Di et al. 2013); and (3) rosmarinic acid is synthesized from caffeic acid/caffeoyl-CoA and 4-dihydroxyphenyllactate (Liu et al. 2019b). The downstream biosynthetic pathways of salvianolic acids remain to be elucidated.

The biosynthesis of flavonoids starts from the condensation and subsequent intramolecular cyclization of one *p*-coumaroyl-CoA molecule and three malonyl-CoA molecules, which are synthesized from the phenylpropanoid pathway and the Krebs cycle, respectively (Deng and Lu 2017). The first step of flavonoid biosynthesis is catalyzed by chalcone synthase. The product (chalcone) of the first step is then converted to different flavonoid molecules under the catalysis of a series of enzymes, such as chalcone isomerase, flavone synthase, flavanone 3-hydroxylase, flavanone 3β-hydroxylase, flavonoid 3′-hydroxylase, flavonoid 3′,5′-hydroxylase, and flavonol synthase. Dihydroflavonol synthesized from the flavonoid pathway may further convert to leucoanthocyanidin under the catalysis of dihydroflavonol 4-reductase, and then to anthocyanidin under the catalysis of anthocyanidin synthase/leucoanthocyanidin dioxygenase. Through genome-wide prediction and molecular cloning, genes encoding these enzymes have been identified in *S. miltiorrhiza* (Deng et al. 2018).

In addition to the pathway and enzyme-encoding genes, the regulatory factors of bioactive compound biosynthesis, such as transcription factors, miRNAs and DNA methylation, have also been intensively studied (Shao and Lu 2013, 2014; Zhang et al. 2014, 2015b, 2016a, 2018; Li and Lu 2014; Xu et al. 2014; Li et al. 2015, 2017b, 2018a, b; Shao et al. 2015). Recent advances on these factors have been reviewed in details in Chaps. 6 and 8 of this book.

1.10 Conclusion

S. miltiorrhiza has been recorded as a medicinal material for more than two thousand years and is a popular and extensively used TCM material in China. The demand for *S. miltiorrhiza* has been rising rapidly. Meanwhile, scientists have made great progress in cultivation, chemical constituents, pharmacological activities, and genes and genomes of *S. miltiorrhiza*. The biosynthetic pathways and regulatory mechanisms of various bioactive compounds are being elucidated. Metabolic engineering of tanshinone and phenolic acid biosynthesis has been conducted (Lee et al. 2008; Kai et al. 2011; Shi et al. 2014, 2016; Yang et al. 2017; Xing et al. 2018). With more and more pharmacological activities and the underlying mechanisms to be elucidated, more and more sequence data to be available, and more and more genes to be identified and functionally analyzed, the economic and academic significance of *S. miltiorrhiza* will be more prominent.

Acknowledgements This work was supported by the National Natural Science Foundation of China (81773836) and the CAMS Innovation Fund for Medical Sciences (CIFMS) (2016-I2M-3-016).

References

Bräuchler C, Meimberg H, Heubl G (2010) Molecular phylogeny of Menthinae (Lamiaceae, Nepetoideae, Mentheae)–Taxonomy, biogeography and conflicts. Mol Phylogenet Evol 55(2):501–523

Cheng TO (2007) Cardiovascular effects of Danshen. Int J Cardiol 121:9–22

Choi JS, Kang HS, Jung HA, Jung JH, Kang SS (2001) A new cyclic phenyllactamide from *Salvia miltiorrhiza*. Fitoterapia 72(1):30–34

Chong CM, Su H, Lu JJ, Wang Y (2019) The effects of bioactive components from the rhizome of *Salvia miltiorrhiza* (Danshen) on the characteristics of Alzheimer's disease. Chin Med 14:19

Cui G, Duan L, Jin B, Qian J, Xue Z, Shen G, Snyder JH, Song J, Chen S, Huang L, Peters RJ, Qi X (2015) Functional divergence of diterpene syntheses in the medicinal plant *Salvia miltiorrhiza*. Plant Physiol 169(3):1607–1618

Cui G, Huang L, Tang X, Zhao J (2011) Candidate genes involved in tanshinone biosynthesis in hairy roots of *Salvia miltiorrhiza* revealed by cDNA microarray. Mol Biol Rep 38(4):2471–2478

Deng Y, Li C, Li H, Lu S (2018) Identification and characterization of flavonoid biosynthetic enzyme genes in *Salvia miltiorrhiza* (Lamiaceae). Molecules 23(6):1467

Deng Y, Lu S (2017) Biosynthesis and regulation of phenylpropanoids in plants. Crit Rev Plant Sci 36(4): 257–290

Di P, Zhang L, Chen J, Tan H, Xiao Y, Dong X, Zhou X, Chen W (2013) ^{13}C tracer reveals phenolic acids biosynthesis in hairy root cultures of *Salvia miltiorrhiza*. ACS Chem Biol 8(7):1537–1548

Don MJ, Shen CC, Lin YL, Syu WJ, Ding YH, Sun CM (2005) Nitrogen-containing compounds from *Salvia miltiorrhiza*. J Nat Prod 68(7):1066–1070

Fang X, Li CY, Yang Y, Cui MY, Chen XY, Yang L (2017) Identification of a novel (-)-5-epieremophilene synthase from *Salvia miltiorrhiza* via transcriptome mining. Front Plant Sci 8:627

Fang ZY, Lin R, Yuan BX, Yang GD, Liu Y, Zhang H (2008) Tanshinone IIA downregulates the CD40 expression and decreases MMP-2 activity on atherosclerosis induced by high fatty diet in rabbit. J Ethnopharmacol 115:217–222

Flora of China Editorial Committee (1994) Folra of China, vol 17. Science Press and Missouri Botanical Garden Press, Beijing, p 212

Gao W, Hillwig ML, Huang L, Cui G, Wang X, Kong J, Yang B, Peters RJ (2009) A functional genomics approach to tanshinone biosynthesis provides stereochemical insights. Org Lett 11:5170–5173

Gao W, Sun HX, Xiao H, Cui G, Hillwig ML, Jackson A, Wang X, Shen Y, Zhao N, Zhang L, Wang XJ, Peters RJ, Huang L (2014) Combining metabolomics and transcriptomics to characterize tanshinone biosynthesis in *Salvia miltiorrhiza*. BMC Genom 15:73

Guo J, Ma X, Cai Y, Ma Y, Zhan Z, Zhou YJ, Liu W, Guan M, Yang J, Cui G, Kang L, Yang L, Shen Y, Tang J, Lin H, Ma X, Jin B, Liu Z, Peters RJ, Zhao ZK, Huang L (2016) Cytochrome P450 promiscuity leads to a bifurcating biosynthetic pathway for tanshinones. New Phytol 210:525–534

Guo J, Zhou YJ, Hillwig ML, Shen Y, Yang L, Wang Y, Zhang X, Liu W, Peters RJ, Chen X, Zhao ZK, Huang L (2013) CYP76AH1 catalyzes turnover of miltiradiene in tanshinones biosynthesis and enables heterologous production of ferruginol in yeasts. Proc Natl Acad Sci U S A 110(29):12108–12113

Han JY, Fan JY, Horie Y, Miura S, Cui DH, Ishii H, Hibi T, Tsuneki H, Kimura I (2008) Ameliorating effects of compounds derived from *Salvia miltiorrhiza* root extract on microcirculatory disturbance and target organ injury by ischemia and reperfusion. Pharmacol Ther 117:280–295

Harley RM, Atkins S, Budantsev AL, Cantino PD, Conn BJ, Grayer R, Harley MM, de Kok R, Krestovskaja T, Morales R, Paton AJ, Ryding O, Upson T (2004) Labiatae. In: Kadereit JW (ed). Flowering plants · Dicotyledons. The families and genera of vascular plants, vol 7. Springer, Berlin, Heidelberg, Germany, pp 167–275

He GQ, Xue RG, Guo CG, Li GQ, Zhao J, Tan JJ (2014) Study on seed germination characteristics of *Salviae miltiorrhizae* Bunge. Seed 33(4):82–85

Ho JH, Hong CY (2011) Salvianolic acids: small compounds with multiple mechanisms for cardiovascular protection. J Biomed Sci 18:30

Hou X, Shao F, Ma Y, Lu S (2013) The phenylalanine ammonia-lyase gene family in *Salvia miltiorrhiza*: genome-wide characterization, molecular cloning and expression analysis. Mol Biol Rep 40(7):4301–4310

Hur KY, Seo HJ, Kang ES, Kim SH, Song S, Kim EH, Lim S, Choi C, Heo JH, Hwang KC, Ahn CW, Cha BS, Jung M, Lee HC (2008) Therapeutic effect of magnesium lithospermate B on neointimal formation after balloon-induced vascular injury. Eur J Pharmacol 586:226–233

Jia C, Ma X, Qin G, Wei Z, Li X (2016) Advances on breeding technology of medicinal plants *Salvia miltiorrhiza* Bge. Med Res Edu 33(2):63–67

Jia Q, Zhu R, Tian Y, Chen B, Li R, Li L, Wang L, Che Y, Zhao D, Mo F, Gao S, Zhang D (2019) *Salvia miltiorrhiza* in diabetes: A review of its pharmacology, phytochemistry, and safety. Phytomedicine 58:152871

Jiang Z, Gao W, Huang L (2019) Tanshinones, critical pharmacological components in *Salvia miltiorrhiza*. Front Pharmacol 10:202

Kai G, Xu H, Zhou C, Liao P, Xiao J, Luo X, You L, Zhang L (2011) Metabolic engineering tanshinone biosynthetic pathway in *Salvia miltiorrhiza* hairy root cultures. Metab Eng 13(3):319–327

Lee CY, Agrawal DC, Wang CS, Yu SM, Chen JJW, Tsay HS (2008) T-DNA activation tagging as a tool to isolate *Salvia miltiorrhiza* transgenic lines for higher yields of tanshinones. Plant Med 74:780–786

Leng J, Chen MH, Zhou ZH, Lu YW, Wen XD, Yang J (2017) Triterpenoids-enriched extract from the aerial parts of *Salvia miltiorrhiza* regulates macrophage polarization and ameliorates insulin resistance in high-fat fed mice. Phytother Res 31(1):100–107

Li B, Cantino PD, Olmstead RG, Bramley GLC, Xiang CL, Ma ZH, Tan YH, Zhang DX (2016a) A large-scale chloroplast phylogeny of the Lamiaceae sheds new light on its subfamilial classification. Sci Rep 6:34343

Li B, Cui G, Shen G, Zhan Z, Huang L, Chen J, Qi X (2017a) Targeted mutagenesis in the medicinal plant *Salvia miltiorrhiza*. Sci Rep 7:43320

Li C, Li D, Li J, Shao F, Lu S (2017b) Characterization of the polyphenol oxidase gene family reveals a novel microRNA involved in posttranscriptional regulation of *PPOs* in *Salvia miltiorrhiza*. Sci Rep 7:44622

Li C, Li D, Shao F, Lu S (2015) Molecular cloning and expression analysis of *WRKY* transcription factor genes in *Salvia miltiorrhiza*. BMC Genom 16:200

Li C, Lu S (2014) Genome-wide characterization and comparative analysis of R2R3-MYB transcription factors shows the complexity of MYB-associated regulatory networks in *Salvia miltiorrhiza*. BMC Genom 15:277

Li CG, Sheng SJ, Pang EC, May B, Xue CC (2011) Plant density-dependent variations in bioactive markers and root yield in Australian-grown *Salvia miltiorrhiza* Bunge. Chem Biodivers 8(4):699–709

Li H, Liu J, Pei T, Bai Z, Han R, Liang Z (2019) Overexpression of *SmANS* enhances anthocyanin accumulation and alters phenolic acids content in *Salvia miltiorrhiza* and *Salvia miltiorrhiza* Bge f. *alba* Plantlets. Int J Mol Sci 20(9):2225

Li J, Li C, Lu S (2018a) Identification and characterization of the cytosine-5 DNA methyltransferase gene family in *Salvia miltiorrhiza*. PeerJ 6:e4461

Li J, Li C, Lu S (2018b) Systematic analysis of DEMETER-like DNA glycosylase genes shows lineage-specific Smi-miR7972 involved in SmDML1 regulation in *Salvia miltiorrhiza*. Sci Rep 8(1):7143

Li XL, Zhan XR, Li Y, Wang RX, Shao AJ (2016b) Biological characteristics of *Salvia miltiorrhiza* seed. Chin J Exp Trad Med Form 22(18):27–30

Li Y, Sun C, Luo HM, Li XW, Niu YY, Chen SL (2010) Transcriptome characterization for *Salvia miltiorrhiza* using 454 GS FLX. Yao Xue Xue Bao 45(4):524–529

Liu M, Ma Y, Du Q, Hou X, Wang M, Lu S (2019a) Functional analysis of polyprenyl diphosphate synthase genes involved in plastoquinone and ubiquinone biosynthesis in *Salvia miltiorrhiza*. Front Plant Sci 10:893

Liu X, Jin M, Zhang M, Li T, Sun S, Zhang J, Dai J, Wang Y (2019b) The application of combined 1H NMR-based metabolomics and transcriptomics techniques to explore phenolic acid biosynthesis in *Salvia miltiorrhiza* Bunge. J Pharm Biomed Anal 172:126–138

Liu Y, Yang SX, Cheng Y, Liu DQ, Zhang Y, Deng KJ, Zheng XL (2015) Production of herbicide-resistant medicinal plant *Salvia miltiorrhiza* transformed with the bar gene. Appl Biochem Biotechnol 177(7):1456–1465

Lu J, Song HP, Li P, Zhou P, Dong X, Chen J (2015) Screening of direct thrombin inhibitors from Radix Salviae Miltiorrhizae by a peak fractionation approach. J Pharm Biomed Anal 109:85–90

Luo H (2015) Liposoluble chemical constituents in danshen. In: Yan X (ed) Dan Shen (*Salvia miltiorrhiza*) in medicine vol 1. Biology and chemistry. People's Medical Publishing House, Beijing, China, and Springer, Dordrecht, Netherlands, pp. 119–209

Luo H, Zhu Y, Song J, Xu L, Sun C, Zhang X, Xu Y, He L, Sun W, Xu H, Wang B, Li X, Li C, Liu J, Chen S (2014) Transcriptional data mining of *Salvia miltiorrhiza* in response to methyl jasmonate to examine the mechanism of bioactive compound biosynthesis and regulation. Physiol Plant 152 (2):241–255

Ma Y, Yuan L, Wu B, Li X, Chen S, Lu S (2012) Genome-wide identification and characterization of novel genes involved in terpenoid biosynthesis in *Salvia miltiorrhiza*. J Exp Bot 63(7):2809–2823

Mei XD, Cao YF, Che YY, Li J, Shang ZP, Zhao WJ, Qiao YJ, Zhang JY (2019) Danshen: a phytochemical and pharmacological overview. Chin J Nat Med 17 (1):59–80

Ngo TM, Tran PT, Hoang LS, Lee JH, Min BS, Kim JA (2019) Diterpenoids isolated from the root of *Salvia miltiorrhiza* and their anti-inflammatory activity. Nat Prod Res. https://doi.org/10.1080/14786419.2019.1596098

Pan XG, He J, Wang XX, Zhang J, Ma BZ, Zhang WK, Xu JK (2018) An unusual racemic C_{12}-norabietane diterpene and a new abietane diterpene alkaloid from *Salvia miltiorrhiza* Bunge. Fitoterapia 125:240–244

Petersen M, Abdullah Y, Benner J, Eberle D, Gehlen K, Hücherig S, Janiak V, Kim KH, Sander M, Weitzel C, Wolters S (2009) Evolution of rosmarinic acid biosynthesis. Phytochemistry 70(15–16):1663–1679

Pharmacopoeia Commission of People's Republic of China (2015) Chinese pharmacopoeia 2015 edition, vol 1. China Medical Science Press, Beijing

Qian L, Liang ZS, Wang JR, Xu WH (2009) Essential oil composition of *Salvia miltiorrhiza* flower. Food Chem 113(2):592–594

Qiu D, Song J, Ma X, Qi J, Zhang Y (2004) Studies on the mass culture of the hairy roots of *Salvia miltiorrhiza* with bioreactor. Mol Plant Breeding 2:699–703

Ravipati AS, Zhang L, Koyyalamudi SR, Jeong SC, Reddy N, Bartlett J, Smith PT, Shanmugam K, Münch G, Wu MJ, Satyanarayanan M, Vysetti B (2012) Antioxidant and anti-inflammatory activities of selected Chinese medicinal plants and their relation with antioxidant content. BMC Complement Altern Med 12:173

Shao F, Lu S (2013) Genome-wide identification, molecular cloning, expression profiling and posttranscriptional regulation analysis of the *Argonaute* gene family in *Salvia miltiorrhiza*, an emerging model medicinal plant. BMC Genom 14:512

Shao F, Lu S (2014) Identification, molecular cloning and expression analysis of five RNA-dependent RNA polymerase genes in *Salvia miltiorrhiza*. PLoS ONE 9(4):e95117

Shao F, Qiu D, Lu S (2015) Comparative analysis of the *Dicer-like* gene family reveals loss of miR162 target site in *SmDCL1* from *Salvia miltiorrhiza*. Sci Rep 5:9891

Sheng SJ, Pang EC, Xue CC, Li CG (2009) Seasonal variations in bioactive marker contents in Australian-grown *Salvia miltiorrhiza* roots. Chem Biodivers 6 (4):551–560

Shi M, Huang F, Deng C, Wang Y, Kai G (2018) Bioactivities, biosynthesis and biotechnological production of phenolic acids in *Salvia miltiorrhiza*. Crit Rev Food Sci Nutr 10:1–12

Shi M, Luo X, Ju G, Li L, Huang S, Zhang T, Wang H, Kai G (2016) Enhanced diterpene tanshinone accumulation and bioactivity of transgenic *Salvia miltiorrhiza* hairy roots by pathway engineering. J Agric Food Chem 64(12):2523–2530

Shi M, Luo X, Ju G, Yu X, Hao X, Huang Q, Xiao J, Cui L, Kai G (2014) Increased accumulation of the cardio-cerebrovascular disease treatment drug tanshinone in *Salvia miltiorrhiza* hairy roots by the enzymes 3-hydroxy-3-methylglutaryl CoA reductase and 1-deoxy-D-xylulose 5-phosphate reductoisomerase. Funct Integr Genomics 14(3):603–615

Song J, Wang Z (2011) RNAi-mediated suppression of the phenylalanine ammonia-lyase gene in *Salvia miltiorrhiza* causes abnormal phenotypes and a reduction in rosmarinic acid biosynthesis. J Plant Res 124 (1):183–192

Song J, Zhang Y, Qi J (1998) Biotechnology of *Salvia miltiorrhiza*. Nat Prod Res Dev 11:86–89

Song JY, Luo HM, Li CF, Sun C, Xu J, Chen SL (2013) *Salvia miltiorrhiza* as medicinal model plant. Yao Xue Xue Bao 48(7):1099–1106

Song JY, Qi JJ, Lei HT, Ren CL, Fu J, Zhang YL (2000) Effect of *Armillaria mellea* elicitor on accumulation of tanshinones in crown gall cultures of *Salvia miltiorrhiza*. Acta Bot Sin 42:316–320

Song Z, Guo L, Liu T, Lin C, Wang J, Li X (2017) Comparative RNA-sequence transcriptome analysis of phenolic acid metabolism in *Salvia miltiorrhiza*, a traditional Chinese medicine model plant. Int J Genomics 2017:9364594

Steinkamp-Fenske K, Bollinger L, Völler N, Xu H, Yao Y, Bauer R, Förstermann U, Li H (2007) Ursolic acid from the Chinese herb danshen (*Salvia miltiorrhiza* L.) upregulates eNOS and downregulates Nox4 expression in human endothelial cells. Atherosclerosis 195(1): e104–111

Su CY, Ming QL, Rahman K, Han T, Qin LP (2015) *Salvia miltiorrhiza*: Traditional medicinal uses, chemistry, and pharmacology. Chin J Nat Med 13(3):163–182

Tung NH, Nakajima K, Uto T, Hai NT, Long DD, Ohta T, Oiso S, Kariyazono H, Shoyama Y (2017) Bioactive triterpenes from the root of *Salvia miltiorrhiza* Bunge. Phytother Res 31(9):1457–1460

Walker JB, Sytsma KJ (2007) Staminal evolution in the genus *Salvia* (Lamiaceae): molecular phylogenetic evidence for multiple origins of the staminal lever. Ann Bot 100(2):375–391

Walker JB, Sytsma KJ, Treutlein J, Wink M (2004) *Salvia* (Lamiaceae) is not monophyletic: implications for the systematics, radiation, and ecological specializations of *Salvia* and tribe Mentheae. Am J Bot 91(7):1115–1125

Wang B, Sun W, Li Q, Li Y, Luo H, Song J, Sun C, Qian J, Zhu Y, Hayward A, Xu H, Chen S (2015) Genome-wide identification of phenolic acid biosynthetic genes in *Salvia miltiorrhiza*. Planta 241(3):711–725

Wang H, Gao X, Zhang B (2005) Tanshinone: an inhibitor of proliferation of vascular smooth muscle cells. J Ethnopharmacol 99:93–98

Wang H, Wei T, Wang X, Zhang L, Yang M, Chen L, Song W, Wang C, Chen C (2018) Transcriptome analyses from mutant *Salvia miltiorrhiza* reveals important roles for *SmGASA4* during plant development. Int J Mol Sci 19(7):E2088

Wang J, Xu J, Gong X, Yang M, Zhang C, Li M (2019) Biosynthesis, chemistry, and pharmacology of polyphenols from Chinese *Salvia* species: a review. Molecules 24:155

Wang K, Luo Q, Cheng H (1998) Study on production of allochtonic formations in callus cells of *Salvia miltiorrhiza* Bge. China J Chin Mater Med 23:592–594

Wang L, Ma R, Liu C, Liu H, Zhu R, Guo S, Tang M, Li Y, Niu J, Fu M, Gao S, Zhang D (2017a) *Salvia miltiorrhiza*: a potential red light to the development of cardiovascular diseases. Curr Pharm Des 23 (7):1077–1097

Wang M, Deng Y, Shao F, Liu M, Pang Y, Li C, Lu S (2017b) *ARGONAUTE* genes in *Salvia miltiorrhiza*: identification, characterization, and genetic transformation. Methods Mol Biol 1640:173–189

Wang QH, Chen AH, Zhang BL (2009) *Salviae miltiorrhiza*: a model organism for Chinese traditional medicine genomic studies. Acta Chin Med Pharmacol 37(4):1–3

Wang Z (2018) Cultivation techniques of medicinal *Salvia miltiorrhiza*. Appl Technol Rural Areas 2018 (02):23–25

Wei T, Deng K, Wang H, Zhang L, Wang C, Song W, Zhang Y, Chen C (2018) Comparative transcriptome analyses reveal potential mechanisms of enhanced drought tolerance in transgenic *Salvia miltiorrhiza* plants expressing *AtDREB1A* from *Arabidopsis*. Int J Mol Sci 19(3):E827

Wei WJ, Zhou PP, Lin CJ, Wang WF, Li Y, Gao K (2017) Diterpenoids from *Salvia miltiorrhiza* and their immune-modulating activity. J Agric Food Chem 65 (29):5985–5993

Wei YK, Wang Q, Huang YB (2015) Species diversity and distribution of *Salvia* (Lamiaceae). Bio Sci 23 (1):3–10

Wenping H, Yuan Z, Jie S, Lijun Z, Zhezhi W (2011) De novo transcriptome sequencing in *Salvia miltiorrhiza* to identify genes involved in the biosynthesis of active ingredients. Genomics 98(4):272–279

Wester P, Claßen-Bockhoff R (2007) Floral diversity and pollen transfer mechanisms in bird-pollinated *Salvia* species. Ann Bot 100(2):401–421

Will M, Claßen-Bockhoff R (2014) Why Africa matters: evolution of Old World *Salvia* (Lamiaceae) in Africa. Ann Bot 114(1):61–83

Will M, Claßen-Bockhoff R (2017) Time to split *Salvia* s. l. (Lamiaceae)—new insights from old World *Salvia* phylogeny. Mol Phylogenet Evol 109:33–58

Xia H, Sun L, Lou H, Rahman MM (2014) Conversion of salvianolic acid B into salvianolic acid A in tissues of Radix *Salviae Miltiorrhizae* using high temperature, high pressure and high humidity. Phytomedicine 21:906–911

Xing B, Liang L, Liu L, Hou Z, Yang D, Yan K, Zhang X, Liang Z (2018) Overexpression of *SmbHLH148* induced biosynthesis of tanshinones as well as phenolic acids in *Salvia miltiorrhiza* hairy roots. Plant Cell Rep 37(12):1681–1692

Xu L (2002) Cultivation techniques of *Salvia miltiorrhiza*. Anhui Agric 10(7):29

Xu X, Jiang Q, Ma X, Ying Q, Shen B, Qian Y, Song H, Wang H (2014) Deep sequencing identifies tissue-specific microRNAs and their target genes involving in the biosynthesis of tanshinones in *Salvia miltiorrhiza*. PLoS ONE 9(11):e111679

Xu YY, Wan RZ, Lin YP, Yang L, Chen Y, Liu CX (2007) Recent advance on research and application of *Salvia miltiorrhiza*. Asian J Pharmacod Pharmacok 7 (2):99–130

Xu Z, Luo H, Ji A, Zhang X, Song J, Chen S (2016a) Global identification of the full-length transcripts and alternative splicing related to phenolic acid biosynthetic genes in *Salvia miltiorrhiza*. Front Plant Sci 7:100

Xu Z, Peters RJ, Weirather J, Luo H, Liao B, Zhang X, Zhu Y, Ji A, Zhang B, Hu S, Au KF, Song J, Chen S (2015) Full-length transcriptome sequences and splice variants obtained by a combination of sequencing platforms applied to different root tissues of *Salvia miltiorrhiza* and tanshinone biosynthesis. Plant J 82 (6):951–961

Xu H, Song J, Luo H, Zhang Y, Li Q, Zhu Y, Xu J, Li Y, Song C, Wang B, Sun W, Shen G, Zhang X, Qian J, Ji A, Xu Z, Luo X, He L, Li C, Sun C, Yan H, Cui G, Li X, Li X, Wei J, Liu J, Wang Y, Hayward A, Nelson D, Ning Z, Peters RJ, Qi X, Chen S (2016b) Analysis of the genome sequence of the medicinal plant *Salvia miltiorrhiza*. Mol Plant 6:949–952

Yan Y, Wang Z (2007) Genetic transformation of the medicinal plant *Salvia miltiorrhiza* by *Agrobacterium* tumefaciens-mediated method. Plant Cell Tiss Org Cult 88(2):175–184

Yang L, Ding G, Lin H, Cheng H, Kong Y, Wei Y, Fang X, Liu R, Wang L, Chen X, Yang C (2013) Transcriptome analysis of medicinal plant *Salvia miltiorrhiza* and identification of genes related to tanshinone biosynthesis. PLoS ONE 8(11):e80464

Yang N, Zhou W, Su J, Wang X, Li L, Wang L, Cao X, Wang Z (2017) Overexpression of *SmMYC2* increases the production of phenolic acids in *Salvia miltiorrhiza*. Front Plant Sci 8:1804

Zeng H, Su S, Xiang X, Sha X, Zhu Z, Wang Y, Guo S, Yan H, Qian D, Duan J (2017) Comparative analysis of the major chemical constituents in *Salvia miltiorrhiza* roots, stems, leaves and flowers during different growth periods by UPLC-TQ-MS/MS and HPLC-ELSD methods. Molecules 22(5):E771

Zhang G, Tian Y, Zhang J, Shu L, Yang S, Wang W, Sheng J, Dong Y, Chen W (2015a) Hybrid *de novo* genome assembly of the Chinese herbal plant danshen (*Salvia miltiorrhiza* Bunge). GigaScience 4:62

Zhang H, Jin W, Zhu X, Liu L, He Z, Yang S, Liang Z, Yan X, He Y, Liu Y (2016a) Identification and characterization of *Salvia miltiorrhizain* miRNAs in response to replanting disease. PLoS ONE 11(8): e0159905

Zhang J, Jin Q, Deng Y, Hou J, Wu W, Guo D (2017) New depsides from the roots of *Salvia miltiorrhiza* and their radical-scavenging capacity and protective effects against H_2O_2-induced H9c2 cells. Fitoterapia 121:46–52

Zhang L (2018) Cultivation techniques for high production of *Salvia miltiorrhiza*. Hebei Agric 2018(03): 9–10

Zhang L, Lu S (2017) Overview of medicinally important diterpenoids derived from plastids. Mini Rev Med Chem 17(12):988–1001

Zhang L, Wu B, Zhao D, Li C, Shao F, Lu S (2014) Genome-wide analysis and molecular dissection of the *SPL* gene family in *Salvia miltiorrhiza*. J Integr Plant Biol 56(1):38–50

Zhang Q, Chang Z, Yang J, Wang Q (2008) Antiatherogenic property of triterpenoids-enriched extract from the aerial parts of *Salvia miltiorrhiza*. Phytother Res 22:1040–1045

Zhang X, Dong J, Liu H, Wang J, Qi Y, Liang Z (2016b) Transcriptome sequencing in response to salicylic acid in *Salvia miltiorrhiza*. PLoS ONE 11(1):e0147849

Zhang X, Luo H, Xu Z, Zhu Y, Ji A, Song J, Chen S (2015b) Genome-wide characterisation and analysis of bHLH transcription factors related to tanshinone biosynthesis in *Salvia miltiorrhiza*. Sci Rep 5:11244

Zhang X, Ma Z, Liang Q, Tang X, Hu D, Liu C, Tan H, Xiao C, Zhang B, Wang Y, Gao Y (2015c) Tanshinone IIA exerts protective effects in a LCA-induced cholestatic liver model associated with participation of pregnane X receptor. J Ethnopharmacol 164:357–367

Zhang Y, Song J, Zhao B, Liu H (1995) Crown gall tissue culture and production of tanshinone in *Salvia miltiorrhiza*. Chin J Biotechnol 11(2):150–152

Zhang Y, Xu Z, Ji A, Luo H, Song J (2018) Genomic survey of *bZIP* transcription factor genes related to tanshinone biosynthesis in *Salvia miltiorrhiza*. Acta Pharm Sin B 8(2):295–305

Zhao GR, Zhang HM, Ye TX, Xiang ZJ, Yuan YJ, Guo ZX, Zhao LB (2008) Characterization of the radical scavenging and antioxidant activities of danshensu and salvianolic acid B. Food Chem Toxicol 46:73–81

Zhao SJ, Zhang JJ, Yang L, Wang ZT, Hu ZB (2011) Determination and biosynthesis of multiple salvianolic acids in hairy roots of *Salvia miltiorrhiza*. Yao Xue Xue Bao 46:1352–1356

Zhou L, Zuo Z, Chow MSS (2005) Danshen: an overview of its chemistry, pharmacology, pharmacokinetics, and clinical use. J Clin Pharmacol 45:1345–1359

Zhou W, Huang Q, Wu X, Zhou Z, Ding M, Shi M, Huang F, Li S, Wang Y, Kai G (2017) Comprehensive transcriptome profiling of *Salvia miltiorrhiza* for discovery of genes associated with the biosynthesis of tanshinones and phenolic acids. Sci Rep 7(1):10554

Salvia miltiorrhiza Resources, Cultivation, and Breeding

2

Chun Sui

Abstract

Salvia miltiorrhiza Bunge is the source species of Danshen listed in Chinese Pharmacopoeia (2015 edition). Currently, the germplasm of *S. miltiorrhiza* and its related species are collected and preserved in several universities, research institutes, and companies where projects on Danshen have been conducted or are being conducted. Attempts have been made to use molecular markers and DNA barcoding in evaluating the genetic diversity and species identification of medicinal Danshen materials. Different cultivation modes, such as direct seeding and root propagation, are used in different production areas in China, and the latter is more popular than the former. From growing seedlings, transplanting, field management, root harvesting, to primary processing and storage, experiments have been conducted by researchers to establish efficient standard operating procedures (SOPs). Elite cultivars are crucial in elevating farming economic values. Bulk selection, system selection, cross-breeding, and space mutation breeding are adopted to obtain new cultivars. With the need for excellent cultivated varieties and the advancement of the application of next-/third-generation sequencing technologies, the preservation, evaluation, and utilization of *S. miltiorrhiza* and its related species will go even further, and new groundbreaking cultivars and breeding techniques will emerge. The nationwide standardization project of seed quality, plantation, reproduction, and crude medicinal material and decoction should promote the Danshen production industry. The release of the genome sequence of *S. miltiorrhiza* and related transcriptome analysis would accelerate germplasm evaluation and utilization and promote the breeding of elite cultivars.

2.1 Introduction

Danshen (*Salvia miltiorrhiza* Bunge) is one of the most important herbs in China (Deng et al. 2016; Guo et al. 2014). It has gradually been known worldwide for its value in treating cardiovascular diseases (Hao and Xiao 2017; Imanshahidi and Hosseinzadeh 2006; Liu et al. 2016). With the increasing research on the molecular biology of medicinal plants, *S. miltiorrhiza* has been deemed a model medicinal plant (Song et al. 2013). The annual yield and average yield per hectare of Danshen in China

C. Sui (✉)
Key Laboratory of Bioactive Substances and Resources Utilization of Chinese Herbal Medicine, Ministry of Education and National Engineering Laboratory for Breeding of Endangered Medicinal Materials, Institute of Medicinal Plant Development (IMPLAD), Chinese Academy of Medical Sciences and Peking Union Medical College, Beijing 100193, China
e-mail: csui@implad.ac.cn

S. Lu (ed.), *The Salvia miltiorrhiza Genome*, Compendium of Plant Genomes,
https://doi.org/10.1007/978-3-030-24716-4_2

have been approximately 20,000 tons and 4500–6000 kg in recent years, respectively. Main production areas have formed with the long-term cultivation in several provinces of China, such as Shandong, Shaanxi, Henan, and Sichuan (Liu et al. 2019). Well-known brands and companies with Danshen have been formed. Few species of *Salvia Lamiaceae* are listed in some local pharmacopoeias of China and used as medicinal Danshen (Zhang 2016; Li and Xiao 2013). Those species are less explored and utilized than *S. miltiorrhiza*. Exploring the germplasm of *S. miltiorrhiza* and its related species will improve the production performance of current cultivated varieties. In this chapter, natural and domestic Danshen resources, their current cultivation and cultivar breeding status, and future challenges in the genomic era are described.

2.2 Danshen Resources

According to the record in Chinese Pharmacopoeia (2015 edition), *S. miltiorrhiza* is the source species of medicinal Danshen. *S. miltiorrhiza* has two varieties; the original variety is *S. miltiorrhiza* var. *miltiorrhiza*, which has oddly pinnate compound leaves, ovate leaflets, or elliptic-ovate to broadly lanceolate leaves; and *S. miltiorrhiza* var. *charbonnelii* (H. Léveillé) C. Y. Wu with simple leaves, which sometimes have three-foliolate, blades, or orbicular-to-suborbicular leaflets. The original variety has been found growing in hillsides, streamsides, and grassy places in forests at 100–1300 m above sea level in the Chinese provinces of Anhui, Hebei, Henan, Hunan, Jiangsu, Shaanxi, Shandong, Shanxi, and Zhejiang. It is also distributed in Japan. The other variety has been found growing in the hillsides and grassy places of Hebei, Henan, Hubei, and Shanxi. The original variety has two forms, namely the original form (i.e., f. *miltiorrhiza*) and white form (i.e., f. *alba* C. Y. Wu et H. W. Li in Addenda 582). The difference between them lies in the color of flowers. The former's flower is purple, and the latter's is white and found in Shandong. Figure 2.1 shows the purple- and white-flowered *S. miltiorrhiza* and the light-purple-flowered breeding line, whose color is between purple and white.

In addition to the *S. miltiorrhiza* recorded in the national pharmacopoeia, other species of *Salvia* were recorded in several local medicinal material standards. For example, *Salvia yunnanensis*, with the common name "Dian Danshen," was recorded in Yunnan Drug Standard and Quality Standard of Traditional Chinese Medicinal Materials in Guizhou Province; *Salvia prezwalskii*, with the common name "Gansu Danshen," was recorded in Chinese Herbal Medicine, Qinghai Drug Standard, and Quality Standard of Traditional Chinese Medicinal Materials in Gansu Province; *S. bowleyana*, which is called Danshen, was recorded in Standard of Traditional Chinese Medicinal Materials in Zhejiang Province; and *S. castanea* f. *tomentosa*, with the common name "Zang Danshen," was approved by the Food and Drug Administration of Tibet Autonomous Region as Danshen (Wang et al. 2016; Zhang 2016). Table 2.1 shows the distribution and action of some *Salvia* species with medicinal usage.

DNA barcoding and molecular markers have been studied for identifying different germplasm resources and exploring genetic diversities (Song et al. 2010; Wang et al. 2011; Wang et al. 2013a; Zhang et al. 2012). *Salvia shandongensis*, *S. miltiorrhiza*, and *S. miltiorrhiza* f. *alba* can be identified by using the plastid psbA-trnH intergenic region (Li et al. 2013). Different regions and their combinations of rbcL, matK, trnlL-F, and psbA-trnH from the chloroplast genome and ITS, ITS1, and ITS2 from the nuclear genome were screened for authenticating the genus *Salvia* using sequences from 27 *Salvia* species. ITS1 is reportedly a powerful barcode for identifying *Salvia* species, especially *S. miltiorrhiza* (Wang et al. 2013a). In addition, the complete chloroplast genome of *S. miltiorrhiza* was sequenced, which is significant for developing new barcodes or markers (Qian et al. 2013).

The contents of tanshinones and phenolic acids, which are the two main groups of the medicinal components of Danshen, vary in different species and environments. The contents of

Fig. 2.1 Purple (**a**, **b**)-, white (**c**, d)-, and light-purple (**e**)-flowered *S. miltiorrhiza*; medicinal root (**f**), and seedlings for transplanting (**g**)

the two groups have been found to be independent variables, and one is usually higher than the other in most species. In addition, tanshinone IIA and salvianolic acid B were selected as the biomarkers in Chinese Pharmacopoeia (2015 edition), whereas the rosmarinic acid content is mostly abundant in the roots of some species (Li and Xiao 2013). Besides their ethnomedicinal usage, *Salvia* plants have an extensive application potential in ornamental and aromatic industries. Currently, the effective protection of wild *Salvia* resources is a great challenge to scientists because it is the essential foundation of future development and utilization (Wu et al. 2011). The main departments in China that collected, evaluated, and preserved Danshen resources are listed in Table 2.2. Communication and opening to the public will promote the application of existing resources.

2.3 Danshen Distribution and Cultivation

2.3.1 Wild and Cultivated Distribution

Danshen has been recognized as an extensively dispersed species since the Song Dynasty and recorded to be distributed in 19 provinces and districts, including Liaoning, Hebei, Henan, Shandong, Shanxi, Jiangsu, Anhui, Zhejiang,

Table 2.1 Some species of *Salvia* with medicinal usage

Original plants	Growing environment	Actions
Salvia miltiorrhiza	Sichuan, Shanxi, Shaanxi; altitude: 120–1300 m	Chest *bi* disorder, heart pain, pain in the epigastrium and abdomen, hypochondriac pain, painful heat *bi* disorder, insomnia caused by vexation, menstrual irregularities, dysmenorrhea, amenorrhea, sore, ulcer, swelling, and pain
Salvia aerea	Southwest Sichuan, northwest Yunnan, northeast and southwest Guizhou; altitude: 2550–3300 m	High fever in warm disease, loss of consciousness. To activate blood, regulate menstruation. For irregular menstruation, kidney deficiency and low back pain
Salvia bowleyana	Zhejiang, Hunan, Jiangxi; altitude: 30–960 m	Promoting blood circulation to dispel blood and stasis, regulate menstruation, and relieve pain. For chest cramps, irregular menstruation
Salvia castana	Sichuan, Yunnan; altitude: 2100–4100 m	Promoting blood circulation to dispel blood, menstrual irregularities. For bruising, insomnia, etc.
Salvia digitaloides	Sichuan, Yunnan; altitude: 2100–3500 m	Qi deficiency with lack of strength, regulate menstruation, and relieve pain. For irregular menstruation, malignant sore
Salvia evanaiana	Sichuan, Yunnan; altitude: 3300-4200 m	Promoting blood circulation to dispel blood, disperse swelling, and relieve pain. For irregular menstruation, bruises
Salvia flava	Northwest Yunnan, southwest Sichuan; altitude: 2500–4000 m	To cool the blood and resolve stasis. For irregular menstruation, congestion, and swelling
Salvia maximowicziana	West Hubei, Sichuan, northeast Yunnan, south Shaanxi; altitude: 1800–3450 m	Dispel wind, remove dampness
Salvia prezwalskii	West and northwest Yunnan; altitude: 2100–4050 m	To activate blood, eliminate stasis, nourish blood, and tranquilize mind. For irregular menstruation, upset, and insomnia
Salvia yunnanensis	East Yunnan, central to west and southwest Sichuan, west Guizhou; altitude: 1800–2600 m	To activate blood, eliminate stasis, tranquilize mind, and calm the heart. For angina pectoris, dysmenorrhea
Salvia plebeia	Sichuan, Guizhou, Yunnan; altitude: under 2000 m	To clear heat remove toxin, remove toxin, cool blood, prostration urination. For sore throat, bronchitis
Salvia chinensis	South Jiangsu, south Anhui, Sichuan; altitude: 120–500 m	To activate blood, eliminate stasis, clear heat, drain dampness. For irregular menstruation, rheumatism, and bone pain
Salvia prionitis	Zhejiang, Anhui, Jiangxi; altitude: 100–800 m	To cool wind and eliminate stasis, clear heat, drain dampness. For colds, fever, vomiting, fetal leakage
Salvia roborowskii	Southwest Gansu, west Sichuan, southwest Qinghai, etc; altitude: 2500–3700 m	To clear the leaver, improve vision. For red eyes and swelling, aponeurosis
Salvia japonica	Zhejiang, south Anhui, Jiangsu; altitude: 220–1100 m	To clear heat, drain dampness, active blood, and regulate menstruation. For jaundice, red squat, wet under the tropics
Salvia cavaleriei var. *simplicifolia*	Sichuan, Hunan, Jiangxi; altitude: 460–2700 m	Qi deficiency with lack of strength, regulate menstruation, and relieve pain. For irregular menstruation, traumatic bleeding
Salvia substolonifera	Zhejiang, Fujian, Hunan; altitude: 40–950 m	To clear heat, resolve phlegm, to tonify the kidney and regulate menstruation. For lung heat cough, menorrhagia, etc.

Table 2.2 Main Chinese departments and their Danshen resources which were preserved and/or created

Institute or any type of organization	Germplasm		Achievement appraisal and evaluations	
	Collected (number)	Selected or created (type and number)	Release time	Achievement title
Institute of Medicinal Plant Development, Chinese Academy of Medical Sciences and Peking Union Medical College (IMPLAD-CAMS and PUMC)	93	27 (hybrid lines)	–	–
	–	17 (white flower Danshen autotetraploid line); 180 (triploid line F1 plant)	2010-3	Breeding and promotion of new triploid varieties of *Salvia miltiorrhiza*
Nankai University (NKU)	48	16 (male sterile line)	2011-12-24	–
Medicinal Plant Research Center, Shandong Academy of Agricultural Sciences (MPRC-SAAS)	More than 10,000 including cultivated, genetically related species, wild types, natural mutants, and spaceflight mutants-	2016-10-10	Research and application of key technologies system of *Salvia miltiorrhiza* resources and quality control	
Zhejiang Sci-Tech University (ZSTU)	–	3 lines	2005	Breeding and industrialization development of white flower types of *Salvia miltiorrhiza*
Laiwu Vegetable Research Institute, Shandong Province (LVRI-SD)	48	–	2012	Identification, evaluation, and innovative utilization of germplasm resources of *Salvia miltiorrhiza*
Institute of Agro-Food Science and Technology, Shandong Academy of Agricultural Sciences (IAST-SAAS)	500 accessions (48 species including 6 varieties and 1 variant)	–	2016	Research and application of key technologies in modern industrial Chain of *Salvia miltiorrhiza*
Sichuan Agricultural University (SAU)	More than 50 species of *Salvia*	–	2015	Metabolic regulation of active components in *Salvia*
Shanghai Chen Shan Botanical Garden	–	4 (excellent variation types)	2011	Research and application of key techniques for high-yield cultivation of genuine medicinal material Salvia miltiorrhiza in Henan
Henan Agricultural University (HAU)	70 (including 32 with diverse characters)	–	2009	Breeding and application of excellent varieties of *S. Miltiorrhiza*

(continued)

Table 2.2 (continued)

Institute or any type of organization	Germplasm		Achievement appraisal and evaluations	
	Collected (number)	Selected or created (type and number)	Release time	Achievement title
Institute of Economic Crops, Hebei Academy of Agricultural and Forestry Sciences	56 (including 12 different types)	1 excellent type	2006	Study on standardized production technology of *Salvia miltiorrhiza*
Shijiazhuang Institute of Herbal and Medicinal Botany				

Note The data on the resources were obtained from achievement appraisal and evaluation reports. Only the first accomplished units are listed. Other contents, except resources and certificated varieties, are excluded

Jiangxi, Fujian, Hubei, Hunan, Guangdong, Guangxi, Shaanxi, Ningxia, Gansu, Sichuan, and Guizhou (Xu 1992). The distribution of wild and cultivated locations in China is shown in Fig. 2.2. The cultivated area is basically identical to a wild distribution.

According to herbal textual research, the *S. miltiorrhiza* product has been gradually developing in China from north to west and then from west to southwest for the past 2000 years. At the earliest, it spread from Tai Mountain and Linyi District of Shandong and Tongbai Mountain of Henan to other areas. In the Song Dynasty, it extended to Shaanxi, Shanxi, and Suizhou of Hubei. In the Ming Dynasty, Suizhou was emphasized. In the Qing Dynasty, cultivated *S. miltiorrhiza* appeared in Sichuan. *S. miltiorrhiza* plants from Anhui and Sichuan were considered the best in the early 1900. After the founding of the People' Republic of China, those from Sichuan were considered genuine (Zhan et al. 2016). Ten Danshen production provinces are generally classified into three categories, namely the first is Shandong producing area, the second is Henan and Shanxi producing area, and the third is Anhui, Hebei, Sichuan, Hubei, Shaanxi, Zhejiang, and Jiansu producing area, according to the comprehensive literature data analysis of the pharmacognostic characteristics and chemical component of crude drug. For Danshen from provinces in the third category, the characteristic values of the category are generally lower than those of the other two categories and only some individual indicators are higher (Liu et al. 2019).

Climatic factors control the geospatial distribution of the active ingredients of Danshen. The main climatic factors include air temperature, precipitation, atmospheric vapor pressure, and sunshine duration. The warming in high-latitude regions can cause the continued northward expansion of planting areas suitable for *S. miltiorrhiza* (Zhang et al. 2019).

2.3.2 Cultivation Procedures

Before the Qing Dynasty, Danshen was mainly wild, and the cultivation of Danshen was recorded in the Qing Dynasty. According to Zhongjiang County History of Medicine, "recorded in Kangxi Records (written in 1715 AD), the production of Danshen in Zhongjiang had reached a certain scale at that time." This indicates that Danshen has been cultivated artificially since the middle of Qing Dynasty (Zhan et al. 2016).

S. miltiorrhiza, a perennial herb, usually grows in temperate climate with sufficient sunlight in air and wet environments. Wild *S. miltiorrhiza* grows at hillsides, streamsides, and forests in China and Japan. For cultivation, an area with annual temperature and relative

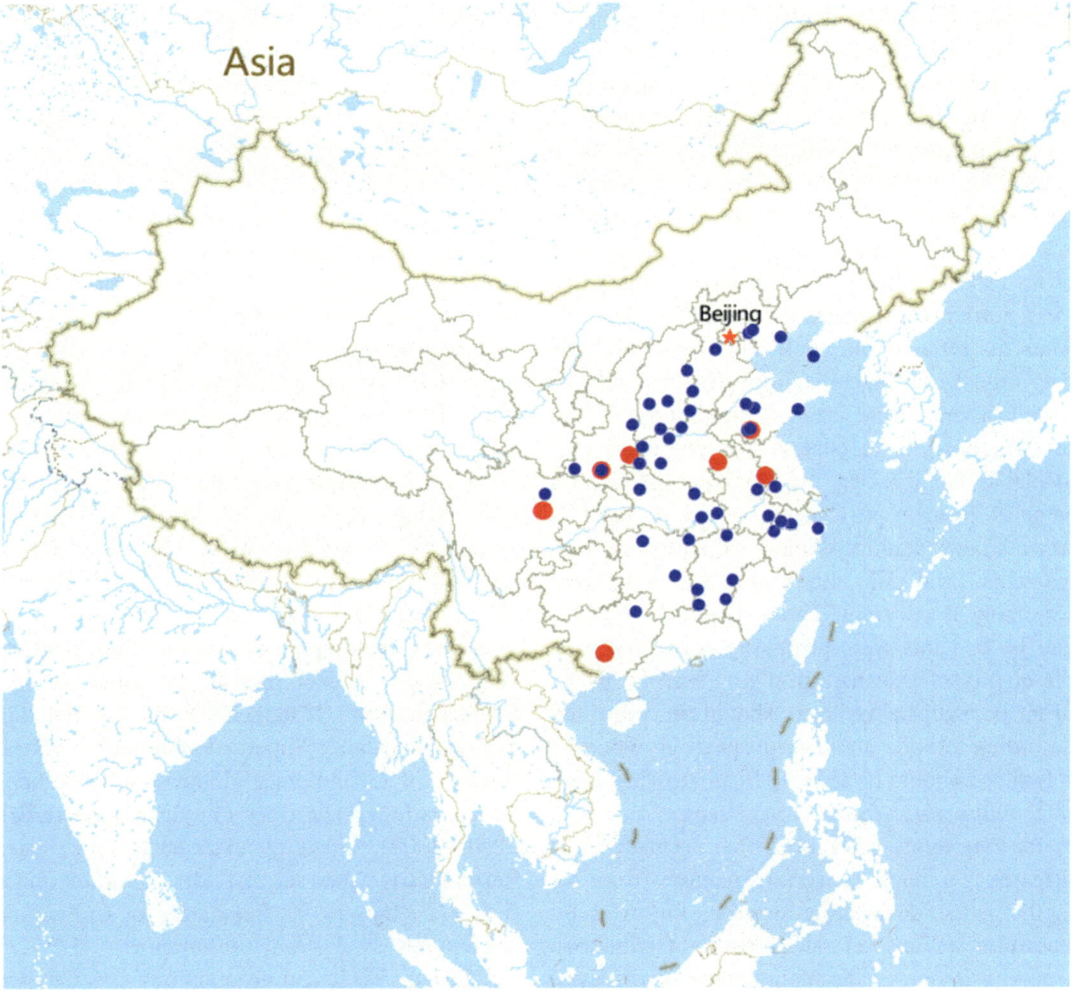

Fig. 2.2 Distribution of wild (blue dots) and cultivated (red dots) *S. miltiorrhiza* in China

humidity averages of 17.1 °C and 77%, respectively, is the most suitable for growing. *S. miltiorrhiza* adapts to the pH value of soil and can grow well in neutral, slightly acidic, and slightly alkaline soil. However, fertile sandy loam is preferred. The degree of soil weathering is closely related to the yield of Danshen, and high terrain and good drainage are vital to its cultivation. Seed sowing and root segment propagation are the two main methods of Danshen cultivation. The optimal germination temperature of seeds is 20–25 °C. At the appropriate temperature, the germination potential and rate are higher under full light than those without light, and germination could be conducted in the wide pH range of 2–10 (Shan et al. 2013). In addition, alternately changing the temperature between 15 °C/25 °C and 20 °C/30 °C promotes germination, while high light intensity inhibits germination (Li et al. 2016). Seeds treated with 0.5–1.0% hydrogen peroxide for 12 h or 600 mg L^{-1} gibberellin for 24 h show elevated germination rate, and the effect of gibberellin is better than that of hydrogen peroxide (Wu and Niu 2017).

Different cultivation modes were compared. Compared with traditional cultivation methods, the plastic film mulching method increases the fresh and dry weights from 14.68 to 48.62%. By maintaining the soil moisture content, the photosynthetic efficiency and yield of Danshen are improved (Yang et al. 2016). However, mild drought could significantly increase the shoot and root biomasses, the accumulation of medicinal active ingredients, and the trace elements in root. Experimentally, the suitable soil moisture content was 55–60%, and the ridge culture mode was suggested (Liu et al. 2011). Generally, the key period of fertilizer demand for *S. miltiorrhiza* is 90–110 days after seedling transplanting when more nutrients are needed for rhizome expansion. Thus, it is the nutrient-efficient period and the best time to fertilize. Additionally, spraying of trace elements could increase the root yield. The investigation of the content and distribution of available elements in the 26 rhizosphere soil samples from 8 provinces growing *S. miltiorrhiza* shows that insufficient N and Mn were available, and all tested soil samples were extremely B deficient. The available N, B, Mn, and Fe contents were positively correlated with the quality of Danshen, implying that the usage of N, B, and Mn fertilizers should be controlled according to different stages of plant growth, and P fertilizer should be reduced in all growing areas of *S. miltiorrhiza* (Shen et al. 2016).

In long-term cultivation, the occurrence of diseases is a major concern of planters. Root rot is the worst devastating disease, followed by Fusarium (Wang et al. 2018a). Continuous cropping is one of the main causes of diseases, including root rot and nematode. Meloidogyne can kill approximately 80% of plants when it occurs. Moreover, continuous cropping inhibits plant development and root elongation. Thus, rotation is usually adopted in Danshen production areas along with soil sogginess prevention and the timely removal of diseased plants.

The roots of *S. miltiorrhiza* are generally harvested in October or November, at least one year after seedling transplant. Differently aged roots have altered bioactive ingredients profiles (Wang et al. 2013b). Harvested roots can be dried by different means, such as drying under the sun, drying in the shade, and drying by baking at different temperatures. Drying in the shade or at 40 °C economically retains the most active components (Wang et al. 2017). In addition, the effects of various temperature drying modes (i.e., first at a low temperature of 30 or 40 °C and then high temperature at 60, 70, or 80 °C), using an air-dry oven on the content of active components, have been investigated. Results revealed that variable temperature drying can significantly affect the contents of active components in roots of *S. miltiorrhiza* f. *alba*. Compared with the traditional process of shade-drying, low-temperature drying can significantly increase the content of water-soluble active components and has a significant promotion effect on liposoluble components, such as tanshinone IIA, cryptotanshinone, and tanshinone I. Most importantly, the variable temperature drying mode can effectively shorten the process of drying and enables the industrialization and standardization of the drying process (Zhou et al. 2017).

Several regulations or standards on Danshen cultivation have been released by China National Standardization Management Committee or Provincial Quality Supervision Bureau, such as Product of Geographical Indication Fangcheng *Salvia miltiorrhiza* Bge. (Yu Danshen) (GB/T 22745-2008), Quality Standards for Seed Reproductive Material of Radix *S. miltiorrhizae* (DB51/T 1044-2010), Standards for Cultivation Techniques of *S. miltiorrhiza* (DB12/T 827-2018), Technical Regulations for Cultivation of *S. miltiorrhiza* (DB34/T 460-2017), and Regulations for Underforest Cultivation Techniques of *S. miltiorrhiza* (DB41/T 1217-2016). In addition, regular companies in China usually have their own technical specifications for the standardized production of *S. miltiorrhiza* Bunge (SOP) to regulate the production process.

2.3.3 Cultivation-Related Research Findings

Rhizosphere bacteria play a vital role in plant nutrition absorption, growth, and disease resistance. High-throughput sequencing technology is used to analyze the rhizosphere bacterial communities of Danshen (Jiang et al. 2019).

One species of Alternaria was found to markedly enhance *S. miltiorrhiza* root growth and active ingredient accumulation under greenhouse and field conditions (Zhou et al. 2018a). Works on this aspect can hopefully provide a reference for improving the soil environment and gains of Danshen cultivation.

Molecular markers were developed based on transcriptome sequences (Deng et al. 2009). A few of genetic linkage maps were constructed (Pan et al. 2016; Liang et al. 2016; Zong et al. 2015). For example, two genetic maps were constructed with near-isogenic lines of sterile male and fertile *S. miltiorrhiza*, and differential fragments and markers with potential functions were found (Liang et al. 2016).

Late embryogenesis abundant (LEA) proteins are a group of proteins associated with tolerance to water-related stress. A LEA gene was cloned from *S. miltiorrhiza* and revealed that SmLEA-transformed *S. miltiorrhiza* plants exhibit faster root elongation, have a lower malondialdehyde concentration, and experience a less rapid rate of water loss with greater superoxide dismutase activity and higher glutathione concentration than the control plants (Wu et al. 2014). A transcription factor gene, CBF1/DREB1B, was introduced into *S. miltiorrhiza*. Transgenic plants display tolerance to drought because of the activation of different downstream DREB/CBF genes by AtDREB1B. Furthermore, no growth inhibition was detected in transgenic plants (Wei et al. 2016).

Using second-generation Illumina sequencing, one gene, SmGASA4, was identified by the comparison of wide type and mutant *S. miltiorrhiza*. The results showed that SmGASA4 is positively regulated by Gibberellin and significantly enhances plant resistance to salt, drought, and paclobutrazol stress, thereby up-regulating the genes involved in salvianolic acid biosynthesis but inhibiting the expression of the genes involved in tanshinone biosynthesis (Wang et al. 2018b). The ability of small RNA to regulate the replanting disease was explored. Five miRNAs were predicted to be involved in plant responses to replanting disease (Zhang et al. 2016).

2.4 Danshen Breeding

The application of bred cultivars with combined virtues that adapt to diverse planting environments mostly symbolizes agricultural modernization and standardization. Continuous efforts are required for cultivar breeding. The *S. miltiorrhiza* germplasm is complex and diverse with rich wild and cultivated populations. Through decades, especially in recent years of breeding practice, some cultivars have been bred and reported (Wang et al. 2012; Wei and Yang 2018). Table 2.3 lists the basic information of cultivars, which have been identified by the provincial authoritative departments in China. The traits of cultivars include root yields, contents of medicinal active components, and stress resistance. Methods, such as system selection, bulk selection, cross-breeding, polyploidy breeding, and space mutation, were employed.

2.4.1 System and Bulk Selection

Given the simplicity and efficiency, bulk and system selections were used at the early stage of the cultivar breeding of *S. miltiorrhiza*. Several cultivars were bred by bulk or system selection (Table 2.3). Usually, in bulk selection, the germplasms collected from different locations are planted in the field, and the spontaneously pollinated seeds are harvested and sowed. The progeny plants were then evaluated on traits, such as botanic characteristics, yield, content of medicinal active components, and stress resistance (Wang et al. 2012). In system selection, low generations of self-pollination plants are crossed within the population in which the genetic consistency and prevention of inbreeding degradation are balanced. Although this kind of method could promote the rapid usage of existing wild or domesticated resources, inherent adverse agronomic traits are usually difficult to avoid, and some improvements are further needed. Especially, when sexual propagation is adopted, the traits of cultivars are impossible to retail.

Table 2.3 Selected or bred cultivars of *S. miltiorrhiza*

Cultivar or line	Characters	Identifier	Certification time (year)	Breeding method	Breeder
Zhongdanyaozhi No. 1	Erect; leaf margin purple; medium-diameter radical type roots; male fertile; 330 kg dried roots per mu in test	Jingpinjianyao2014030	2014	Cross-breeding between DPT101-4-2 male sterile line from Kushan Shandong and DT807-5-1 self-bred line from Shangluo Shaanxi	IMPLAD (Institute of Medicinal Plant Development
Zhongdanyaozhi No. 2	Semi-erect; leaf margin green; medium-diameter radical type roots; male fertile; 318 kg dried roots per mu in test	Jingpinjianyao2014031	2014	Cross-breeding between DPT101-4-2 male sterile line from Kushan Shandong and DT804-17-1 self-bred line from Shangluo Shaanxi	Institute of Medicinal Plant Development (IMPLAD)
Beidan No. 1	Buds purple, flowers purple, roots fine and high lignification; high tanshinone II A content; strong reproduction and germination abilities; 280 kg dried roots per mu in test.	Jingpinjianyao2009008	2009	System selection from Hebei landrace	IMPLAD (Institute of Medicinal Plant Development)
Jidan No. 1	Lower plant; wrinkle leaf surface; bright red root; not flowering at the first year; high resistance to root rot; 1123.5 kg fresh roots per mu in test; tanshinone II A 0.32%, salvianolic acid B 11.6%	JiS-SV-PF-012-2012	2012	System selection	Institute of Cash Crops, Hebei Academy of Agriculture and Forestry Sciences (ICC-HAAFS)
Jidan No. 2	Purple red root with white inside; wrinkle leaf surface; resistance to root rot; 486.35 kg dry roots per mu in test; tanshinone II A 0.26%, salvianolic acid B 7.02%	JiS-SV-PF-013-2012	2012	System selection	Institute of Cash Crops, Hebei Academy of Agriculture and Forestry Sciences (ICC-HAAFS)
Jidan No. 3	Flat leaf surface; bright red root; resistance to root rot; 393.9 kg dry roots per mu in test; tanshinone II A 0.24%, salvianolic acid B 7.33%	JiS-SV-PF-019-2012	2012	System selection	Institute of Cash Crops, Hebei Academy of Agriculture and Forestry Sciences (ICC-HAAFS)

(continued)

Table 2.3 (continued)

Cultivar or line	Characters	Identifier	Certification time (year)	Breeding method	Breeder
Danza No. 1	Flat leaf surface; bright red root; 375.3 kg dry roots per mu in test; tanshinone II A 0.25%, salvianolic acid B 8.10%	JiS-SV-SM-030-2014	2014	Cross-breeding with Jidan No. 2 (D0540) as female parent and tissue cultured Danshen plant (D0501) as male parent	Institute of Cash Crops, Hebei Academy of Agriculture and Forestry Sciences (ICC-HAAFS)
Danza No. 2	Winkle leaf surface; bright red root; 371.9 kg dry roots per mu in test; tanshinone II A 0.29%, salvianolic acid B 11.7%	JiS-SV-PF-031-2014	2014	Cross-breeding with Jidan No. 2 (D0540) as female parent and Jidan No. 1 as male parent	Institute of Cash Crops, Hebei Academy of Agriculture and Forestry Sciences (ICC-HAAFS)
Ludanshen No. 1	Erect; Flowering late; 355.8 kg dried roots per mu in demonstration and extension project in hillside wasteland; tanshinone II A 0.52%, salvianolic acid B 8.2%	–	2011	Bulk selection with landrace of Mengyin Shandong	Institute of Agro-Food Science and Technology, Shandong Academy of Agricultural Sciences (IAST-SAAS)
Luyuandanshen No. 1	Compact, erect, and tall; fewer branch roots, suitable for processing slices; 524 kg dried roots per mu in production test; tanshinone II A 0.39%, salvianolic acid B 7.3%	Lunongshenzi2013061hao	2013	Space mutation breeding with landrace seeds of Lv County Shandong	Institute of Agro-Food Science and Technology, Shandong Academy of Agricultural Sciences (IAST-SAAS)
Chuandanshen No. 1	Creeping plant; coreless root section; 895.61 kg fresh roots per mu in production test; tanshinone II A 0.25%, salvianolic acid B 8.0%	Chuanshenyao2011002	2011	System selection	Sichuan Agricultural University (SAU)
Zhongdan No. 1	Erect quadrilateral and multi-branched stem; slender cores in the center of root; 227.8 kg dried roots per mu in production test; tanshinone II A 0.43%, salvianolic acid B 10.6%	Chuanshenyao2012003	2012	System selection from Zhongjiang landrace	Sichuan Academy of Chinese Medicine Sciences (SACMS)
Hangdan No. 1	Dwarf plant; Winkle leaf surface; more branch root number; 560 kg dried roots per mu in production test; tanshinone II A 0.56%, salvianolic acid B 6.81%	XPD010-2012	2012	Selected from seedlings of Bozhou market	Hunan University of Traditional Chinese Medicine (HUCM)

(continued)

Table 2.3 (continued)

Cultivar or line	Characters	Identifier	Certification time (year)	Breeding method	Breeder
Danshen No. 1	Seedless; less fibrous root; about 1700 kg fresh roots per mu in production test; tanshinone II A 0.43%, salvianolic acid B 5.4%	Wanpinjiandengzidi1306013	2013	Selected from wild-type fine individual	Horticultural Research Institute of Anhui Academy of Agricultural Sciences (HRI-AAAS); Anhui Guotai Pharmaceutical Co., Ltd.
Tiandan No. 1	About 102 g Dry root per plant; no root rot disease found and wilt disease average incidence rate <0.16%; about 960 kg fresh roots per mu; tanshinone II A > 0.48%, salvianolic acid B > 7.5%	Shanyaocaidengzi2011001	2011	System selection from plant population with purple flower	Shaanxi Tianshili Plant Pharmaceutical Co., Ltd. and Northwest University of Agriculture and Forestry Science and Technology (NUAF)
Tiandan No. 2	278.45 kg dried roots per mu in variety comparison test; tanshinone II A 0.56%, salvianolic acid B 7.4%	Shanyaocaidengzi2011002	2011	Cross-breeding with male sterile line SH-B-18 as female parent and A-107 as male parent	Northwest University of Agriculture and Forestry Science and Technology (NUAF) and Shaanxi Tianshili Plant Pharmaceutical Co., Ltd.
Shannongdanshen No. 1	Compact plant; erect stem; High resistance to root knot nematode, waterlogging and barrenness; 435 kg dried roots per mu in production test; tanshinone II A 0.42%, salvianolic acid B 9.5%	Lunongshen2015081	2015	System selection from Tai mountain wild-type population	Shandong Agricultural University (SAU)

2.4.2 Cross-breeding

Cross-breeding is one of the effective methods of creating elite cultivars. Based on the successful experience in crop cross-breeding, researchers have conducted related works on Danshen. Before 2005, a male sterile individual plant was found and cultivated as a line for cross-breeding (Shu et al. 2005). The sterility characteristics were explored as the basis of cross-breeding (Song et al. 2009; Zhong et al. 2010). GGE-biplot was applied to analyze the stability of yield and quality traits of hybrid lines (Chen et al. 2017). Two hybrid cultivars of *S. miltiorrhiza* were reported with high-yield and active compounds content derived from male sterile and self-pollinating lines (Chen et al. 2016). Selecting a continuous cropping tolerant cultivar is one of the effective countermeasures to continuous cropping. Hybrids were obtained by crossing between high-generation self-pollinating materials with artificial male removal and assisted pollination. Seven lines with good performance in a three-year continuous cropping soil were screened from 50 hybrid lines (Ni et al. 2017).

2.4.3 Polyploidy Breeding

Polyploidy breeding is one of the effective methods of creating new varieties and has been widely used in many crops and vegetables. Owing to chromosome addition, the vegetative organs of polyploid plants often exhibit "giantism," showing larger roots, stems, leaves, flowers, and fruits than normal diploid plants. The medicinal parts of many medicinal plants are roots or whole overground vegetative organs. Thus, polyploidy varieties imply the increase of yield and profit for most of medicinal plant species. Danshen is usually diploid with $2n = 2x = 16$, although a few natural mutants in wild-type and cultivated populations were found with $2n = 4x = 32$ (Zhao et al. 2006). Chromosome doubling has been shown to enhance biomass and the production of some medicinal active components (Chen et al. 2018). Triploid, autotetraploid, and allotetraploid *S. miltiorrhiza* have been created in some laboratories (Chen and Li 2009; Fang et al. 2011; Li and Chen 2012; Liu et al. 2009). Meanwhile, the utilization of polyploidy is rare in Danshen production.

2.4.4 Space Mutation

The botanic characteristics, medicinal quality, and chemical fingerprint of the plants derived from the dried seeds of *S. miltiorrhiza* carried by the Shenzhou 7 spaceship of China in 2008 for 68 h were investigated. The outer space environment brought inheritable variations, which are a rich germplasm bank for the new cultivar selection and breeding of *S. miltiorrhiza* (Yang et al. 2013).

A total of 28 lines of second-generation *S. miltiorrhiza* responding to a spaceflight environment were surveyed on their root yields and the active constituents' characterizations, which varied remarkably among the 28 lines. Compared with untreated plants, the lines with much increased yield and active constituents could be selected among 28 lines. The positive relationships lie between the contents of tanshinone I and tanshinone IIA ($r = 0.790$, $p < 0.01$), and rosmarinic acid and salvianolic acid B ($r = 0.728$, $p < 0.01$; Peng et al. 2014).

2.5 Perspectives

Generally, Danshen, as one of the most important Chinese traditional herbs and a model medicinal plant, has received significant attention and work. Rich wild and cultural resources contribute to the wide development and utilization prospect of Danshen. Not only can Danshen be used as medicine, but also for perfume and gardening. Currently, field management in China, including seeds and seedlings quality control, is being standardized from different levels of government regulations and industry specifications. Large specialized plantations which linked to related drug companies are being encouraged to construct. The breeding of cultivars has made progress.

Based on the analysis of the current situation, a few of key issues on Danshen resources, cultivation, and breeding should be considered in future work. To date, few resources of Danshen have been explored. Many wild and cultivated germplasms of *S. miltiorrhiza* and related species are unevaluated and underexplored. The communication among departments that preserve and study Danshen resources must be strengthened. A big room for cultivar improvement exists with such rich resources. Even for existing cultivars, the performance in distinct growing environments and adaptable field management techniques must be determined further. This will broaden the planting area and elevate the applied values of existing cultivars. Good agriculture practice and the related traceability system of medicinal material production is a long-term continuous optimization process. Owing to the traditional usage of wild resources, the acceptance of bred medicinal cultivars, especially those obtained by interspecies cross or transgenic-related methods is a highly controversial topic. Technically, root processing for decoction requires few thin branching rootlets. A high content of active components may be attractive for extraction. In such situation, different breeding goals may be set first, such as the use for decoction or extraction, and the high content of salvianolic acid or tanshinone.

With the release of the *S. miltiorrhiza* genome and transcriptome sequencing results, more functional genes will be identified. Newly emerging biological technologies, such as gene-editing technology, will be used. The genes involved in the biosynthetic pathways of active medicinal components have been edited using the CRISPR/Cas9 tool (Li et al. 2017; Zhou et al. 2018b). The resources of Danshen will be intensively explored, and the effective production and high quality of Danshen will benefit the health of the people of China and around the world.

Acknowledgements This work was supported by the National Natural Science Foundation of China (Grant No. 81773847) and the CAMS Innovation Fund for Medical Sciences (CIFMS) (Grant No. 2016-I2M-2-003).

References

Chen EG, Tsai KL, Chung HH, Chen JT (2018) Chromosome doubling-enhanced biomass and dihydrotanshinone I production in *Salvia miltiorrhiza*, a traditional Chinese medicinal plant. Molecules 23(12): E3106

Chen L, Li XL (2009) Induction and identification of Autotetraploid *Salvia miltiorrhiza*. Chin Tradit Herbal Drugs 40(12):1995–1997

Chen M, Wei J, Jin Y, Sui C, Yang CM (2017) GGE biplots in stability and adaptation evaluation for hybrid F1 *Salvia miltiorrhiza* lines. Mod Chin Med 19(6): 809–814, 838

Chen M, Yang CM, Sui C, Jin Y, Wei JH (2016) Zhongdanyaozhi No. 1 and Zhongdanyaozhi No. 2 are hybrid cultivars of *Salvia miltiorrhiza* with high yield and active compounds content. PLoS One 11(9): e0162691

Deng AP, Guo LP, Zhan ZL, Huang LQ (2016) Decipherment of ancient literature about Danshen. China J Chin Mater Med 41(22):4274–4279

Deng KJ, Zhang Y, Xiong BQ, Peng JH, Zhang T, Zhao XN, Ren ZL (2009) Identification, characterization and utilization of simple sequence repeat markers derived from *Salvia miltiorrhiza* expressed sequence tags. Acta Pharm Sin 44(10):1165–1172

Fang CP, Shan CG, Wang WT, Chen QL, Ni DP, Wang ZF (2011) Preliminary study on induction and identification of polyploidy of Shandong *Salvia miltiorrhiza* buds treated by colchicine. Agr Sci Tech 12(9):1338–1341

Guo Y, Li Y, Xue L, Severino RP, Gao S, Niu J, Qin LP, Zhang D, Brömme D (2014) *Salvia miltiorrhiza*: an ancient Chinese herbal medicine as a source for anti-osteoporotic drugs. J Ethnopharmacol 155(3): 1401–1416

Hao DC, Xiao PG (2017) Impact of drug metabolism/pharmacokinetics and their relevance upon *Salvia* based drug discovery. Curr Drug Metab 18(12):1071–1084

Imanshahidi M, Hosseinzadeh H (2006) The pharmacological effects of *Salvia* species on the central nervous system. Phytother Res 20(6):427–437

Jiang JY, Wang TL, Chi XL, Lyu CG, Jiang DQ, Wang S, Kang CZ, Zhang Y, Huang LQ, Guo LP, Sun K (2019) Community structure and diversity of soil bacteria in rhizosphere of *Salvia miltiorrhiza* and *Salvia miltiorrhiza* f. *alba* based on pyrosequencing. China J Chin Mater Med 44(8):1545–1551

Li B, Cui G, Shen G, Zhan Z, Huang L, Chen J, Qi X (2017) Targeted mutagenesis in the medicinal plant *salvia miltiorrhiza*. Sci Rep 7:43320

Li MH, Xiao PG (2013) Chapter 14—investigation of Danshen and related species. In: Zhang JT, Du GH (eds) The chemistry, pharmacology and clinical therapeutic effects of Danshen. Chemical Industry Press, Beijing, pp 227–241

Li XJ, Han JP, Li JX, Chen XC, Zhang LF, Li J, Gu ZW, Zhang YQ (2013) Identification of *Salvia shandongensis* new species based on sequences of the plastid *psbA-trnH* intergenic region. Acta Pharm Sin 48 (8):1338–1344

Li XL, Chen L (2012) Breeding for triploids of *Salvia miltiorrhiza* and its sustainable utilization. Chin Tradit Herb Drugs 43(2):375–379

Li XL, Zhan XR, Li Y, Wang RX, Shao AJ (2016) Biological characteristics of *Salvia miltiorrhiza* seed. Chin J Exp Tradit Med Formulae 22(18):27–30

Liang WJ, Lu CY, Shu ZM, Zhou ZL, Ye JY, Zhang YJ, Liang ZS, Guo HB (2016) Screening of different AFLP fragments between near-isogenic lines of male sterile and fertile *Salvia miltiorrhiza* and their comparison analysis. China J Chin Mater Med 41 (8):1415–1421

Liu CY, Chu JY, Chiang JH, Yen HR, Hsu CH (2016) Utilization and prescription patterns of traditional Chinese medicine for patients with hepatitis C in Taiwan: a population-based study. BMC Complement Altern Med 16(1):397

Liu D, Guo L, Huang L, Jin H, Wu L, Zeng Y, Zhang J, Yang Y (2011) Effects of soil water content on seedlings growth and active ingredients of *Salvia miltiorrhiza*. China J Chin Mater Med 36(3):321–325

Liu F, Li Y, Li M, Du Y, Yang Y (2019) Systematic evaluation and analysis on genuine regional drug *Salvia miltiorrhiza*. J Liaoning Univ TCM 21(1):119–125

Liu JF, Gao SL, Huang HP, Liu L, Li XY (2009) Allotetraploid induction and identification of *Salvia miltiorrhiza* Bunge and its hybridization breeding. Pharm Biotechnol 16(3):260–264

Ni DP, Zhu YW, Xue J, Tang Y, Shan CG, Wang ZF (2017) Evaluation on tolerance to continuous cropping of new hybrid lines of *Salvia miltiorrhiza* Bge. Chin Med J Res Prac 31(1):20–24

Pan YL, Zhao Q, Song ZQ, Wang JH (2016) Construction of a genetic linkage map in *Salvia miltiorrhiza*. J Chin Med mater 39(7):1443–1445

Peng L, Ru M, Wang K, Li B, Wang Y, Xia P, Liang Z (2014) Spaceflight environment-induced variation in root yield and active constituents of *Salvia miltiorrhiza*. Planta Med 80(12):1029–1035

Qian J, Song J, Gao H, Zhu Y, Xu J, Pang X, Yao H, Sun C, Li X, Li C, Liu J, Xu H, Chen S (2013) The complete chloroplast genome sequence of the medicinal plant *Salvia miltiorrhiza*. PLoS ONE 8(2):e57607

Salvia miltiorrhiza Bunge in Flora of China. http://foc.iplant.cn/content.aspx?TaxonId=200020230

Shan CG, Zhang JH, Wang GC, Zhang F, Ni DP, Zhu JB, Wang ZF (2013) Study on the seed characteristic of *Salvia miltiorrhiza*. Modern Chin Med 15(8):680–684

Shen XF, Zhang Q, Yan ZY, Yang XJ, Guo XH, Chen X, Wan DG (2016) Correlation analysis between effectiveness of element in rhizonsphere soil and quality of Danshen (*Salvia miltiorrhiza*). China J Chin Mater Med 41(7):1212–1217

Shu ZM, Liang ZS, Wei XR, Ye ZL (2005) A male sterile line of *Salvia miltiorrhiza* and its breeding method. Patent CN 1586119 A

Song JY, Luo HM, Li CF, Sun C, Xu J, Chen SL (2013) *Salvia miltiorrhiza* as medicinal model plant. Acta Pharm Sin 48(7):1099–1106

Song Z, Li X, Wang H, Wang J (2010) Genetic diversity and population structure of *Salvia miltiorrhiza* Bge in China revealed by ISSR and SRAP. Genetica 138 (2):241–249

Song ZQ, Wang JH, Wang HG, Zhao FJ, Hao LW (2009) Studies of the floral biology, breeding characters of *Salvia miltiorrhiza*. Acta Hortic Sin 36(6):905–910

Wang C, Li J, Zhang YQ (2012) Advances in germplasm resources and breeding of excellent varieties of *Salvia miltiorrhiza*. Mod Chin Med 14(4):37–42

Wang H, Wei T, Wang X, Zhang L, Yang M, Chen L, Song W, Wang C, Chen C (2018b) Transcriptome analyses from mutant *Salvia miltiorrhiza* reveals important roles for SmGASA4 during plant development. Int J Mol Sci 19(7) pii: E2088

Wang M, Li J, Zhang L, Yang RW, Ding CB, Zhou YH, Yin ZQ (2011) Genetic diversity among *Salvia miltiorrhiza* Bunge and related species using morphological traits and RAPD markers. J Med Plants Res 5 (13):2687–2694

Wang M, Zhao HX, Wang L, Wang T, Yang RW, Wang XL, Zhou YH, Ding CB, Zhang L (2013a) Potential use of DNA barcoding for the identification of *Salvia* based on cpDNA and nrDNA sequences. Gene 10(528):206–215

Wang TL, Guan W, Sun K, Wang S, Chi XL, Guo LP (2018a) Progress in researches on pathogens, epidemiology and integrated control of diseases on *Salvia miltiorrhiza* in China. China J Chin Mater Med 43 (11):2402–2406

Wang XY, Li QM, Wu P, Guo JX, Chen TZ, Xiao T (2017) Effects of different processing methods on the medicinal constituents of *Salvia miltiorrhiza*. Tradit Chin med 40(4):831–833

Wang Y, Peng H, Shen Y, Zhao R, Huang L (2013b) The profiling of bioactive ingredients of differently aged *Salvia miltiorrhiza* roots. Microsc Res Tech 76 (9):947–954

Wang YL, Hao JD, Li MH (2016) Herbal textural and original plants research on medicines from *Salvia* in China. China J Chin Mater Med 41(6):1140–1143

Wei JH, Yang CM (eds) (2018) Compilation of new bred varieties of Chinese medicinal plants. China Agricultural Science and Technology Publishing House, Beijing

Wei T, Deng K, Gao Y, Liu Y, Yang M, Zhang L, Zheng X, Wang C, Song W, Chen C, Zhang Y (2016) Arabidopsis DREB1B in transgenic *Salvia miltiorrhiza* increased tolerance to drought stress without stunting growth. Plant Physiol Biochem 104:17–28

Wu WL, Zhang L, Yang ZJ, Yang RW, Ding CB, Zhou YH (2011) Evaluation of endangered levels and conservation priority classes on *Salvia* L. in Sichun

Southwestern China. J Zhejiang Univ (Agric & Life Sci) 37:162–168

Wu Y, Liu C, Kuang J, Ge Q, Zhang Y, Wang Z (2014) Overexpression of SmLEA enhances salt and drought tolerance in *Escherichia coli* and *Salvia miltiorrhiza*. Protoplasma 251(5):1191–1199

Wu ZH, Niu JP (2017) Effects of different treatments on seed germination of *Salvia miltiorrhiza*. Heilongjiang Agri Sci 8:87–90

Xu GJ (ed) (1992) Study on the varieties and quality of common Chinese medicinal materials. Fujian Science and Technology Publishing House, Fuzhou, China, p p140

Yang F, Liu W, Wei YY, Zhou J, Wang X, Geng YL, Li FS (2016) Effect of the cultivation pattern on photosynthesis and yield of Salvia miltiorrhiza J Chin Med Mat 39(4):704–707

Yang XG, Chen B, Liu TS, Gong LM, Zeng XY (2013) Botany characteristics, quality evaluation and chemistry fingerprint of *Salvia miltiorrhiza* Bge. after spaceflight. Chin J Tradit Chin Med Pharm 28(10): 2932–2937

Zhan ZL, Deng AP, Peng HS, Zhang XB, Guo LP, Huang LQ (2016) Study of genuineness based on changes of ancient herbal origin–taking *Astragalus membranaceus* and *Salvia miltiorrhiza* as example. China J Chin Mater Med 41(17):3202–3208

Zhang C, Yang D, Liang Z, Liu J, Yan K, Zhu Y, Yang S (2019) Climatic factors control the geospatial distribution of active ingredients in *Salvia miltiorrhiza* Bunge in China. Sci Rep 9(1):904

Zhang H, Jin W, Zhu X, Liu L, He Z, Yang S, Liang Z, Yan X, He Y, Liu Y (2016) Identification and characterization of *Salvia miltiorrhizain* miRNAs in response to replanting disease. PLoS ONE 11(8): e0159905

Zhang L (ed) (2016) Plant research on *Salvia miltiorrhiza* and its related species resources. Science Press, Beijing, pp 3–18

Zhang L, Zhao HX, Fang X, Wang M, Ding CB, Yang RW, Yin ZQ, Xie XL, Zhou YH, Wan DG (2012) Genetic diversity among *Salvia miltiorrhiza* Bunge and related species inferred from nrDNA ITS sequences. Turk J Biol 35(3):319–326

Zhao HX, Zhang L, Fan X, Yang RW, Ding CB, Zhou YH (2006) Studies on chromosome numbers of *Salvia miltiorrhiza*, *S. flava* and *S. evansiana*. China J Chin Mater Med 31(22): 1847–1849

Zhong G, Zhang L, Yang R, Ding C (2010) Pollen viability and stigma receptivity of *Salvia miltiorrhiza* and its relative. China J Chin Mater Med 35(6):686–689

Zhou BQ, Lv HH, Yang F, Liu W, Geng YL, Wang X, Wang T, Qu ZY (2017) Further promotion effect of variable temperature drying on effective components in roots of *Salvia miltiorrhiza* f. *alba*. China J Chin Mater Med 42(10): 1883–1893

Zhou LS, Tang K, Guo SX (2018a) The plant growth-promoting fungus (PGPF) *Alternaria* sp. *A13* markedly enhances *Salvia miltiorrhiza* root growth and active ingredient accumulation under greenhouse and field conditions. Int J Mol Sci 19(1): pii: E270

Zhou Z, Tan H, Li Q, Chen J, Gao S, Wang Y, Chen W, Zhang L (2018b) CRISPR/Cas9-mediated efficient targeted mutagenesis of RAS in *salvia miltiorrhiza*. Phytochemistry 148:63–70

Zong CK, Song ZQ, Chen HM, Liu C, Wang JH, Guo LL, Liu T, Pan YL (2015) Construction of the first genetic linkage map of *Salvia miltiorrhiza* Bge. using SSR, SRAP and ISSR markers. Acta Pharm Sin 50(3): 360–366

Molecular Maps and Mapping of Genes and QTLs of *Salvia miltiorrhiza*

3

Xingfeng Li, Jianhua Wang and Zhenqiao Song

Abstract

Salvia miltiorrhiza is an important medicinal crop in traditional Chinese medicine (TCM). Knowledge of its genetic foundation is limited. The first genetic linkage map of *S. miltiorrhiza* was constructed in 94 F_1 individuals from an intraspecific cross, consisting of 53 SSR, 38 SRAP, and 2 ISSR loci in eight linkage groups. Another genetic map was constructed based on 111 genomic SSRs markers. Moreover, specific length amplified fragment sequencing (SLAF-seq) is a recently developed high-throughput strategy for large-scale SNP, which was used to construct a high-density genetic map for *S. miltiorrhiza*. The results not only provide a platform for mapping quantitative trait loci but also offer a critical new tool for *S. miltiorrhiza* genome and comparative genomics as well as a valuable reference for TCM studies. A total of six QTL have been identified through interval mapping. It includes three salvianolic acid B (SAB) content-related QTL on linkage group LG7, two lithospermic acid (LA) content markers on LG2 and LG4, and one rosmarinic acid (RA) content-linked QTL on LG7. These QTL could be used in marker-assisted breeding for phenolic acid contents of *S. miltiorrhiza*. The results provide important basis for QTL mapping, map-based cloning, and association studies for commercially important traits in *S. miltiorrhiza*.

X. Li · J. Wang · Z. Song (✉)
State Key Laboratory of Crop Biology, Agronomy College, Shandong Agricultural University, Taian 271018, China
e-mail: szqsdau@163.com

X. Li
e-mail: lixf@sdau.edu.cn

J. Wang
e-mail: jhwangjh@163.com

3.1 Introduction

Salvia miltiorrhiza Bunge, also known as Danshen in Chinese, is a typical herb plant used for traditional Chinese medicine (TCM). This plant has been extensively used for thousands of years to treat various diseases, particularly coronary heart disease and cerebrovascular diseases, in China and Japan and recently in the USA and many European countries (Zhou et al. 2005). Currently, more than 200 compounds have been isolated, of which diterpenoid quinones and hydrophilic phenolic acids are the major constituents (Ma et al. 2015). Moreover, recent studies have found several new bioactivities of Danshen constituents, such as antioxidant, antitumor, and protective effects on the kidney and liver. It suggests the potential for new applications. More than 320 Danshen preparations are produced by different pharmaceutical manufacturers.

S. Lu (ed.), *The Salvia miltiorrhiza Genome*, Compendium of Plant Genomes,
https://doi.org/10.1007/978-3-030-24716-4_3

There is constantly increasing demand for *S. miltiorrhiza* because of its varied and diverse pharmacologic properties. Currently, the annual demand for Danshen in China is approximately 10 million kg. Interest in biotechnology research on Danshen is increasing in many research groups. A large number of genes involved in the biosynthesis of phenolics and terpenoids have been identified through either molecular cloning or transcriptome-wide analysis (Gao et al. 2014; Luo et al. 2014; Xu et al. 2015). However, the genetic background and regulators of these two biosynthetic pathways, especially in the later steps, remain unknown. Danshen is of considerable research interest, and its superior genetic characteristics, such as its modest genome size, low number of chromosomes, vitality, short generation cycle, and mature tissue culture technology, have resulted in Danshen becoming an ideal model plant among TCM (Song et al. 2013).

At present, almost all of the Danshen preparations are primarily obtained by extraction from plants. An important and urgent task is to focus on improvements in Danshen for optimizing desirable traits, e.g., effective components, resistance, and yield. According to the literature on *S. miltiorrhiza*, several molecular markers, including amplified fragment length polymorphisms (AFLPs) (Yang et al. 2012), sequence-related amplified polymorphisms (SRAPs) (Song et al. 2010), EST-simple sequence repeats (SSRs), and inter-SSRs (ISSRs), have been used to analyze the genetic diversity of Danshen. Previous studies have shown that *S. miltiorrhiza* is a cross-pollinated plant with high differentiation of its germplasm (Song et al. 2010). These results provide an important basis for further construction of genetic maps with the aim of selecting parents and markers. Genetic linkage maps, particularly high-density genetic maps, are valuable tools in meeting the requirement of high-throughput superior trait selection among various germplasms, including plants and animals.

Secondary metabolites are usually synthesized in specific organs or tissues of plants with a relatively low level of contents, which were generally considered as complex agronomic traits, and controlled by multiple genes known as quantitative trait loci (QTL). QTL analysis based on linkage maps is useful for screening closely linked molecular markers of complex traits. It can effectively elucidate the genetic basis of complex agronomic traits, estimate gene actions and genetic parameters, and further accelerate the breeding progress through marker-assistant selection (MAS) in different species (Shehzad and Okuno 2016).

Since the construction of genetic map of *Artemisia annua* L. and the identification of QTL affecting yields of artemisinin were reported (Graham et al. 2010), progresses have been obtained for various plant species, such as *Bupleurum chinense* DC., *Siraitia grosvenorii*, *Eucommia ulmoides*, and *Dendrobium officinale* (Zhan et al. 2010; Liu et al. 2011; Li et al. 2015; Lu et al. 2018). Although this opens a new door for genetic breeding researches of medicinal plant, genetic researches about active compound contents of medicinal or aromatic plants only conducted in a few species. The studies on medicinal or aromatic plants are behind far from other crops.

3.2 Molecular Linkage Map Construction

3.2.1 Map Strategy and Mapping Populations

An appropriate mapping population, which generally includes RIL, DH, F_2 or backcrossed progeny, is very important in the construction of genetic maps (Grattapaglia and Sederoff 1994). However, similar to most medicinal species, it is very difficult to obtain a typical family-based population in *S. miltiorrhiza* due to its high heterozygosity resulting from a long history of natural cross-pollination and inbreeding depression. The double pseudo-testcross strategy was first proposed by Grattapaglia and Sederoff (1994) and was successfully applied to construct a genetic map of forest trees. In the pseudo-testcross, an F_1 progeny is developed as a mapping population by hybridizing two unrelated and highly heterozygous individuals, where gene segregation patterns can be interpreted as a backcross. This strategy has been widely used in plant species that lack appropriate pedigrees (Myburg et al. 2003; Feng et al. 2013; Sudarshini et al. 2014).

Because of its cross-pollinated habit and lack of basic genetic and breeding study, it is difficult to construct the F_2 and RIL populations in *S. miltiorrhiza*. In many fruit and forest trees with cross-pollination and asexual reproduction, the F_1 mapping populations were used for genetic study of interested agronomic traits based on the principle of 'pseudocrossing', such as *Mangifera indica* L. (Luo et al. 2016) and *Sesamum indicum* L. (Zhang et al. 2013a, b). Because both parents were heterozygous, the genotype among different F_1 progeny was different, which provides a feasible method for most medicinal plants. Many medicinal plants have a cross-pollinating habit. There is no doubt that this tendency provides favorable conditions for the selection of mapping parents. If there is a poor genetic basis and the lack of a typical family-based population, the pseudo-testcross is the most promising method for creating genetic maps of medicinal plants at the present time.

Study route for genetic mapping and QTL location of *S. miltiorrhiza* showed on Fig. 3.1. The first step is to select female and male parents with significant different characteristics, which is the guarantee for the analysis of QTL for important trait. A considerable difference between the parents was required to generate large polymorphic markers. The next step is to create F_1 interspecific hybrid population, which usually contains at least 100 F_1 lines. The third step is the main part of work, which includes two parallel elements. One is DNA extraction, genotyping by various molecular markers and further genetic map construction. Another is to investigate agronomic traits. The fourth step is association of agronomic traits with genotyping by software to identify the location, number of loci, and genetic effect of QTL. The final aim is to identify molecular markers closely linked to important trait, providing important basis for marker-assisted selection (MAS) and map-based gene cloning.

3.2.2 F_1 Genotyping, Data Analysis and Map Construction Method

The marker analysis was carried out according to the isolation types of parents and F_1 population. Marker segregation types can be divided into three categories: parental heterozygous markers, maternal-specific markers, and paternal-specific markers, including ab × cd, ef × eg, hk × hk, lm × ll, and nn × np (Fig. 3.2). There are four different amplification bands in the ab × cd segregation type with ab derived from maternal parent and cd from paternal parent, which produce four types including ac, ad, bc, and bd with same ratio of 1:1:1:1 in their F_1 population. For the ef × eg type, the "e" amplification band was a common one from two parents, but "f" and "g" were the unique markers from maternal parent and paternal parent, respectively. There will appear four types including ee, eg, fe, and fg in F_1 population with the ratio of 1:1:1:1. The two parents have the same bands type in the hk × hk type, and the F_1 population show hh:hk:kk with the ratio of 1:2:1. There are two different amplification bands in lm × ll type or nn × np type, with "l" or "n" as a common band of two parents, "m" and "p" as unique band in maternal parent and paternal parent, respectively. Therefore, their F_1 population displays two types with same ratio of 1:1.

Markers with blurred or missing data are labeled as "–". According to different types of markers, the data were analyzed by statistical analysis, and the obtained data were tested for segregation ratios using the Chi-squared (χ^2) test. The distorted markers ($p < 0.05$) were assigned a "*" suffix. JoinMap 4.0 software was employed to establish the genetic linkage map at likelihood of odds (LOD) score at least 3.0 with a

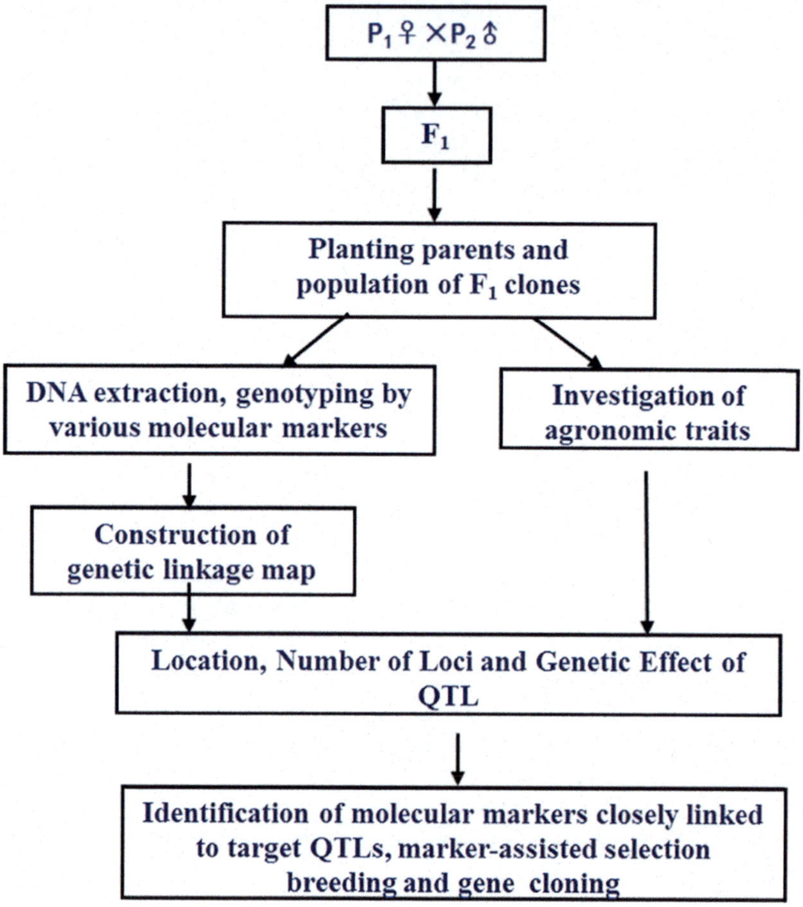

Fig. 3.1 Study route for genetic mapping and QTL location of *S. miltiorrhiza*

cross-pollinator (CP) population type. Map distances were calculated using Kosambi's mapping function. MapChart 2.2 was used to draw the genetic map (Cui et al. 2015).

3.2.3 Genetic Linkage Map Construction of *S. miltiorrhiza*

In our group, we selected the lines ZH74, ZH23, and ZH105 as female parents and BH18 as male parent (Fig. 3.3) to create three F_1 interspecific hybrid population of *S. miltiorrhiza* for marker genotyping and the construction of genetic linkage maps using the double pseudo-testcross mapping strategy. A total of four genetic linkage maps were constructed using the above three F_1 populations.

3.2.3.1 The First Linkage Genetic Map and Another Map Based on Integration of Multiple Markers

The first genetic linkage map of *S. miltiorrhiza* was constructed in 94 F_1 individuals from crossing ZH74 and BH18 by using simple sequence repeat (SSR), sequence-related amplified polymorphism (SRAP) and inter-simple sequence repeat (ISSR) markers. A total of 93 marker loci in the linkage map, consisting of 53 SSR, 38 SRAP, and 2 ISSR loci, were made up of eight linkage groups, covering a total length of 400.1 cm with an average distance of 4.3 cm per marker. The length of linkage groups varied from 3.3 to 132 cm and each of them included 2–23 markers, separately (Zong et al. 2015).

Another map was constructed in 138 F_1 individuals from crossing ZH105 and BH18

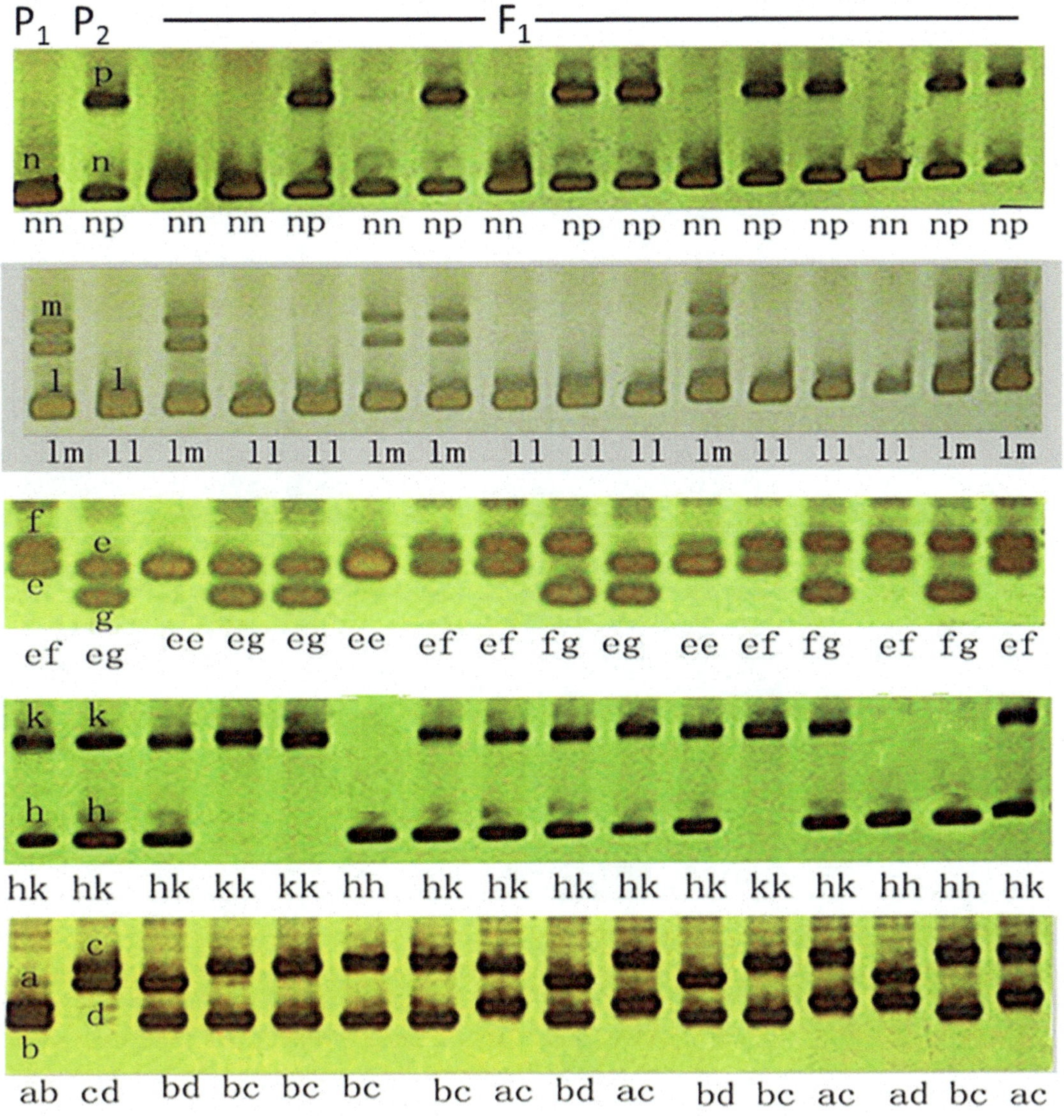

Fig. 3.2 Five segregation types in F_1 population of *S. miltiorrhiza*

based on integration of multiple markers. A set of 503 SSR primers, 100 SRAP primers, and 8 ISSR primers developed from *S. miltiorrhiza* were screened in the two parents. A total of 340 primers yielded amplification effectively, with 206 primers possess polymorphism. It contained 158 SSR primers, 46 SRAP primers, and 2 ISSR primers with the percentage to be 31.4, 46.0, and 25.0%, respectively. It included 9 linkage groups, contained 111 loci, and covered genome 857.4 cM with a mean marker interval of 7.7 cM. The length of linkage groups varied from 41.6 to 149.7 cM and each of them included 2–27 markers, respectively. Seventeen point one percent partially segregated markers distributed in the map were mainly located on LG1, LG2, and LG3 linkage groups.

3.2.3.2 The Genomic SSR Linkage Genetic Map

Genome SSR was developed by simplified genome sequencing of female parent ZH23.

Fig. 3.3 Lines ZH74, ZH23, and ZH105 as female parents and BH18 as male parent for creating three F_1 interspecific hybrid population of *S. miltiorrhiza*

A total of 665 pairs of SSR primers were synthesized and used to analyze the polymorphism between ZH23 and BH18, of which 568 pairs of primers generated clear bands. A total of 138 pairs of genomic SSR primers that amplified 151 loci were polymorphic and stable among the parents and 206 F_1 lines. Among 138 pairs of polymorphic SSR primers, a total of five separation types were produced. The separation types of ab × cd, ef × eg, hk × hk, lm × ll, and nn np involved 1, 13, 0, 48, and 49 primer pairs, respectively. One hundred fifty-one polymorphic loci were subjected to the χ^2 test to determine if they met the expected segregation ratio given by the JoinMap4.0 software program. The LOD value was selected to 3.0, and finally, 111 SSR markers were used to construct the genetic linkage map.

The second genetic map of *S. miltiorrhiza* was constructed based on genomic SSRs. Simple sequence repeat (SSR) is one of the most popular and versatile marker type and is considered as an ideal co-dominant marker for plant genetic mapping (Morgantae and Olivieri 2010). It has several prior characteristics, such as neutrality, abundance and polymorphism, high information content, and mostly single-gene inheritance. Totally, 111 SSR markers were finally assigned into eight linkage groups, which corresponded to the haploid chromosome number of *S. miltiorrhiza* ($2n$ = 16). The genetic distance spanned 397.5 cM in this maps accounting for about 92.7% of the entire genome of *S. miltiorrhiza.* The length of individual linkage group ranged from 24.8 cM (LG4) to 102.0 cM (LG1) with the number of markers on each group ranged from 6 to 24. Average distance between markers was 3.6 cM, with intervals between loci ranged from 1.8 to 9.7 cM (Feng et al. 2019).

3.2.3.3 The High-Density Linkage Genetic Map

To construct a high-density genetic map, specific length amplified fragment sequencing (SLAF-seq) was used for the rapid discovery of SNPs in the F_1 population. Analysis of the F_1 mapping population indicated that 83,154 SLAFs were generated, with an average depth of 6.17-fold for each offspring. Among the 151,035 high-quality SLAFs, 62,834 were polymorphic, resulting in a polymorphism rate of 41.60%. Of the 62,834 polymorphic SLAFs, 47,701 were classified into eight segregation patterns. For the F_1 population, five segregation patterns, including ab × cd, ef × eg, hk × hk, lm × ll, and nn × np, were used for genetic map construction, and a total of 5198 SLAFs were used for map construction (Fig. 3.4) (Liu et al. 2016).

Then, the high-density genetic map was constructed. After completing the data preparation, 5164 of the 5198 SLAFs were mapped onto the genetic map (i.e., a ratio of 99.34%). There were a total of 2966 BH18 (male) markers, 3038 ZH74 (female) markers, and 5164 SLAFs (7554 SNPs), which fell into 8 LGs, for the integrated map. The coverage of the markers was 75.67-fold in the female parent, 91.20-fold in the male parent, and 10.36-fold in each F_1 individual (on average). The final map was 1516.43 cM in length, with an average inter-marker distance of 0.29 cM (Table 3.1). According to the formula for genetic linkage maps, the coverage ratio of the total length to the expected length was 99.83%, which is considerably stronger compared to the value of 84.4% obtained for the first map. The current linkage map covers nearly the entire genome with a resolution of 0.29 cM (Liu et al. 2016).

3.3 Quantitative Trait Loci Analysis of Important Agronomic Traits in *S. miltiorrhiza*

3.3.1 Quantitative Trait Loci (QTL) Mapping Method

Quantitative trait loci (QTL) mapping was performed using the MapQTL 6.0 software based on the interval mapping (IM) algorithm with a significance level of 0.05, 1000 times of permutation and walking speed of 1 cM. Quantitative trait loci (QTL) identified in this way were marked by the marker with the highest LOD score in the corresponding QTL region.

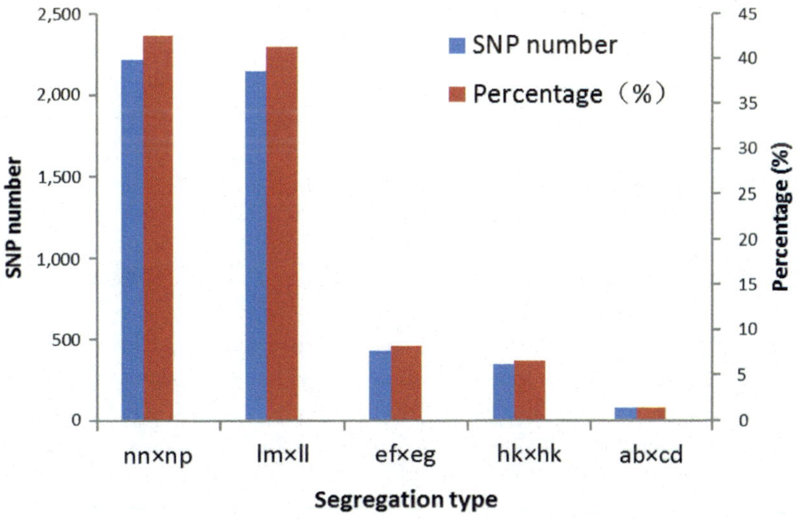

Fig. 3.4 SNP number and percentage of five segregation types in F_1 population of *S. miltiorrhiza*

Table 3.1 Genetic map for eight linkage groups (LGs)

	Total marker			Total distance (cM)			Max gap (cM)		
	Female map	Male map	Integrated map	Female map	Male map	Integrated map	Female map	Male map	Integrated map
LG1	347	316	559	141.153	107.028	132.85	15.79	6.68	7.88
LG2	341	371	613	186.544	180.181	190.39	17.24	13.01	7.88
LG3	426	430	747	195.187	216.418	213.66	21.86	9.12	9.6
LG4	378	390	652	183.249	175.127	181.17	10.38	18.74	9.11
LG5	279	266	478	153.085	108.194	130.64	13.01	26.95	10.45
LG6	404	372	675	196.998	202.857	207.41	114.00	18.44	11.53
LG7	352	353	610	115.325	170.161	163.048	6.68	20.27	19.22
LG8	511	468	830	293.022	297.244	297.25	26.95	17.24	20.67
Total	3038	2966	5164	1464.563	1457.21	1516.43	26.95	26.95	20.67

3.3.2 QTL Analysis for Phenolic Acid Content

The main peaks in the chromatograms of rosmarinic acid (RA), salvianolic acid (SAB), and lithospermic acid (LA) were identified by HPLC, based on their retention time and UV absorptions. Measurement of three phenolic acid contents showed significant differences between two parents. The variation coefficients of these traits varied from 33.64% (SAB content) to 53.03% (RA content), which showed a significant variation among F_1 individuals. Skewness and peakedness test indicated that these traits were segregated continuously and controlled by multiple genes, suggesting their suitability for QTL analysis.

Six QTL were identified for three phenolic acids content traits with phenotypic variance explained from 10.8 to 15.9%. Among them, 4 QTL was anchored on LG7, 1 QTL on LG2, and 1 QTL on LG4, respectively. Three QTL for SAB content were detected in a continuous interval between DSSR-540 and DSSR-98 markers on LG7 with a length of 27.2 cM (Fig. 3.5), in which, three continuous QTL explained 13.8, 12.9, and 13.6% of the phenotypic variation, respectively. Two QTL for LA content were located on LG2 and LG4, explained 10.8 and 15.9% of the phenotypic variation, respectively. These two QTL were localized with a single marker DSSR-140 and DSSR-77, respectively. Only one QTL for RA content was detected on LG7, which accounted for 12.4% of the phenotypic variation, and this locus was in the same region of QTL for SAB content.

3.3.3 QTL Analysis for Root Yield

Various characters in ZH105/BH18 F_1 generation groups showed continuous distribution of the root traits of *S. miltiorrhiza*. These characters conformed the quantitative trait loci for genetic and normal distribution. Twenty-six loci were detected in the ten phenotypic traits of *S. miltiorrhiza* by the interval mapping method. Root diameter, root range, root fresh weight, and root number were detected QTL loci, explaining the phenotypic variation rate raged from 8.0 to 13.7%.

3.4 Discussion

Most of the genetic maps constructed in medicinal plants mainly use the universal markers (ISSR, SRAP, and AFLP), which indicates that the related genetic research of medicinal plants is still in its infancy, far behind that of food crops and other crops. The genetic map of Danshen, even though it is an emerging model medicinal

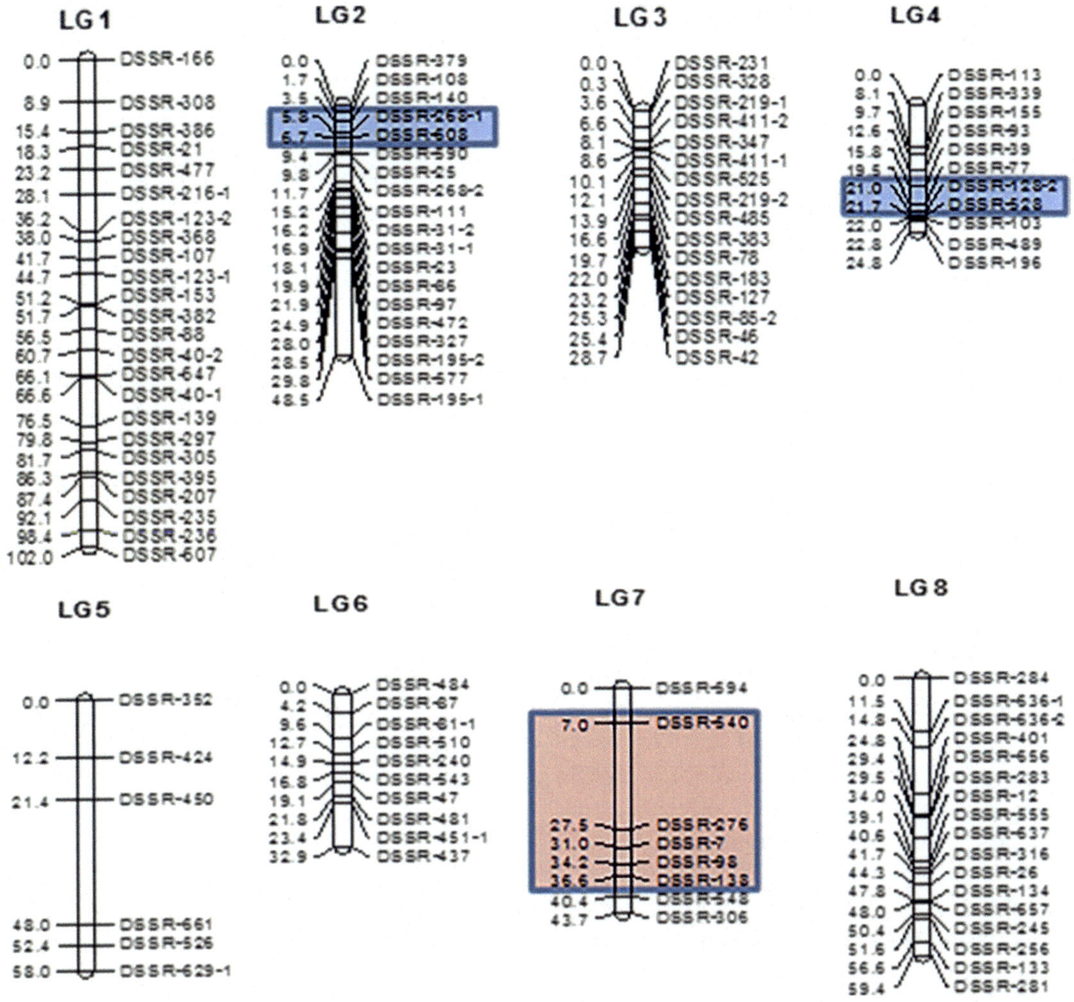

Fig. 3.5 Genetic map of the F_1 population (SNDS1 × BH18) based on the simple sequence repeat (SSR) markers and distribution of quantitative trait loci (QTL) effecting phenolic acid concentration. Genetic linkage map of *S. miltiorrhiza* presented as eight pseudo-chromosomes. The names of loci are shown on the right, and the positions of the loci are shown, in Kosambi centiMorgan (cM), on the left. QTL are shown on the light blue boxes on LG2 and LG4 for two loci of lithospermic acid content. The light red box showed an important cluster consisting of four QTL, one for rosmarinic acid and three for salvianolic acid B content

plant, only recently begun to be constructed. The first map, which was constructed using SRAPs, ISSRs, and SSRs in the F_1 population, includes 94 loci with an average interval distance of 4.3 cM. However, this unsaturated genetic map has limited future application. SNPs are the most abundant and stable form of genetic variation in most genomes and have become the marker type of choice in many genetic studies. Subsequently, a high-density genetic map of *S. miltiorrhiza* was constructed, which contained 5164 high-quality SLAFs and spanned 1516.43 cM, with an average marker interval of 0.29 cM. Due to different parents and different population, the genetic maps usually have poor versatility. In integration, the genomic SSR was essential because of the characteristics of co-dominance, easy to score, simple banding, and convenient identification (Wu et al. 2016; McCallum et al. 2016). Species-specific SSR markers need to be further developed and widely used in TCM. The methods used in this study for genetic mapping and for the development of markers provide a valuable reference for other medicinal plants.

Quantitative trait loci analysis based on linkage map can effectively elucidate the genetic basis of complex agronomic traits (Ma et al. 2014). A large number of QTL have been mapped in many crops related to various important agronomic traits, such as quantity (Tharanya et al. 2018), quality (Vallejo et al. 2015), plant adaptation and tolerance to stresses (Pilet-nayel et al. 2017), and some of which had been successfully utilized in breeding programs (Singh and Miklas 2015).

Phenolic acids are one of the main active components in *S. miltiorrhiza* root. Thus, improvement of phenolic acid contents has been an important objective of its genetic and breeding research. In our group, QTL analyses of phenolic acid contents in *S. miltiorrhiza* were firstly reported. It opened a new window for future MAS breeding of phenolic acids and identification of genes associated with phenolic acids biosynthesis.

Quantitative trait loci for correlated traits often position on the same chromosomal regions. Consistently, QTL for rosmarinic acid (RA) and three QTL for salvianolic acid (SAB) content were located within the same region. This region can be considered as further target region for fine mapping and marker-assisted selection in *S. miltiorrhiza* breeding programs. Besides, SAB content also had a significantly high correlation with the other two secondary compounds contents. Selection of SAB content will have positive effects on other traits. The same phenomenon of overlaps in QTL regions was also reported in many plant species, such as *Pyrus* L. (Zhang et al. 2013a, b), *Glycine max* L. Merr. (Akond et al. 2014), and *Cynodon dactylon* var. dactylon (Guo et al. 2017a, b). The above findings may be due to two genes controlling different traits. These QTL are likely to be either within the same region or in adjacent locations on a chromosome or a single gene tended to have pleiotropic effect (Zhang et al. 2013a, b). A limited number of QTL for phenolic acid contents were detected due to the genetic maps with low density. Improvement in the marker density would increase the identification of QTL number. It is suggested that the density of the genetic map is a key to perform QTL fine mapping of interested agronomic traits.

In future breeding of *S. miltiorrhiza*, attention should be paid to the continuous recording and dynamic analysis of agronomic traits for many years. It will greatly improve the predictability and accuracy of the selection of good genotypes for quantitative traits and will effectively improve the MAS efficiency. In addition, expanding mapping population, increasing the number of markers in genetic map, constructing ultra-high density map, and integrating and locating physical map can effectively improve the establishment of closely linked QTL and lay an important foundation for molecular breeding of *S. miltiorrhiza*.

Acknowledgements This work was supported by funds from the National Natural Science Foundation of China (81872949), the Natural Science Foundation of Shandong province of China (ZR2019HM081), and the key research and development plan of Shandong Province (2017CXGC1302).

References

Akond M, Liu SM, Boney M, Kantartzi SK, Meksem K, Bellaloui N, Lightfoot DA, Kassem MA (2014) Identification of quantitative trait loci (QTL) underlying protein, oil, and five major fatty acids contents in soybean. Am J Plant Sci 5(1):158–167

Cui JJ, Cheng JW, Wang GP, Tang X, Wu ZM, Lin MB, Li LF, Hu KL (2015) QTL analysis of three flower-related traits based on an interspecific genetic map of *Luffa*. Euphytica 202(1):45–54

Feng YY, Guo LL, Jin H, Lin CC, Zhou CH, Fang XS, Wang JH, Song ZQ (2019) Quantitative trait loci analysis of phenolic acids contents in *Salvia miltiorrhiza* based on genomic simple sequence repeat markers. Industrial crops and products 133:365–372

Feng S, Zhao H, Lu J, Liu J, Shen B, Wang H (2013) Preliminary genetic linkage maps of chinese herb *Dendrobium nobile* and *D. moniliforme*. J Genet 92 (2):205–212

Gao W, Sun HX, Xiao H, Cui G, Hillwig ML, Jackson A, Wang X, ShenY Zhao N, Zhang LX, Wang XJ, Peters RJ, Huang LQ (2014) Combining metabolomics and transcriptomics to characterize tanshinone biosynthesis in *Salvia miltiorrhiza*. BMC Genom 15 (1):1–14

Graham IA, Besser K, Blumer S, Braniga, CA, Czechowski T, Elias L, Guterman I, Harvey D, Isaac PG, Khan AM, Larson TR, Li Y, Pawson T, Penfield T, Rae AM, Rathbone DA, Reid S, Ross J, Smallwood MF, Segura V, Townsend T, Vyas D, Winzer T, Bowles D (2010) The genetic map of *Artemisia annua* L. identifies loci affecting yield of the antimalarial drug artemisinin. Science 327(5963): 328–331

Grattapaglia D, Sederoff R (1994) Genetic linkage maps of *Eucalyptus grandis* and *Eucalyptus urophylla* using a pseudo-testcross: mapping strategy and RAPD markers. Genetics 137:1121–1137

Guo Q, Guo LL, Zhang LX, Zhang LX, Ma HL, Guo DL, Hou XG (2017a) Construction of a genetic linkage map in tree peony (*Paeonia* Sect. *Moutan*) using simple sequence repeat (SSR) markers. Sci Hortic 219:294–301

Guo Y, Wu Y, Anderson JA, Moss JQ, Zhu L, Fu J (2017b) SSR marker development, linkage mapping, and QTL analysis for establishment rate in common Bermuda grass. Plant Genome 10(1):1–11

Li Y, Wei JK, Li ZQ, Lei AN, Liu MH (2015) Quantitative trait locus analysis of leaf morphological characters, yield-related traits, and secondary metabolite contents in *Eucommia ulmoides*. Genet Mol Res 14(4):17871–17884

Liu LH, Ma XJ, Wei JH, Qin JM, Mo CM (2011) The first genetic linkage map of Luohanguo (*Siraitia grosvenorii*) based on ISSR and SRAP markers. Genome 54 (1):19–25

Liu T, Guo LL, Pan YL, Zhao Q, Wang JH, Song ZQ (2016) Construction of the first high-density genetic linkage map of *Salvia miltiorrhiza* using specific length amplified fragment (SLAF) sequencing. Sci Rep 6:24070

Lu JJ, Liu YY, Xu J, Mei ZW, Shi YJ, Liu PL, He JB, Wang XT, Meng YJ, Feng SG, Shen CJ, Wang HZ (2018) High-density genetic map construction and stem total polysaccharide content-related QTL exploration for Chinese Endemic *Dendrobium* (Orchidaceae). Front Plant Sci 9:398

Luo C, Shu B, Yao QS, Wu HX, Xu WT, Wang SB (2016) Construction of a high-density genetic map based on large-scale marker development in mango using specific-locus amplified fragment sequencing (SLAF-seq). Front Plant Sci 7:1310

Luo H, Zhu Y, Song J, Xu L, Sun C, Zhang X, Xu Y, He L, Sun W, Xu H, Wang B, Li X, Li C, Liu J, Chen S (2014) Transcriptional data mining of *Salvia miltiorrhiza* in response to methyl jasmonate to examine the mechanism of bioactive compound biosynthesis and regulation. Physiol Plant 152(2):241–255

Ma JQ, Yao MZ, Ma CL, Wang XC, Jin JQ, Wang XM, Chen L (2014) Construction of a SSR-based genetic map and identification of QTLs for catechins content in tea plant (*Camellia sinensis*). Plos One 9(3):e93131

Ma PD, Liu JY, Osbourn A, Donga J, Liang ZS (2015) Regulation and metabolic engineering of tanshinone biosynthesis. RSC Adv 5:18137–18144

Mccallum S, Graham J, Jorgensen L, Rowland LJ, Bassil NV, Hancock JF, Wheeler EJ, Vining K, Poland JA, Olmstead JW, Buck E, Wiedow C, Jackson E, Brown A, Hackett CA (2016) Construction of a SNP and SSR linkage map in autotetraploid blueberry using genotyping by sequencing. Mol Breed 36(4):1–24

Morgantae M, Olivieri AM (2010) PCR-amplified microsatellites as markers in plant genetics. Plant J 3 (1):175–182

Myburg AA, Griffin AR, Sederoff RR, Whetten RW (2003) Comparative genetic linkage maps of *Eucalyptus grandis*, *Eucalyptus globulus* and their F1 hybrid based on a double pseudo-back cross mapping approach. Theor Appl Genet 107:1028–1042

Pilet-Nayel ML, Moury B, Caffier V, Montarry J, Kerlan MC, Fournet S, Durel CE, Delourme R (2017) Quantitative resistance to plant pathogens in pyramiding strategies for durable crop protection. Front Plant Sci 8:1838

Shehzad T, Okuno K (2016) QTL mapping for yield and yield-contributing traits in sorghum (*Sorghum bicolor* (L.) Moench) with genome-based SSR markers. Euphytica 203(1):1–15

Singh SP, Miklas PN (2015) Breeding common bean for resistance to common blight: a review. Crop Sci 55 (3):971

Song JY, Luo HM, Li CF, Sun C, Chen SL (2013) *Salvia miltiorrhiza* as medicinal model plant. Yao Xue Xue Bao 48(7):1099–1106

Song ZQ, Li XF, Wang HG, Wang JH (2010) Genetic diversity and population structure of *Salvia miltiorrhiza* Bge in China revealed by ISSR and SRAP. Genetica 138:241–249

Sudarshini KV, Padmakar B, Patil MS, Lakshman R, Aswath C (2014) The genetic linkage maps of *Anthurium* species based on RAPD, ISSR and SRAP marker. Sci Hortic 178:132–137

Tharanya M, Kholova J, Sivasakthi K, Seghal D, Hash CT, Raj B, Srivastav, RK, Baddam R, Thirunalasundari T, Yadav R, Vadez V (2018) Quantitative trait loci (QTL) for water use and crop production traits co-locate with major QTL for tolerance to water deficit in a fine-mapping population of pearl millet (*Pennisetum glaucum* L.R.Br.). Theor Appl Genet 131(7):1509–1529

Vallejo VA, Tychonievich J, Lin WK, Wangchu L, Barry CS, Warner RM (2015) Identification of QTL for crop timing and quality traits in an interspecific *Petunia* population. Mol Breed 35(1):2

Wu QH, Chen YX, Lin F, Zhou SH, Chen JJ, Zhao XJ, Zhang D, Ouyang SH, Wang ZZ, Li D, Wang GX, Zhang DY, Yuan CG, Wang LX, You MS, Han J, Liu ZY (2016) QTL mapping of flag leaf traits in common wheat using an integrated high-density SSR and SNP genetic linkage map. Euphytica 208(2):337–351

Xu Z, Peters RJ, Weirather J, Luo H, Liao B, Zhang X, Zhu Y, Ji A, Zhang B, Hu S, Au KF, Song J, Chen S (2015) Full-length transcriptome sequences and splice variants obtained by a combination of sequencing platforms applied to different root tissues of *Salvia miltiorrhiza* and tanshinone biosynthesis. Plant J 82:951–961

Yang D, Ma P, Liang X, Liang Z, Zhang M, Shen S, Liu H, Liu Y (2012) Metabolic profiles and cDNA-AFLP analysis of *Salvia miltiorrhiza* and *Salvia castanea Diel* f. *tomentosa* Stib. PLoS One 7: e29678

Zhan QQ, Sui C, Wei JH, Fan SC, Zhang J (2010) Construction of genetic linkage map of *Bupleurum Chinense* DC. Using ISSR and SSR markers. Yao Xue Xue Bao 45(4):517–529

Zhang RP, Wu J, Li XG, Khan, MA, Chen H, Korban SS, Zhang SL (2013a) An AFLP, SRAP, and SSR genetic linkage map and identification of QTL for fruit traits in Pear (*Pyrus* L.). Plant Mol Biol Rep 31(3):678–687

Zhang Y, Wang L, Xin H, Li D, Ma C, Ding X, Hong W (2013b) Construction of a high-density genetic map for sesame based on large scale marker development by specific length amplified fragment (SLAF) sequencing. BMC Plant Biol 13(1):141

Zhou L, Zuo Z, Chow MSS (2005) Danshen: an overview of its chemistry, pharmacology, pharmacokinetics, and clinical use. J Clin Pharmacol 45:1345–1359

Zong CK, Song ZQ, Chen HM, Liu C, Wang JH, Guo LL, Liu T, Pan YL (2015) Construction of the first genetic linkage map of *Salvia miltiorrhiza* Bge. Using SSR, SRAP and ISSR markers. Yao Xue Xue Bao 50 (3):360–366

The Genome of *Salvia miltiorrhiza*

4

Zhichao Xu

Abstract

Salvia miltiorrhiza is one of the most commonly used medicinal plants that belong to the *Salvia* genus of the Lamiaceae family. It is highly valued for its dried root or rhizome in traditional Chinese medicine (TCM), which is better known for its use in the treatment of cardiovascular diseases and its anti-oxidative function. This species is a source of bioactive natural products, particularly diterpenoid tanshinones and phenolic acids. The *S. miltiorrhiza* genome and transcriptome data provide important basis for the molecular mechanism of active compound biosynthesis and regulation. So far, the 538 Mb draft genome with 54.44% repeat elements has been decoded, with 30,478 predicted genes. Four potential TPS/CYP gene clusters that are probably involved in the biosynthesis of diterpenoids have been identified. In addition, a combination of the next-generation sequencing (NGS) and the single-molecule real-time (SMRT) sequencing has been used to analyze various root tissues, particularly the periderm. The results provide a relatively complete view of the *S. miltiorrhiza* transcriptome and further insight into tanshinone biosynthesis.

Z. Xu (✉)
Institute of Medicinal Plant Development, Chinese Academy of Medical Sciences, Beijing 100193, China
e-mail: zcxu@implad.ac.cn

4.1 Introduction

Salvia miltiorrhiza Bunge, belongs to the *Salvia* genus of the Lamiaceae family, is considered a potential model medicinal plant of traditional Chinese medicine (TCM) research due to its significant medicinal value, relatively small genome, short life cycle, efficient transgenic system and undemanding tissue culture (Cheng 2006; Song et al. 2013; Zhou et al. 2005). The dried root or rhizome of *S. miltiorrhiza*, named Danshen or Tanshen in Chinese, is one of the most commonly used herbs in traditional Chinese medicine (TCM) (Pharmacopoeia Commission of People's Republic of China 2015). Danshen is better known for its use in the treatment of cardiovascular diseases and its anti-oxidative function. The main active constituents of *S. miltiorrhiza* are lipophilic diterpenoid components and hydrophilic phenolic acids (Dong et al. 2011; Ma et al. 2015; Zhou et al. 2005). Phytochemical analysis and bioactivity assays suggested that the hydrophilic phenolic acids, such as rosmarinic, lithospermic and salvianolic acids, and the reddish diterpenoid tanshinones, including tanshinone I, tanshinone IIA and cryptotanshinone, had antitumor, anti-inflammatory, anti-atherosclerotic and antioxidant activities through various therapeutic pathways. More than 90 lipophilic diterpenoid components and 40 hydrophilic phenolic acids have been isolated and identified from *S. miltiorrhiza*, including tanshinone I, tanshinone IIA,

S. Lu (ed.), *The Salvia miltiorrhiza Genome*, Compendium of Plant Genomes,
https://doi.org/10.1007/978-3-030-24716-4_4

cryptotanshinone, dihydrotanshinone, salvianolic acid A, salvianolic acid B, rosmarinic acid, lithospermic acid and dihydroxyphenyllactic acid. Clarifying the biosynthetic pathway and regulatory mechanisms of the active constituents will provide the theoretical basis to cultivate TCM and produce innovative medicines (Xu et al. 2016c; Xin et al. 2019; Chen et al. 2015).

Recently, the biosynthesis of active compounds of *S. miltiorrhiza* has drawn wide attention (Cui et al. 2015; Guo et al. 2013; Kai et al. 2011; Ma et al. 2015). Many key enzymes involved in the biosynthetic pathways have been cloned and identified, such as enzymes of the 2-C-methyl-D-erythritol 4-phosphate (MEP) and the mevalonate (MVA) pathways involved in tanshinones biosynthesis (Ma et al. 2012), and the phenylpropanoid and tyrosine-derived pathways related to phenolic acid biosynthesis (Wang et al. 2015; Xu et al. 2016b). With the remarkable advances in high throughput sequencing, the whole genome, chloroplast genome and transcriptome of *S. miltiorrhiza* have been gradually improved (Qian et al. 2013; Xu et al. 2015; Luo et al. 2014). An early report used unsequenced cDNAs from hairy root cultures to construct a microarray. Differential expression correlated with either culture time/development or induction (both of which are associated with tanshinone accumulation) was utilized to highlight cDNAs for sequencing (Li et al. 2010). Tanshinones are labdane-related diterpenoids (Gao et al. 2014), whose biosynthesis requires a copalyl diphosphate synthase (CPS) and a cyclase related to kaurene synthases involved in gibberellin phytohormone metabolism. The latter is often termed kaurene synthase-like (KSL). Functional characterization of the two inducible diterpene synthases found in the microarray study (SmCPS1 and SmKSL1) led to identification of the resulting diterpene olefin precursor of tanshinones, miltiradiene (Gao et al. 2009). Later, next-generation sequencing (NGS)-based RNA-Seq analysis of similarly induced hairy root cultures led to the identification and functional characterization of a cytochrome P450 (CYP) involved in tanshinone biosynthesis (Guo et al. 2013). The P450 is known as CYP76AH. It carries out the initial hydroxylation of aromatized miltiradiene to form ferruginol. Recent works suggest that genes for synthesis of the same metabolites tend to arrange in tandem array. It facilitates the elucidation of these pathways by gene cluster analysis. *S. miltiorrhiza* whole-genome sequencing can greatly help in identification of genes involved in the tanshinone biosynthetic pathway (Xu et al. 2016a).

4.2 The Sequencing and Assembly of *S. miltiorrhiza* Genome

4.2.1 Sequencing of the *S. miltiorrhiza* 99-3 Genome (Xu et al. 2016a)

The genome of *S. miltiorrhiza* line 99-3 was estimated to be 615 ± 12 M in size by flow cytometric analysis. It is noteworthy that no obvious peak was observed in the *k*-mer distribution of the Illumina reads, highlighting complexity of this genome. The young leaves of *S. miltiorrhiza* were used to extract genomic DNA and total RNA for genome and transcriptome sequencing, respectively. For Illumina sequencing, PE and MP libraries with multiple insert sizes ranging from 350 bp to 8 kb were prepared following the manufacturer's instructions and sequenced on an Illumina HiSeq 2000 platform (Illumina, USA). For 454 sequencing, shotgun libraries were prepared following the manufacturer's instructions and sequenced on Roche/454 platform (Roche, Swiss). For PacBio sequencing, SMRTbell DNA template libraries with an insert size of 8–10 kb were prepared in accordance with the manufacturer's protocol and sequenced on a PacBio®RS platform (Pacific Biosciences, USA). Firstly, 158.2 Gb short reads were produced for different pair-end and mate-pair libraries (350, 400, 500, 700, 3000, 5000, 8000 bp) using Illumina HiSeq 2000 platform (Table 4.1). The Trimmomatic was employed to remove the adapters and low-quality reads of Illumina raw data. Then, in order to improve the assembly quality, 8.19 Gb long reads with the average length of 3740 bp were

Table 4.1 Statistics of raw sequencing data using different platforms for *S. miltiorrhiza* (Xu et al. 2016a)

Platforms	Length of insert reads (bp)	Sequencing read length (bp)	Total raw data (Gb)	Coverage
Illumina HiSeq	350	100	96.50	156.9×
Illumina HiSeq	400	100	69.33	112.7×
Illumina HiSeq	500	100	90.40	147.0×
Illumina HiSeq	700	100	58.98	95.9×
Illumina HiSeq	3000	100	36.64	59.6×
Illumina HiSeq	5000	100	35.88	58.3×
Illumina HiSeq	8000	100	8.51	13.8×
PacBio RS	8000–10,000	3740	8.19	13.3×
Roche 454	500	364	8.65	14.1×

generated by the PacBio RS platform. Because of the low quality of PacBio reads with about 0.85, 8.65 Gb Roche/454 high-quality reads were harvested for the error correction of long reads.

4.2.2 Genome Assembly of the *S. miltiorrhiza*

Totally, 158.2 Gb Illumina reads were used for de novo assembly using Phusion2. This resulted in a draft assembly of 558 Mb with contig N50 of 2.47 Kb. In addition, SOAPdenovo assembler gave similar assembly metrics, implied intrinsic complexity of *S. miltiorrhiza*. The PacBio RS long reads were used to improve the assembly performance, and the length of about 4 Gb PacBio high-quality reads showed longer than 2 Kb, after the base error correction with 454 data using CAP3 and BLASR. The corrected long reads were assembled with Celera v7.0. The resulted contigs were combined with Roche/454 data for re-assembly. Illumina reads were then mapped onto these contigs, and SNPs and small indels in homozygotes, which presumably resulted from sequencing chemistry bias of Roche/454 platform, were used for error correction of the original contigs. After scaffolding with mate-pair data, a final assembly of 538 Mb genome was derived, with contig and scaffold N50 of 12.38 and 51.02 Kb, respectively (Table 4.2).

4.2.3 Genome Annotation and Repeat Element Analysis

Structural repeat annotation of the *S. miltiorrhiza* genome was performed by the RepeatModeler (http://www.repeatmasker.org/RepeatModeler/; version 1.0.9) package. Two de novo repeat finding programs, RECON and RepeatScout, were employed for identifying and classifying repeat elements. Repbase (version 21.12) was also used, and consensus classification of TEs was performed according to the default parameters. Finally, RepeatMasker (version 4.0.6) was used to count and mask the TE sequences. RMBlast, a modified version of the National Center for Biotechnology Information (NCBI) Blast program, was used for all-by-all alignments by RepeatModeler and RepeatMasker. The results showed that repeat elements accounted for 54.44% of the *S. miltiorrhiza* genome, twice that of *Sesamum indicum* (28.5%). Among them, the long terminal repeats spanned 18.03% of the *S. miltiorrhiza* genome, and 55.58% of the repeats were unclassified (30.26% of the genome) (Table 4.3).

The repeat elements were masked using RepeatMasker, and then the masked genome were performed gene prediction using MAKER package (http://www.yandell-lab.org/software/maker.html) (Cantarel et al. 2008). Augustus

Table 4.2 Summary statistics of de novo genome assembly (Xu et al. 2016a)

		Illumina	PacBio + 454 + Illumina
Contig	Number	310,742	60,349
	Average length (Kb)	1.64	8.69
	N50 length (Kb)	2.47	12.38
	Total length (Mb)	510	524
Scaffold	Number	223,112	21,045
	Average length (Kb)	2.50	25.56
	N50 length (Kb)	9.38	51.02
	Total length (Mb)	558	538

Table 4.3 Statistics of repeat element in *S. miltiorrhiza* genome (Xu et al. 2016a)

	Repeat elements
Total interspersed repeats	292,782,059 bp (54.44%)
SINEs	9312 (0.00%)
LINEs	15,849,707 (2.95%)
LTR elements	96,960,659 (18.03%)
DNA elements	17,208,679 (3.20%)
Unclassified	162,753,702 (30.26%)
Satellites	66,389 (0.01%)
Simple repeats	9,070,284 (1.69%)
Low complexity	1,305,919 (0.24%)

was employed for Ab initio gene prediction using default parameters, and the *Arabidopsis* genome was selected as model reference. Homology-based annotation was based on the following: (1) BLASTN alignment of RNA transcripts derived from trinity-assembled transcriptomes from root, shoot and leaf materials (http://trinityrnaseq.sourceforge.net) and (2) BLASTX searches of 35,838 available proteins (trimmed proteins from mitochondria and chloroplast) from *Lamiales* species. A total of 30,478 putative genes in *S. miltiorrhiza* genome were predicted, with an average transcript length of 2825 bp. Most of the genes (27,793; 91.19%) had homologs in the Nr database (with *E*-value less than 10^{-5} using BLASTP), and more than half (56.60%) could be assigned to KEGG pathways.

A total of 1602 transcription factor (TF) genes were identified. It includes genes encoding 171 APETALA2, 139 basic helix-loop-helix, 291 MYB and 78 WRKY TFs. These genes might be involved in plant developmental regulation. TPSs and CYP are essential enzymes for biosynthesis of plant terpenoids. TPSs are involved in synthesizing the hydrocarbon backbone of monoterpenes, sesquiterpenes, diterpenes and isoprene. CYPs are mono-oxygenases that catalyze various reactions, albeit largely hydroxylations, in various secondary metabolite production pathways. Total 82 TPSs and 456 CYPs were annotated in the assembled *S. miltiorrhiza* genome. Some of them have been known to participate in the biosynthesis of phenolic acids and tanshinones.

4.2.4 Genome Family Evolution of *S. miltiorrhiza*

The amino acid sequences of eight vascular plants, including *S. miltiorrhiza*, tomato, bladderwort, grape, sesame, rice, *Arabidopsis* and potato, were clustered using OrthoMCL (V1.4)

(Li et al. 2003). A core set of genes from 8018 gene families were found to present in all eight plants, while *S. miltiorrhiza* contained 956 unique gene families. Interestingly, 19.2% of these *S. miltiorrhiza* specific genes are involved in plant–pathogen interactions (KEGG Pathway Entry Ko04626), amino sugar and nucleotide sugar metabolism (Ko00520), amino acid biosynthesis (Ko01230) and plant hormone signal transduction (Ko04075). Among the genome families analysis, 1824 one-to-one single-copy orthologous genes were used to construct a phylogenetic tree for eight selected plant genomes. Divergence times between potato–tomato (7.2–7.4 MYA) and monocot–eudicot (128.7–234.4 MYA) were used as references for time calibration. The phylogenetic analysis indicated that *S. miltiorrhiza* was most closely related to sesame. The divergence time between them was approximately 67 million years ago (MYA).

In addition, we recently sequenced the genome of *Scutellaria baicalensis* and *Scutellaria barbata*. Comparative analysis showed that two *Scutellaria* species were most closely related to *S. miltiorrhiza* with an estimated divergence time of 51.35 MYA. The *Ks* distribution of *S. miltiorrhiza* showed that no recent whole-genome duplication (WGD) events occurred after the speciation of *S. miltiorrhiza*; however, the distribution of *Ks* values peaked around 1.02, 0.89 and 0.86 in *S. miltiorrhiza*, *S. baicalensis* and *S. barbata*, respectively. An ancient WGD event, occurred 72.04–85.44 MYA, was shared in *S. miltiorrhiza*, *S. baicalensis* and *S. barbata*. It suggests that the At-α event might be the Lamiaceae-specific WGD event.

Gene family evolution among eight plant species was analyzed by CAFÉ (De Bie et al. 2006). The results indicated that the gene families underwent significant expansion ($P < 0.01$) in the *S. miltiorrhiza* genome were mainly involved in stilbenoid, diarylheptanoid and gingerol biosynthesis (Ko00945), terpenoid biosynthesis (Ko00902) and steroid biosynthesis (Ko00100). It is consistent with the high production of tanshinones and phenolic acids in this herbal plant.

4.2.5 Gene Clusters Related to Tanshinone Biosynthesis

Analysis of the *S. miltiorrhiza* genome identified 82 terpene synthase genes (TPS) and 437 CYP450s. Physical clustering of TPSs and CYPs showed that four TPS/CYP pairs, including CPS1, CPS2, CPS5 and CPS7, were found in the draft *S. miltiorrhiza* genome. Among them, *SmCPS1* and *SmCPS2* are involved in tanshinone biosynthesis in roots and leaves, respectively, while *SmCPS5* is required for gibberellin phytohormone metabolism (Cui et al. 2015). *CYP76AH1*, *CYP76AH3* and *CYP76AK1* have been reported to be involved in tanshinone biosynthesis (Guo et al. 2016, 2013). Interestingly, *SmCPS1* and *SmCPS2* are clustered with the members from the CYP76AH subfamily, respectively. The *SmCPS1* clustered with two CYP450s, *CYP76AH12* and *CYP76AH13*; and *SmCPS2* clustered with *CYP76AH1*, *CYP76AH3* and *CYP76AH28*. However, the assembled scaffolds are rather short (N50 of 51.02 Kb), so the gene cluster analysis need to be further confirmed via the assembly of high-quality genome.

4.3 The Chloroplast Genome of *S. miltiorrhiza*

Qian et al. (2013) reported the complete chloroplast (cp) genome sequence of *S. miltiorrhiza*. *Salvia miltiorrhiza* cp genome is similar to the typical angiosperm cp genomes. It has 151,328 bp in length and exhibits a typical quadripartite structure of the large (LSC, 82,695 bp) and small (SSC, 17,555 bp) single-copy regions, separated by a pair of inverted repeats (IRs, 25,539 bp). The genome contains 114 annotated unique genes, including 80 protein-coding genes, 30 tRNAs and four rRNAs. Chloroplast genome comparison of *S. miltiorrhiza* and three other *Lamiales* chloroplast genomes showed a high degree of sequence similarity and a relatively high divergence of intergenic spacers. Phylogenetic analysis demonstrated a sister relationship between *S. miltiorrhiza* and *S. indicum*. The results of phylogenetic analysis of chloroplast genome are consistent with the evolutionary of single-copy genes from whole genome.

4.4 The Full-Length Transcriptome Based on *S. miltiorrhiza* Genome

4.4.1 Full-Length Transcriptome Analysis

Xu et al. (2015) reported the full-length transcriptome sequences and splice variants obtained by a combination of sequencing platforms applied to different root tissues of *S. miltiorrhiza* and tanshinone biosynthesis. Since medicinally active constituents of *S. miltiorrhiza* mainly exist in the root, two experiments to identify transcript isoforms and quantify them using two sequencing platforms were carried out. First, nine mRNA samples of three root tissues (periderm, phloem and xylem; triplicate) were subjected to 2 × 100 pair-end sequencing using the HiSeq 2500 platform, and then, 489,309,772 reads were produced. Second, full-length cDNAs of nine pooled poly(A) RNA samples were normalized and subjected to single-molecule long-read survey using the PacBio platform. In total, 1,202,336 raw reads (4.8 billion bases) of PacBio *RS* were generated. After filtering using the RS_Subreads.1 of PacBio software, 796,011 subreads representing 4.3 billion bases were obtained. Next, the RS_IsoSeq.1 protocols were performed to classify, cluster and map the sequence reads to reference genome. A total of 70,761 consensus CCS reads and 725,250 subreads, which could not produce CCS reads because of insufficient sequencing, were obtained. Finally, 223,368 full-length reads were obtained by detecting the poly(A), 5′ primers and 3′ primers.

All of the long reads were mapped against the *S. miltiorrhiza* genome, and 96% of the reads were identified as one or more mappings using BLAST. Although 70,761 CCS reads compared with the reference genome and showed constant quality, 90% of long reads could not undergo CCS. To resolve the high error rates of non-CCS reads, 636,855 long reads were corrected to high constant using about 50 M NGS reads as input data via LSC software (Au et al. 2012).

The transcripts assembled from Illumina short reads by Trinity could not cover entire RNA molecules, although the coverage was sufficiently high (about 200×) (Grabherr et al. 2011). Approximately, 61% of transcripts assembled from NGS reads were less than 600 bp, but only 4% of transcripts assembled from PacBio reads were less than 600 bp. Meanwhile, the mean full-length read lengths from different libraries (<1 kb, 1–2 kb, 2–3 kb and >3 kb) produced by PacBio were 923, 1283, 2026 and 3020 bp, respectively. After removing redundant sequences to all TGS subreads using cd-hit-est ($c = 0.90$), 160,468 non-redundant reads were produced and the mean read length was 2059 bp.

The SpliceMap (Au et al. 2010) was used to analyze 60,584,058 Illumina short reads from velamen tissue of *S. miltiorrhiza*, and then, 113,613 junctions were detected, 110,715 of which were chosen after nUM filtering with approximately 95% high sensitivity. A total of 1,313,216 long reads combined corrected long reads that have lost some flanking sequences and original subreads as input long data were subjected to IDP with detected junctions to detect and predict gene isoforms (Au et al. 2013). To identify alternative splicing in *S. miltiorrhiza*, multiexon genes were mainly focused. Approximately, 84% of long reads (1,109,011) covered at least one junction detected using SpliceMap. Although there are 30,478 annotated gene loci and 26,064 multiexon genes in the *S. miltiorrhiza* genome, the total number of multiexon gene loci identified by cufflinks(Trapnell et al. 2012) in the root is only 10,245 with the expression of FPKM > 10 (fragments per kilobase of exon model per million mapped reads). In total, 4035 gene isoforms compared with annotated multiexon gene transcripts with an FPKM > 10 were directly detected using IDP with a 36% detection rate. According to the spliced alignment of long reads, 16,241 gene isoforms were predicted, which expanded the sensitivity of isoform identification up to 65%.

4.4.2 Identification of Alternatively Spliced Isoforms Using Full-Length Transcriptome Data

To confirm the distribution of different alternatively spliced types in *S. miltiorrhiza*, all of the junctions and isoforms detected using SpliceMap and IDP were further analyzed. A total of 12,264 identified junctions contained annotated junctions and novel junctions, which represented alternatively spliced (AS) isoforms of known genes. Twenty-one percent and four percent of the alternative splicing were resulted from intron retention events and exon skipping events, respectively. Eighteen percent and thirty-nine percent of the junctions were characterized as alternative 5′ splice site events and alternative 3′ splice site events, respectively.

After mapping all of the detected and predicted isoforms to gene loci, 10,323 gene loci were covered. About 40% of detected gene loci expressing alternatively spliced isoforms were characterized. Among them, 25 and 10% of gene loci harbored two and three isoforms, respectively. Another 5% of gene loci were identified to be more than or equal to four isoforms. Two gene loci, including the StAR (steroidogenic acute regulatory protein)-related protein of SMil_00002871 and transporter from the MFS family (major facilitator family) of SMil_00012576, could express seven isoforms. According to the expression statistics of different isoforms using IDP and cufflinks, only one or two of all isoforms from 94% of the gene loci were dominantly expressed.

The complexity of alternative splicing events played a potential role in tanshinone biosynthesis. Six gene loci from 41 expressed terpenoid synthases and three gene loci from 33 candidate tanshinone biosynthesis-related CYP450s underwent alternative splicing. Compared with 40% AS events in *S. miltiorrhiza*, the terpenoid synthases and CYP450s showed lower alternative splicing efficiency. It indicates that the high conservatism of evolution and function in terpenoid synthases and CYP450s can result in high constitutive splicing in *S. miltiorrhiza*. Moreover, based on full-length transcriptome and gene co-expression analysis, candidate CYP450s involved in the catalysis of miltiradiene and ferruginol to tanshinones were identified. Further study of candidate genes with transgenics and biochemical reaction might clearly decode tanshinone biosynthesis.

4.5 Conclusion

Genome survey showed the high heterozygosity of *S. miltiorrhiza*. In order to produce a high-quality genome, the PacBio long reads and Illumina short reads were used to perform the de novo assembly of *S. miltiorrhiza* genome. The resulting draft genome of *S. miltiorrhiza* provides new insights for the biosynthesis of active compounds, such as tanshinones and phenolic acids. Based on genome annotation, CPS, KSL, candidate CYP450s, SDRs and 2OGDs were identified and determined to be involved in tanshinone biosynthesis (Xu and Song 2017). In addition, four TPS/CYP clusters, including CPS1, CPS2, CPS5 and CPS7, were found in the draft *S. miltiorrhiza* genome. It provides important information for analysis of tanshinone biosynthesis. However, the assembly statistics of *S. miltiorrhiza* genome need to be improved for discovering gene clusters of tanshinone biosynthesis. Following the significant progress of sequencing platforms, we will try to sequence and de novo assemble the *S. miltiorrhiza* genome using new PacBio Sequel or Oxford Nanopore platforms, which produce the average read length over 10 kb. In addition, the Hi-C technology could be used to construct the chromosomes for high-quality genome assembly (Belton et al. 2012; Burton et al. 2013; Servant et al. 2015). Furthermore, the chloroplast genome of *S. miltiorrhiza* was assembled to trace the phylogenetic relationship with the closely related plants, such as sesame.

The study in full-length transcriptome in *S. miltiorrhiza* is the first attempt to directly sequence the full-length transcriptome of plants. It provides good model for the identification of full-length transcripts and AS events via hybrid sequencing. In previous studies, applying NGS analysis to identify novel introns and splicing variants indicated that up to 60% of multiexonic genes underwent alternative splicing events in different plants, such as *Arabidopsis thaliana* (Marquez et al. 2012), *Glycine max* (Shen et al. 2014), *Brachypodium distachyon* (Walters et al. 2013) and *Oryza sativa* (Reddy et al. 2013). Although the NGS data have shown impressive performance in the detection of spliced junctions using SpliceMap or Tophat, the length of assemble transcripts limited the detection of isoforms. Combined PacBio long reads and NGS short reads could sensitively detect and predict the isoforms and alternative splicing events in the medicinal plant *S. miltiorrhiza* (Xu et al. 2015). Therefore, with the progress of sequencing technology and bioinformatics, high-quality omics of *S. miltiorrhiza*, including genome and transcriptome, will be further elucidated.

Acknowledgements This work was supported by the National Natural Science Foundation of China (31700264).

References

Au KF, Jiang H, Lin L, Xing Y, Wong WH (2010) Detection of splice junctions from paired-end RNA-seq data by SpliceMap. Nucleic Acids Res 38 (14):4570–4578

Au KF, Sebastiano V, Afshar PT, Durruthy JD, Lee L, Williams BA, van Bakel H, Schadt EE, Reijo-Pera RA, Underwood JG, Wong WH (2013) Characterization of the human ESC transcriptome by hybrid sequencing. Proc Natl Acad Sci USA 110(50):E4821–E4830

Au KF, Underwood JG, Lee L, Wong WH (2012) Improving PacBio long read accuracy by short read alignment. PLoS ONE 7(10):e46679

Belton JM, McCord RP, Gibcus JH, Naumova N, Zhan Y, Dekker J (2012) Hi-C: a comprehensive technique to capture the conformation of genomes. Methods 58 (3):268–276

Burton JN, Adey A, Patwardhan RP, Qiu R, Kitzman JO, Shendure J (2013) Chromosome-scale scaffolding of de novo genome assemblies based on chromatin interactions. Nat Biotechnol 31(12):1119–1125

Cantarel BL, Korf I, Robb SM, Parra G, Ross E, Moore B, Holt C, Sanchez Alvarado A, Yandell M (2008) MAKER: an easy-to-use annotation pipeline designed for emerging model organism genomes. Genome Res 18(1):188–196

Chen SL, Song JY, Sun C, Xu J, Zhu YJ, Verpoorte R, Fan TP (2015) Herbal genomics: examining the biology of traditional medicines. Science 347(6219 Suppl):S27–S28

Cheng TO (2006) Danshen: a popular chinese cardiac herbal drug. J Am Coll Cardiol 47(7):1498

Cui G, Duan L, Jin B, Qian J, Xue Z, Shen G, Snyder JH, Song J, Chen S, Huang L, Peters RJ, Qi X (2015) Functional divergence of diterpene syntheses in the medicinal plant *Salvia miltiorrhiza*. Plant Physiol 169 (3):1607–1618

De Bie T, Cristianini N, Demuth JP, Hahn MW (2006) CAFE: a computational tool for the study of gene family evolution. Bioinformatics 22(10):1269–1271

Dong Y, Morris-Natschke SL, Lee KH (2011) Biosynthesis, total syntheses, and antitumor activity of tanshinones and their analogs as potential therapeutic agents. Nat Prod Rep 28(3):529–542

Gao W, Hillwig ML, Huang L, Cui G, Wang X, Kong J, Yang B, Peters RJ (2009) A functional genomics approach to tanshinone biosynthesis provides stereochemical insights. Org Lett 11(22):5170–5173

Gao W, Sun HX, Xiao H, Cui G, Hillwig ML, Jackson A, Wang X, Shen Y, Zhao N, Zhang L, Wang XJ, Peters RJ, Huang L (2014) Combining metabolomics and transcriptomics to characterize tanshinone biosynthesis in *Salvia miltiorrhiza*. BMC Genom 15:73

Grabherr MG, Haas BJ, Yassour M, Levin JZ, Thompson DA, Amit I, Adiconis X, Fan L, Raychowdhury R, Zeng Q, Chen Z, Mauceli E, Hacohen N, Gnirke A, Rhind N, di Palma F, Birren BW, Nusbaum C, Lindblad-Toh K, Friedman N, Regev A (2011) Full-length transcriptome assembly from RNA-Seq data without a reference genome. Nat Biotechnol 29 (7):644–652

Guo J, Ma X, Cai Y, Ma Y, Zhan Z, Zhou YJ, Liu W, Guan M, Yang J, Cui G, Kang L, Yang L, Shen Y, Tang J, Lin H, Jin B, Liu Z, Peters RJ, Zhao ZK, Huang L (2016) Cytochrome P450 promiscuity leads to a bifurcating biosynthetic pathway for tanshinones. New Phytol 210(2):525–534

Guo J, Zhou YJ, Hillwig ML, Shen Y, Yang L, Wang Y, Zhang X, Liu W, Peters RJ, Chen X, Zhao ZK, Huang L (2013) CYP76AH1 catalyzes turnover of miltiradiene in tanshinones biosynthesis and enables heterologous production of ferruginol in yeasts. Proc Natl Acad Sci USA 110(29):12108–12113

Kai G, Xu H, Zhou C, Liao P, Xiao J, Luo X, You L, Zhang L (2011) Metabolic engineering tanshinone biosynthetic pathway in *Salvia miltiorrhiza* hairy root cultures. Metab Eng 13(3):319–327

Li L, Stoeckert CJ Jr, Roos DS (2003) OrthoMCL: identification of ortholog groups for eukaryotic genomes. Genome Res 13(9):2178–2189

Li Y, Sun C, Luo HM, Li XW, Niu YY, Chen SL (2010) Transcriptome characterization for *Salvia miltiorrhiza* using 454 GS FLX. Yao Xue Xue Bao 45(4):524–529

Luo H, Zhu Y, Song J, Xu L, Sun C, Zhang X, Xu Y, He L, Sun W, Xu H, Wang B, Li X, Li C, Liu J, Chen S (2014) Transcriptional data mining of *Salvia miltiorrhiza* in response to methyl jasmonate to examine the mechanism of bioactive compound biosynthesis and regulation. Physiol Plant 152(2):241–255

Ma XH, Ma Y, Tang JF, He YL, Liu YC, Ma XJ, Shen Y, Cui GH, Lin HX, Rong QX, Guo J, Huang LQ (2015) The biosynthetic pathways of tanshinones and phenolic acids in *Salvia miltiorrhiza*. Molecules 20 (9):16235–16254

Ma Y, Yuan L, Wu B, Li X, Chen S, Lu S (2012) Genome-wide identification and characterization of novel genes involved in terpenoid biosynthesis in *Salvia miltiorrhiza*. J Exp Bot 63(7):2809–2823

Marquez Y, Brown JW, Simpson C, Barta A, Kalyna M (2012) Transcriptome survey reveals increased complexity of the alternative splicing landscape in *Arabidopsis*. Genome Res 22(6):1184–1195

Pharmacopoeia Commission of People's Republic of China (2015) Chinese Pharmacopoeia 2015 edition Volume 1. China Medical Science Press, Beijing

Qian J, Song J, Gao H, Zhu Y, Xu J, Pang X, Yao H, Sun C, Li X, Li C, Liu J, Xu H, Chen S (2013) The complete chloroplast genome sequence of the medicinal plant *Salvia miltiorrhiza*. PLoS ONE 8(2):e57607

Reddy AS, Marquez Y, Kalyna M, Barta A (2013) Complexity of the alternative splicing landscape in plants. Plant Cell 25(10):3657–3683

Servant N, Varoquaux N, Lajoie BR, Viara E, Chen CJ, Vert JP, Heard E, Dekker J, Barillot E (2015) HiC-Pro: an optimized and flexible pipeline for Hi-C data processing. Genome Biol 16:259

Shen Y, Zhou Z, Wang Z, Li W, Fang C, Wu M, Ma Y, Liu T, Kong LA, Peng DL, Tian Z (2014) Global dissection of alternative splicing in paleopolyploid soybean. Plant Cell 26(3):996–1008

Song JY, Luo HM, Li CF, Sun C, Xu J, Chen SL (2013) *Salvia miltiorrhiza* as medicinal model plant. Yao Xue Xue Bao 48(7):1099–1106

Trapnell C, Roberts A, Goff L, Pertea G, Kim D, Kelley DR, Pimentel H, Salzberg SL, Rinn JL, Pachter L (2012) Differential gene and transcript expression analysis of RNA-seq experiments with TopHat and Cufflinks. Nat Protoc 7(3):562–578

Walters B, Lum G, Sablok G, Min XJ (2013) Genome-wide landscape of alternative splicing events in *Brachypodium distachyon*. DNA Res 20(2): 163–171

Wang B, Sun W, Li Q, Li Y, Luo H, Song J, Sun C, Qian J, Zhu Y, Hayward A, Xu H, Chen S (2015) Genome-wide identification of phenolic acid biosynthetic genes in *Salvia miltiorrhiza*. Planta 241(3): 711–725

Xin T, Zhang Y, Pu X, Gao R, Xu Z, Song J (2019) Trends in herbgenomics. Sci China Life Sci 62 (3):288–308

Xu Z, Peters RJ, Weirather J, Luo H, Liao B, Zhang X, Zhu Y, Ji A, Zhang B, Hu S, Au KF, Song J, Chen S (2015) Full-length transcriptome sequences and splice variants obtained by a combination of sequencing platforms applied to different root tissues of *Salvia miltiorrhiza* and tanshinone biosynthesis. Plant J 82 (6):951–961

Xu H, Song J, Luo H, Zhang Y, Li Q, Zhu Y, Xu J, Li Y, Song C, Wang B, Sun W, Shen G, Zhang X, Qian J, Ji A, Xu Z, Luo X, He L, Li C, Sun C, Yan H, Cui G, Li X, Li X, Wei J, Liu J, Wang Y, Hayward A, Nelson D, Ning Z, Peters RJ, Qi X, Chen S (2016a) Analysis of the genome sequence of the medicinal plant *Salvia miltiorrhiza*. Mol Plant 9(6):949–952

Xu Z, Luo H, Ji A, Zhang X, Song J, Chen S (2016b) Global identification of the full-length transcripts and alternative splicing related to phenolic acid biosynthetic genes in *Salvia miltiorrhiza*. Front Plant Sci 7:100

Xu ZC, Ji AJ, Zhang X, Song JY, Chen SL (2016c) Biosynthesis and regulation of active compounds in medicinal model plant *Salvia miltiorrhiza*. Chin Herbal Med 8(1):3–11

Xu Z, Song J (2017) The 2-oxoglutarate-dependent dioxygenase superfamily participates in tanshinone production in *Salvia miltiorrhiza*. J Exp Bot 68(9): 2299–2308

Zhou L, Zuo Z, Chow MS (2005) Danshen: an overview of its chemistry, pharmacology, pharmacokinetics, and clinical use. J Clin Pharmacol 45(12):1345–1359

The Chloroplast and Mitochondrial Genomes of *Salvia miltiorrhiza*

5

Haimei Chen and Chang Liu

Abstract

Salvia miltiorrhiza is one of the most widely used medicinal plants. Here, we discuss research progresses made on its chloroplast and mitochondrial genomes. In these studies, strand-specific RNA-Seq and single-molecule real-time (SMRT) sequencing analyses were conducted. Then, the RNA-Seq reads were mapped to the genome assembly to determine the relative expression levels of genes, DNA modifications and RNA editing events. For the chloroplast, the expression levels of all 80 protein-coding genes and 136 putative anti-sense and intergenic noncoding RNA (ncRNA) genes were detected. A total of 2687 putative modification sites were identified. Further analysis identified two DNA modification motifs: "TATANNNATNA" and "WNYANTGAW". For the mitochondrial genome, the assembled genome has been validated extensively. A pipeline was developed to predict the RNA editing events using REDItools. A total of 1123 editing sites were identified, including 225 "C" to "U" sites in the protein-coding regions. The nucleotides on both strands at 115 of the 225 sites had undergone RNA editing, which were called symmetrical RNA editing (SRE). Taken together, a complex interplay among DNA transcriptome, modifications in *S. miltiorrhiza* plastome has been reported. In addition, symmetrical RNA editing events have been identified in its mitochondrial genome.

H. Chen · C. Liu (✉)
Key Laboratory of Bioactive Substances and Resources Utilization of Chinese Herbal Medicine, Ministry of Education, Institute of Medicinal Plant Development, Chinese Academy of Medical Sciences and Peking Union Medical College, No. 151 Malianwa North Road, Haidian District, Beijing 100193, People's Republic of China
e-mail: cliu@implad.ac.cn

H. Chen
e-mail: hmchen@implad.ac.cn

5.1 Introduction

Salvia miltiorrhiza Bunge (Danshen in Chinese) is a deciduous perennial flowering plant in the family Lamiaceae from the order Lamiales. It is one of the mostly used traditional Chinese medicinal herbs, widely cultivated in China (Zhong et al. 2009). The dried roots of *S. miltiorrhiza*, commonly known as "Chinese sage" or "red sage" in western countries, are widely used in the treatment of several diseases, including but not limited to cardiovascular, cerebrovascular and hyperlipidemia diseases (Wang 2010; Zhou et al. 2005; Cheng 2007; Chan et al. 2010). To date, more than 70 compounds have been isolated and structurally identified from the root of *S. miltiorrhiza* (Li et al. 2008, 2009). These compounds can be divided into two major groups: the hydrophilic phenolic acids, including

S. Lu (ed.), *The Salvia miltiorrhiza Genome*, Compendium of Plant Genomes,
https://doi.org/10.1007/978-3-030-24716-4_5

rosmarinic, lithospermic and salvianolic acids; and the lipophilic components, including diterpenoids and tanshinones (Wang 2010; Li et al. 2008). In addition to the significant medicinal value described above, *S. miltiorrhiza* is exemplary for its relatively small genome size (~600 Mb), short life cycle and genetic transformability (Ma et al. 2012; Lee et al. 2008; Yan and Wang 2007; Xiao et al. 2011). These characteristics make *S. miltiorrhiza* an attractive model to investigate the mechanism of medicinal plant secondary metabolism.

Chloroplasts and mitochondria are all integral part of plant cells. Chloroplasts and mitochondria both contain their own genomes, which interact with each other as well as those from the nucleus. Recent studies have shown that chloroplasts are ideal hosts for the expression of genes related to the production of secondary metabolites (Verma and Daniell 2007). A detailed understanding of the gene content of the chloroplast genome (plastome), and the expression and regulation of these genes is a prerequisite for the development of effective chloroplast-based transgenic systems. For these purposes, detailed studies of the transcriptome and methylome of the plastome of *S. miltiorrhiza* have been carried out. Similarly, mitochondria play important roles in many cellular metabolic pathways and protein syntheses, illustrating its gene content, gene expression and regulation will improve our understanding regarding the interaction among the nucleus and the mitochondria. In this chapter, we discuss the recent research progress made regarding the chloroplast and mitochondrial genome of *S. miltiorrhiza*.

5.2 Transcriptome of *S. miltiorrhiza* Chloroplasts

To date, two draft plastome sequences of *S. miltiorrhiza* have been reported (Xu et al. 2016; Zhang et al. 2015). Strand-specific RNA-Seq and SMRT technologies were used to characterize the genome, transcriptome and DNA modifications of the *S. miltiorrhiza* plastome.

Chloroplast genes are of prokaryotic origin. Therefore, a protocol designed for prokaryotic transcriptome can be used to analyze that of *S. miltiorrhiza*. Using Tophat (Trapnell et al. 2012), 553,014 (1.0%) of the reads were mapped to the *S. miltiorrhiza* plastome. The overall results are summarized in Fig. 5.1, which include the following information: the putative DNA modification sites on the positive strand (a) and negative strand (b); the expression level of noncoding genes on the positive strand (c) and negative strand (d); the expression level of coding genes on the positive strand (e) and negative strand (f); and the identified polycistrons on the positive strands (g) and negative strands (h).

Except for 14 out of 18 tRNA genes, all genes were expressed, including 80 protein-coding genes, 4 rRNA genes and 16 tRNA genes (Table 5.1). By contrast, 16 tRNA genes were expressed, 9 of which were embedded in polycistrons (Table 5.1) and 3 have large introns. The expressed four tRNA genes (*trnF-GAA*, *trnN-GUU*, *trnS-GGA* and *trnQ-UUG*) were embedded in long noncoding transcripts. These expression levels of these genes were quantified using the cuffdiff software. The most highly expressed genes were the rRNA 23S and rRNA 16S, as well as *psbA*, *psbB*, *psbC*, *psbD* and *rbcL*. This result is consistent with the involvement of rRNA in the protein synthesis as well as with the participation of *psb* genes and *rbcL* in photosynthesis, two of the most important biosynthesis pathways. Two tRNA genes, namely *trnA-UGC* and *trnI-UUC*, were highly expressed at levels comparable to those of rRNA genes and *psb* genes. The *trnA-UGC* and *trnI-GAU* transcripts accounted for 60.60 and 38.18% of all tRNA transcripts, respectively, even though alanine and isoleucine only account for 2/30 (6.7%) of the amino acids in the *S. miltiorrhiza* plastome. These two tRNA genes are located in polycistron pc17, which contains the rRNA genes. Their expression levels are consistent with those of other genes in the same polycistron. Most genes from plastomes are transcribed in polycistrons (Stern et al. 1997). In total, there are 19 polycistronic transcripts containing 71 genes, which include 58 protein-coding genes, 9 tRNA and all 4 rRNA genes.

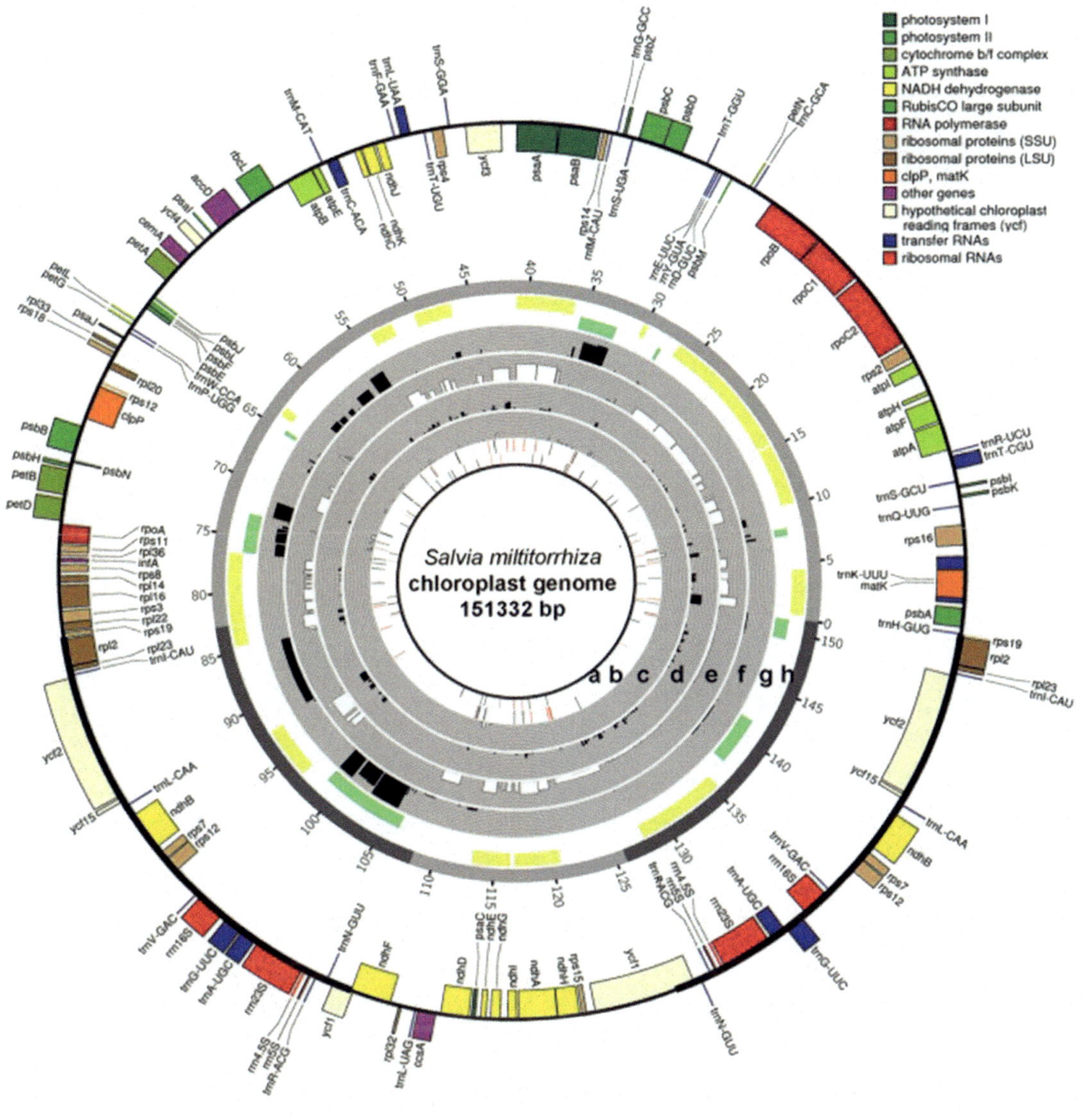

Fig. 5.1 An ideogram showing the transcriptome and DNA modifications of the *S. miltiorrhiza* plastome (Chen et al. 2014b). The predicted genes are drawn on the outermost circle. Those drawn outside the circle are transcribed in a clockwise direction, whereas those drawn inside the circle are transcribed in a counterclockwise direction. The distribution of DNA modification motif 1 on the positive strand (**a**) and the negative strand (**b**) is shown as bars. The red bar indicates motifs that are significantly modified. The expression levels of the noncoding genes on the positive (**c**) and negative strands (**d**); the coding genes on the positive (**e**) and negative strand (**f**) are shown in black and white arcs on grey background, the polycistrons on the positive (**g**) and negative strands (**h**) are shown as green and yellow arcs on white background. The heights of the arcs in (**c**) to (**h**) represent the expression levels of the corresponding transcripts

Noncoding RNAs (ncRNAs) are RNAs that do not encode proteins, but have multiple functions. A number of studies have elucidated the regulatory roles of chloroplast ncRNA. For the first example, an *ndhB* asRNA covers two editing sites of the *ndhB* gene and a group II intron splice acceptor site, and might play an important role in RNA maturation or stability (Georg et al. 2010). For the second example, two different antisense RNAs of *psbT* gene were found to form

Table 5.1 Expression levels of all genes and polycistrons determined using RNA-Seq analysis

Gene name[a]	Start	End	Strand	Arbitrary log abundance	Gene name	Start	End	Strand	Arbitrary log abundance
1 *psb*A	392	1450	N[b]	12.93	*psb*B	71218	72744	P	14.08
1 *trn*K-UUU	1672	4266	N	6.90	13 *psb*T	72921	73028	P	1.85
1 *mat*K	1972	3534	N	5.46	*psb*N	73089	73220	N	5.46
*rps*16	4834	5944	N	5.85	13 psb*H*	73308	73547	P	5.56
2 *psb*K	7380	7559	P[b]	8.21	13 *pet*B	73673	75022	P	8.88
2 *psb*I	7905	8066	P	5.18	13 pet*D*	75216	76418	P	7.88
*trn*G-UCC	8923	9677	P	3.85	14 *rpo*A	76595	77602	N	5.46
3 *atp*A	10042	11565	N	9.67	14 *rps*11	77674	78090	N	6.44
3 *atp*F	11661	12913	N	7.96	14 *rpl*36	78200	78313	N	3.85
3 *atp*H	13187	13432	N	8.74	14 *inf*A	78409	78642	N	5.18
3 *atp*I	14403	15146	N	5.66	14 *rps*8	78769	79173	N	5.54
4 *rps*2	15374	16084	N	7.02	14 *rpl*14	79363	79731	N	5.31
4 *rpo*C2	16293	20480	N	6.98	14 *rpl*16	79863	81143	N	7.06
4 *rpo*C1	20638	23451	N	7.69	14 rps*3*	81279	81941	N	6.66
4 *rpo*B	23478	26690	N	7.06	14 *rpl*22	81926	82393	N	6.31
5 trn*C-GCA*	27834	27904	P	1.70	14 *rps*19	82463	82741	N	3.85
5 *pet*N	28061	28156	P	2.85	15 *rpl*2	82801	84283	N	9.38
*psb*M	29133	29237	N	4.85	2nd copy	149748	151230	P	–
6 *trn*D-GUC	29758	29831	N	1.85	15 *rpl*23	84302	84583	N	7.06
6 *trn*Y-GUA	29942	30025	N	1.85	2nd copy	149448	149729	P	–
6 *trn*E-UUC	30097	30169	N	1.70	*ycf*2	84911	91762	P	7.38
7 *psb*D	32159	33220	P	14.92	2nd copy	142269	149120	N	–
7 *psb*C	33168	34589	P	13.64	16 *trn*L-CAA	92362	92442	N	1.70
7 *psb*Z	35262	35450	P	8.62	2nd copy	141589	141669	P	–
8 *rps*14	36188	36490	N	10.09	16 *ndh*B	93002	95155	N	12.80
8 *psa*B	36613	38817	N	12.50	2nd copy	138876	141029	P	–
8 *psa*A	38843	41095	N	12.58	16 *rps*7	95484	95951	N	12.59
*ycf*3	41886	43817	N	9.64	2nd copy	138080	138547	P	–
*trn*S-GGA	44634	44720	P	3.44	16*3′*-rps*12*	96005	96785	N	10.72
*rps*4	44997	45602	N	6.66	2nd copy	137246	138026	P	–
9 *trn*L-UAA	46728	47267	P	5.02	17 *rrn*16S	98690	100180	P	18.34
9 trn*F-GAA*	47566	47638	P	2.85	2nd copy	133851	135341	N	–
9 *ndh*J	48171	48647	N	6.94	17 *trn*I-GAU	100479	101490	P	13.16
9 *ndh*K	48758	49435	N	9.43	2nd copy	132541	133552	N	–
9 *ndh*C	49487	49849	N	8.90	17 *trn*A-UGC	101555	102422	P	13.82
*trn*V-UAC	50799	51447	N	5.66	2nd copy	131609	132476	N	–
10 *atp*E	51904	52305	N	5.66	17 *rrn*23S	102581	105397	P	18.79
10 *atp*B	52302	53798	N	9.76	2nd copy	128634	131450	N	–

(continued)

Table 5.1 (continued)

Gene name[a]	Start	End	Strand	Arbitrary log abundance	Gene name	Start	End	Strand	Arbitrary log abundance
*rbc*L	54563	56017	P	13.52	17 *rrn*4.5S	105497	105599	P	7.56
*acc*D	56708	58195	P	8.57	2nd copy	128432	128534	N	–
*psa*I	58658	58768	P	1.85	17 *rrn*5S	105824	105954	P	2.85
*ycf*4	59224	59778	P	5.31	2nd copy	128077	128207	N	–
*cem*A	60326	61015	P	5.44	17 *trn*R-ACG	106188	106261	P	3.85
*pet*A	61221	62183	P	6.66	2nd copy	127770	127843	N	–
11 *psb*J	63245	63367	N	4.44	*ndh*F	108206	110422	N	7.85
11 *psb*L	63497	63613	N	6.56	*rpl*32	110875	111045	P	1.85
11 *psb*F	63637	63756	N	7.56	*ccs*A	111907	112884	P	2.85
11 *psb*E	63771	64022	N	9.60	18 ndh*D*	113116	114630	N	9.29
12 *pet*L	64829	64924	P	1.70	18 *psa*C	114766	115011	N	5.31
12 *pet*G	65103	65216	P	1.85	18 *ndh*E	115262	115567	N	6.76
*psa*J	65939	66073	P	5.66	18 *ndh*G	115783	116313	N	7.88
*rps*18	66890	67195	P	2.85	19 *ndh*I	116689	117195	N	7.35
*rpl*20	67427	67813	N	4.18	19 *ndh*A	117276	119352	N	9.72
*5′-rps*12	68623	68736	N	1.85	19 *ndh*H	119354	120535	N	7.31
*clp*P	68860	70769	N	6.71	*ycf*1	121333	126849	N	4.66

This table was reproduced from our previous publication (Chen et al. 2014b)
[a] The number in front of each gene name indicate the polycistron number
[b] P: positive strand; N: negative strand [c] 2nd copy: the 2nd copy of the same gene on the other IR region

double-stranded RNA/RNA hybrids, which results in translational inactivation of the *psbT* mRNA. The hybrid was further hypothesized to provide protection against nucleolytic degradation of mRNA during photo-oxydative stress conditions (Zghidi-Abouzid et al. 2011). A systematic analysis was performed to identify ncRNA in the plastome of *S. miltiorrhiza*. Non-coding RNA (ncRNA) was defined as any transcribed RNA transcripts that do not encode proteins and have a length >100 bp. 136 ncRNA transcripts were identified based on these criteria. These ncRNAs are classified into two groups, namely intergenic ncRNA (Fig. 5.2a) and antisense ncRNA (asRNA, Fig. 5.2b), based on their positions relative to the cRNA genes. Intergenic ncRNA are those located between two transcripts or polycistrons and its distance to the start or end position of each transcript was at least 100 bp (Fig. 5.2a). Three types of ncRNAs, namely A1, A2 and A3, were found. A special case of intergenic ncRNA expresses from both strands and was defined as bilateral ncRNA (A4, Fig. 5.2a).

5.3 Methylation of *S. miltiorrhiza* Plastome

DNA can undergo a large variety of functionally important modifications (He et al. 2011). The most common modification, DNA methylation, involves the addition of a methyl group to nucleotides cytosine or adenine by methyl transferase (MTase). The most common types of methylation are N6-methyladenine (m6A), N4-methylcytosine (m4C) and 5-methylcytosine (m5C) (Clark et al. 2012). In plants, DNA methylation is essential for growth and development, and it affects gene expression, genomic

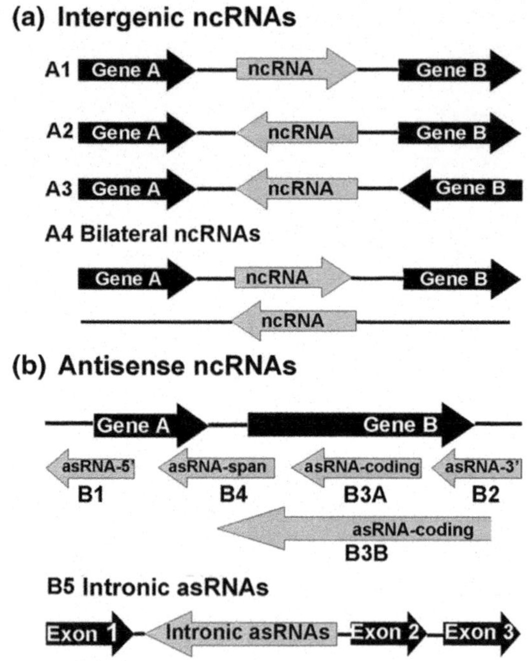

Fig. 5.2 Schematic representations of various types of ncRNA identified in *S. miltiorrhiza* (Chen et al. 2014b). Protein-coding genes and their exons are presented in black, whereas ncRNA are presented in grey. The arrows indicate the direction of the transcripts. **a** Intergenic ncRNA, which are ncRNA expressed in the intergenic region. Four different types, namely A1, A2 and A3, A4 are shown. **b** Antisense ncRNA, which are ncRNA transcribed from the antisense strand of protein-coding genes. This ncRNA overlaps with either the 5′ (B1) end or 3′ (B2) end of the coding region, is embedded in (B3A) or spans (B3B) one entire protein-coding gene, or spans two protein-coding genes (B4). Intronic asRNA (B5) is transcribed entirely from an intron of a coding gene

imprinting, heterochromatin assembly and protection of the genome against migrating transposable elements (Martienssen and Colot 2001; Henderson and Jacobsen 2007). Identification of potential DNA modification sites in the *S. miltiorrhiza* plastome would provide valuable information on the epigenetic regulation of its gene expression.

SMRT technology can provide detailed kinetic information of the DNA polymerase during DNA replication through the detection of the fluorescent signal during the period when the DNA polymerase is "holding" a nucleotide. These form fluorescent "pulses" as DNA polymerase continue incorporating nucleotides into the elongating DNA strand. The interpulse duration (IPD) is the time difference between two "pulses". IPD ratio is calculated as [IPD at a site at a condition under study]/[IPD at the same site at the control condition]. Based on the IPD ratio, the DNA modification status of the base can be inferred (Clark et al. 2012). Using the DNA modification analysis module of SMRT Portal 1.3.2, we identified 2687 ($p < 0.01$) putatively modified bases throughout the plastome (Fig. 5.3a). Among these, 90.3% IPD ratio ranged from −2 to 2. Two DNA motifs using the motif analysis module were identified, namely "TATANNNATNA" (Fig. 5.3b) and "WNYANTGAW" (Fig. 5.3c), whose percentage of putatively modified sites was 35/97 (36.1%) and 91/369 (24.7%), respectively (Table 5.2).

5.4 Basic Information for *S. miltiorrhiza* Mitochondrial Genome

Mitochondria are membrane-bound organelles found in most eukaryotic cells and play important roles in energy conversion, the storage of calcium ions and other metabolic tasks, such as regulation of the membrane potential, steroid synthesis and so on (Henze and Martin 2003). The plant mitochondrial genomes (mitogenomes) have been known to be highly polymorphic. As a result, validation of any genome assembly is of greater importance for mitochondrial genomes. Three methods were used to validate the assembly of the mitogenome sequence of *S. miltiorrhiza*. Firstly, short reads generated from the same species and sequenced using the Illumina platform were mapped to the reference genome. The entire assembly was covered with short reads at an average coverage of 1390×. Secondly, long reads generated from the same species and sequenced with the PacBio RS II platform were mapped to the reference genome (Chen et al. 2014a). The average coverage was 30× and only 0.72% of the reference genome had zero coverage. Last, the predicted CDS sequences were extracted from the reference genome and translated to the protein

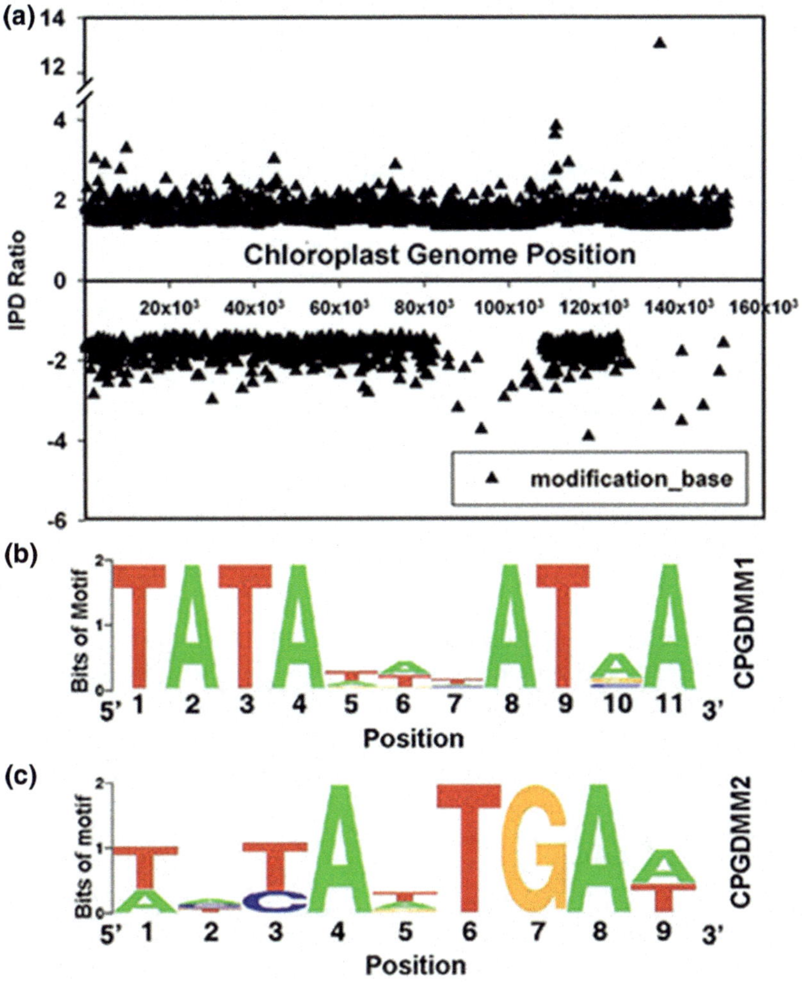

Fig. 5.3 Identification of putative DNA modification sites and motifs in the *S. miltiorrhiza* plastome using SMRT technology (Chen et al. 2014b). **a** Distribution of interpulse duration (IPD) ratio across the genome. "m" indicates that the corresponding IPD ratio ("m") has a $p < 0.05$, which suggests that the base at this position is modified. **b** Sequence LOGO for putative DNA modification motif 1 (DMM1). **c** Sequence LOGO for putative DNA modification motif 2 (DMM2)

Table 5.2 Putative DNA modification motifs associated with base modifications

DMM id	Motif sequence	Total no. of motifs in genome	No. of motifs modified	Fraction of motifs modified	Mean IPD ratio	Mean coverage	Position modified
CPGDMM1	TATANNNATNA	97	35	0.36	1.71	66.00	−1
CPGDMM2	WNYANTGAW	369	91	0.25	1.77	84.71	−3

This table was reproduced from our previous publication (Chen et al. 2014b)

sequences. The protein sequences were then used to search the public databases to identify their homologous sequences, which were then subjected to multiple sequence alignment. The alignments of the mitochondrial proteins of *S. miltiorrhiza* and their homologs show that no significant discrepancy was observed. Results from these three lines of analyses suggest that the reference genome is of high quality, particular for the regions encoding proteins, can be used for downstream analysis.

5.5 RNA Editing of *S. miltiorrhiza* Mitogenome

RNA editing is an important mechanism to increase the diversities of transcriptomes and proteomes in eukaryotic organisms through post-transcriptional modifications of mRNA sequences (Gott 2003; Mallela and Nishikura 2012). The phenomenon has been observed in the nuclei of higher eukaryotes (Gott 2003; Gott and Emeson 2000) and plant mitochondria (Giege and Brennicke 1999; Handa 2003; Mower and Palmer 2006; Notsu et al. 2002; Picardi et al. 2010) and plastids (Hoch et al. 1991). The modifications of mRNA sequences include the insertion, deletion and substitution of nucleotides; among them, base substitution is the most frequently observed. RNA editing can occur in the coding and noncoding regions. Most RNA editing sites were observed at the first or second position of a codon, thereby directly altering the coded amino acid (Takenaka et al. 2013). Particularly, RNA editing can create a start or stop codon (Brennicke et al. 1999; Takenaka et al. 2008) or remove a stop codon, thereby producing proteins having different sizes (Brennicke et al. 1999; Grewe et al. 2009). In contrast, RNA editing in the introns and untranslated regions has been found to regulate the stability of particular mRNA molecules (Castandet et al. 2010).

To determine if RNA editing events also occur in the mitochondria of *S. miltiorrhiza*, we carried out an extensive analysis (Wu et al. 2017). Briefly, the sequence for the mitogenome of *S. miltiorrhiza* was downloaded from RefSeq (NC_023209). The RNA-Seq data of the flower, leaf and root samples (SRR1043988, SRR1045051 and SRR1020591) were generated in our previous study (Chen et al. 2014a) and can be downloaded from GenBank SRA database (http://www.bemd.ncbi.nlm.nih.gov/sra). The cleaned reads from the three tissues were mapped to the mitogenome by bowtie2 (version 2.2.1) with mismatch = 7.

The results showed that 569,489 reads from the leaf, 221,727 reads from the flower, and 52,485 reads from the root were successfully mapped (Chen et al. 2014b). These mapped reads were grouped based on the strands to which they were mapped by samtools. RNA editing sites from each strand were identified by REDItools. A total of 1123 sites were found to have undergone RNA editing, and these sites will be called RNA editing site for simplicity. Among them, 575 sites were found on the forward strand and 548 were found on the reverse strand.

Some of the observed nucleotide changes in mRNAs might result from the polymorphism at these sites in the genome DNA. As a result, possible SNP sites in the genome need to be identified and excluded them from the analyses. A low-coverage sequencing ($\sim 40\times$) was conducted with total genomic DNA. The resulting data were subjected to putative SNP site identification in the mitogenome. A total of 63 SNPs and 32 indels were identified; 24 of these sites overlapped with the predicted RNA editing sites and 7 sites overlapped with SRE (see below) sites. These sites were excluded from the following analyses. Among the 1123 putative editing sites, 24 sites were excluded because they overlapped with the predicted high-quality SNP sites. The schematic representation of all predicted RNA editing sites is shown in Fig. 5.4. As shown, the outermost circle showed the location of the annotated genes on the *S. miltiorrhiza* mitogenome as constructed with OrganellarGenomeDRAW(4.1). The inner circles were drawn by custom Perl scripts. The following information are shown in Fig. 5.4: the predicted RNA editing sites on the positive and negative strands; the types of RNA editing sites; the editing frequency of each site; and the 225 "C to U" predicted editing events in the CDS regions.

We then predicted RNA editing sites using data of each tissue. All RNA editing sites found in each tissue. 89 and 97 RNA editing sites on the positive and negative strands were found across all three tissue types. And, 126 and 124 RNA editing sites on the positive and negative

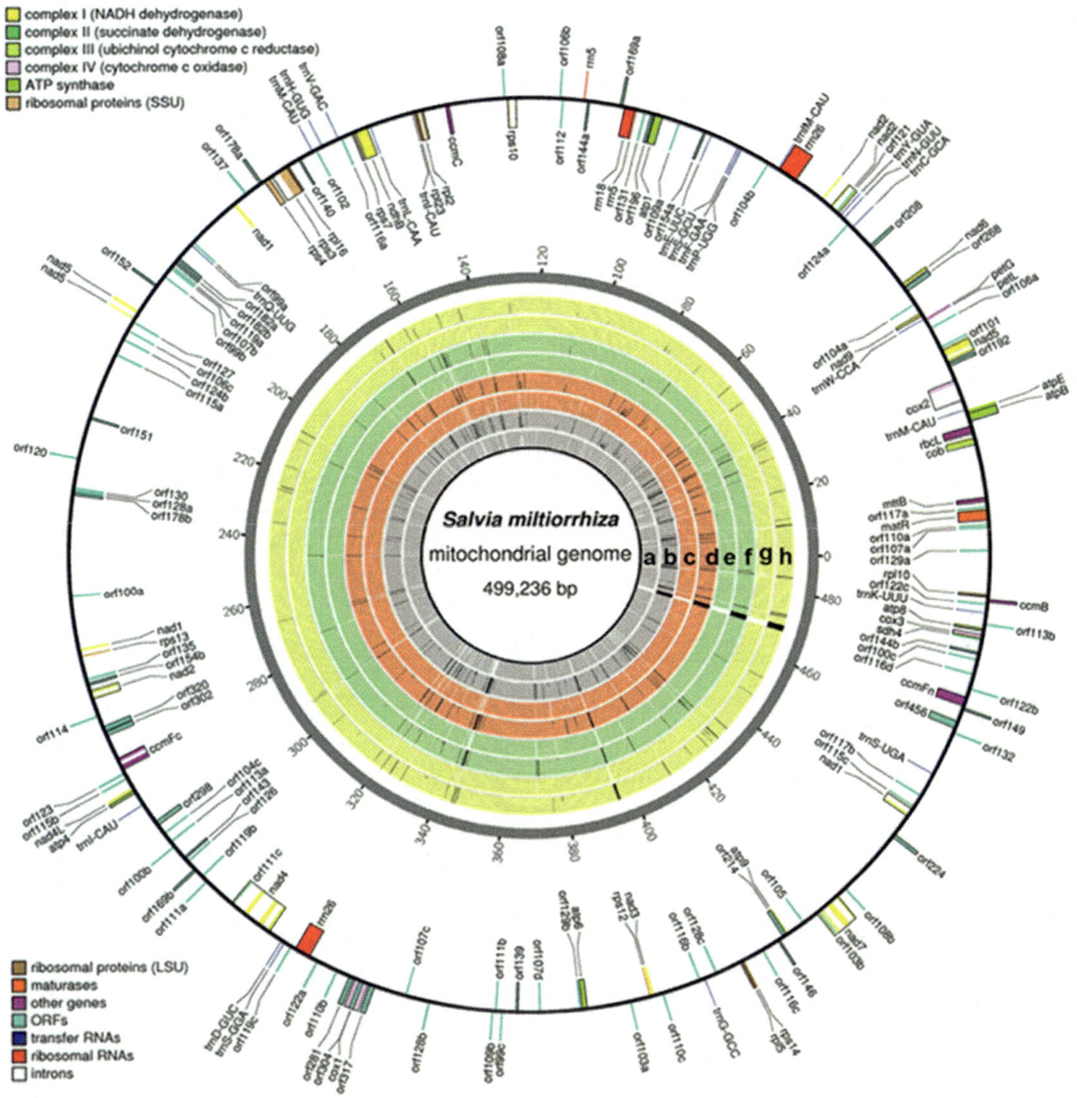

Fig. 5.4 An ideogram showing the location, types and frequencies of RNA editing on the mitogenome of *S. miltiorrhiza* (Wu et al. 2017). The predicted genes are shown on the outermost circle. Genes outside the circle are on the positive strand and oriented clockwise. Genes inside the circle are on the negative strand and oriented anticlockwise. The eight rings correspond to: all RNA editing events found on the positive (**a**) and negative strands (**b**), events found in the flower sample on the positive (**c**) and negative strands (**d**), events found in the leaf sample on the positive (**e**) and negative strands (**f**), and events found in the root sample on the positive (**g**) and negative strands (**h**). The height of the bar represents the relative frequency

strands were only found in the flower tissue, more unique RNA editing sites found in the leaf and root tissues, suggesting that RNA editing events might occur more frequently in the flower tissue than in the other two tissues.

A RNA editing site that was found missing in a particular type of tissue might be due to two reasons. Firstly, no reads were found in the region containing the RNA editing sites. Alternatively, there were reads found in the regions containing the RNA editing sites. However, the coverage and/or the editing frequency did not pass our threshold. Many RNA editing sites failed to be predicted could result from either the failure to pass the threshold or the missing of read coverage. These results highlight the impact of read coverage on the prediction of RNA editing sites.

We filtered out the sites that have undergone more than one type of RNA editing here. As a result, the sites reported here are associated with only one type of RNA editing. Twelve types of RNA editing events (Fig. 5.5) were identified; these events included all possible intra-base transitions: "A to C", "A to G", "A to U", "C to A", "C to G", "C to U", "G to A", "G to C", "G to U", "U to A", "U to C" and "U to G". Overall, the "C to U" transition is the most abundant type, which represents 30.26 and 41.42% of all RNA editing sites on the forward and reverse strands, respectively (Fig. 5.5a). The complementary "G to A" is the second-most abundant type, which represents 25.39 and 17.70% of all RNA editing sites on the forward and reverse strands, respectively. In terms of the editing frequency, the majority of the sites have frequencies between 91 and 100%, including 59.13% on the forward strand and 9.85% on the reverse strand (Fig. 5.5b).

In the CDS regions, the "C to U" transition is also the most abundant type, which represents 64.17 and 14.29% of all events on the forward and reverse strands, respectively (Fig. 5.5c). Similarly, the complementary "G to A" is the

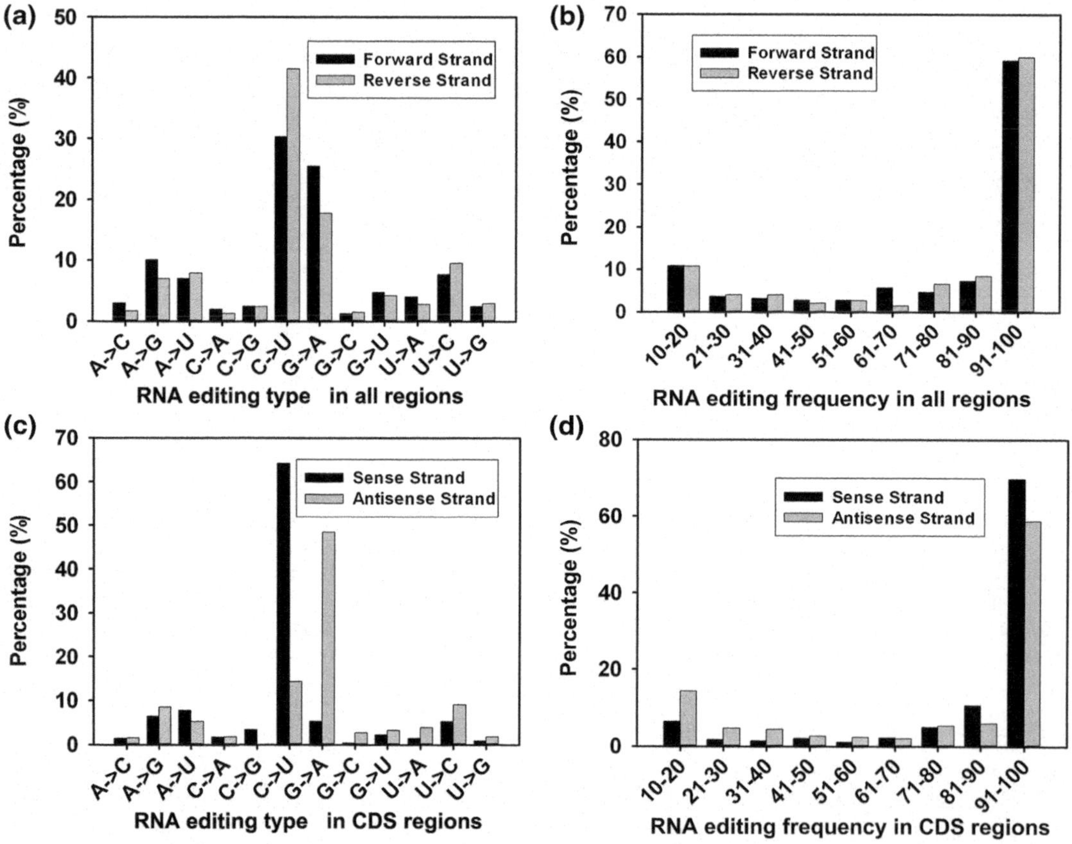

Fig. 5.5 Distribution of RNA editing events on the mitogenome of *S. miltiorrhiza* (Wu et al. 2017). The types and frequencies of RNA editing sites across the entire genome are shown in Panels **a** and **b**, respectively. The types and frequencies of RNA editing sites in the CDS regions are shown in Panels **c** and **d**, respectively

second-most abundant type, which represents 5.28% and 48.40% of all RNA editing sites on the forward and reverse strands, respectively. In terms of the editing frequency, the majority of sites have frequencies between 91 and 100%, including 69.72% on the forward strand and 58.60% on the reverse strand (Fig. 5.5d). Interestingly, an obvious bias of "C to U" sites exists on the strands, such that more "C to U" sites are on the forward strand; this trend probably reflects the fact that more genes are coded on the forward strand.

Sixteen predicted RNA editing sites were randomly selected for experimental validation: cox2-544, cox2-557, cox2-632, cox3-419, cox3-422, nad1-889, nad1-919, nad5-835, nad6-463, nad6-569, rps3-1471, rps-1525, rpl12-196, rpl16-146, orf214-305 and orf456-493. Compared with those observed for the products amplified from the control (gDNA), the predicted "C to U" conversions are visible in the chromatograms at the predicted RNA editing sites for the products amplified from the corresponding cDNA samples, confirming the RNA editing of these sites. For sites such as rps3-1471 and rps3-1525, two overlapping peaks were seen at the predicted RNA editing sites; these peaks might result from the incomplete RNA editing at these sites for all mRNA molecules.

The use of strand-specific RNA-Seq data allowed us to distinguish RNA editing events observed on particular strands. Carefully examining the identified RNA editing sites in the protein-coding regions revealed that the nucleotides on both strands at 115 sites were edited. We designated these RNA editing events as symmetrical RNA editing (SRE). To validate these predicted strand-specific editing events, the cDNA was reverse transcribed with forward and reverse primers specific for four predicted SRE sites: rps3-1525, nad5-835, orf214-305 and rpl16-146. The transcribed products were then subjected to PCR amplification and Sanger DNA sequencing as described. Compared with those amplified from gDNA, the "C to U" editing events on the sense strand (the strand coding for proteins) and "G to A" editing events on the antisense strand were simultaneously observed, agreeing with the predictions.

To determine the potential functional effects of these RNA editing events, the RNA and protein sequences resulting from the 225 "C to U" type of RNA editing were subjected to structure prediction. Some significant changes in the RNA and protein tertiary structures were observed. For example, the RNA secondary structure of the Rps3 gene was changed after editing; the large loop (highlighted with the arrowhead) was divided into two smaller loops (Fig. 5.6a). The tertiary structure of the Rps3 protein was also altered (Fig. 5.6b); part of the domain was separated from the others, thereby suggesting the creation of a protein product with novel functions.

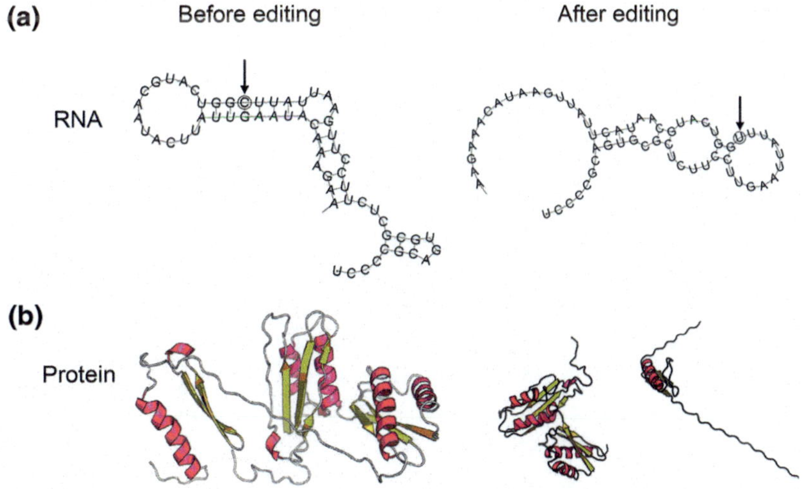

Fig. 5.6 Potential effects of RNA editing at Rps3-1558 on the structures of the mRNA (Wu et al. 2017) (**a**) and protein (**b**) of the *Rps3* gene

5.6 Conclusion and Perspectives

This chapter first reviewed one study taking advantage of two advanced technologies, RNA-Seq and SMRT, to identify asRNA and DNA modifications and then characterize their effects on gene expression in the *S. miltiorrhiza* plastome. RNA-Seq is a revolutionary tool for transcriptomic analysis (Wang et al. 2009). It has several advantages over other transcriptomic methods such as tiling microarray and cDNA or EST sequencing (Martin and Wang 2011), namely independence from genomic sequence, low background noise, wide dynamic range to quantify gene expression level, and high throughput, among others. Moreover, strand-specific RNA-Seq methods provide information on the orientation of transcripts, which is valuable for transcriptome annotation particularly for regions with overlapping transcription from opposite directions (Sultan et al. 2012). SMRT is a novel technology for detecting DNA methylation while simultaneously determining the context of the corresponding DNA sequence (Clark et al. 2012).

The SMRT technology allows the collection of kinetic data for the enzyme during the incorporation of each dNTP into the DNA strand. Significant changes in kinetic parameters such as IPD ratio should be observed when the DNA polymerase encounters m6A, m5C or 5-hmC on the template strand. These distinct kinetic signatures allow the identification of the type and position of the base modification in the DNA template (Clark et al. 2012). The SMRT technology is likely to enhance DNA modification studies on samples that were not previously accessible for this type of research, such as organelle genomes. However, the validation of these prediction results is a challenge.

Mapping the RNA-Seq reads to the genome assembly allowed the determination of the relative expression levels of 80 protein-coding genes. In addition, the expression levels of 19 polycistronic transcription units and 136 putative antisense and intergenic noncoding RNA (ncRNA) genes were also detected. Comparison of the abundance of protein-coding transcripts (cRNA) with and without overlapping antisense ncRNAs (asRNA) suggests that the presence of asRNA is associated with increased cRNA abundance ($p < 0.05$), 2687 potential DNA modification sites and two potential DNA modification motifs were predicted.

Then, the identification of RNA editing sites from stand-specific RNA-Seq data was reviewed for *S. miltiorrhiza* in this chapter (Wu et al. 2017). The predictions were validated using results from reverse transcription, RT-PCR amplification and Sanger sequencing experiments. Putative sequence motifs were characterized. In total, 1123 editing sites, including 225 "C to U" sites in the protein-coding regions were identified. Fourteen of sixteen (87.5%) sites were validated. Three putative DNA motifs were identified around the predicted sites. The nucleotides on both strands at 115 of the 225 sites had undergone SRE. Four of six these SRE sites (66.7%) were experimentally confirmed. These studies have identified complex mechanisms for gene expression regulation in the chloroplast and mitocondria of *S. miltiorrhiza*. Future work needs to setup the appropriate experimental system to validate the predictions and also to illustrate the potential functions of the modifications described here.

Acknowledgements This work has been supported by Chinese Academy of Medical Sciences, Innovation Funds for Medical Sciences (CIFMS) (2016-I2M-3-016, 2017-I2M-1-013); National Natural Science Foundation of China (81872966).

References

Brennicke A, Marchfelder A, Binder S (1999) RNA editing. FEMS Microbiol Rev 23:297–316

Castandet B, Choury D, Begu D, Jordana X, Araya A (2010) Intron RNA editing is essential for splicing in plant mitochondria. Nucleic Acids Res 38(20):7112–7121

Chan E, Tan M, Xin J, Sudarsanam S, Johnson DE (2010) Interactions between traditional Chinese medicines and Western therapeutics. Curr Opin Drug Discov Dev 13(1):50–65

Chen H, Wu B, Nelson DR, Wu K, Liu C (2014a) Computational identification and systematic classification of novel cytochrome P450 genes in *Salvia miltiorrhiza*. PLoS ONE 9(12):e115149

Chen H, Zhang J, Yuan G, Liu C (2014b) Complex interplay among DNA modification, noncoding RNA expression and protein-coding RNA expression in *Salvia miltiorrhiza* plastome. PLoS ONE 9(6):e99314

Cheng T (2007) Cardiovascular effects of Danshen. Int J Cardiol 121(1):9–22

Clark TA, Murray IA, Morgan RD, Kislyuk AO, Spittle KE, Boitano M, Fomenkov A, Roberts RJ, Koriach J (2012) Characterization of DNA methyltransferase specificities using single-molecule, real-time DNA sequencing. Nucleic Acids Res 40(4):e29

Georg J, Honsel A, Voss B, Rennenberg H, Hess WR (2010) A long antisense RNA in plant chloroplasts. New Phytol 186(3):615–622

Giege P, Brennicke A (1999) RNA editing in *Arabidopsis* mitochondria effects 441 C to U changes in ORFs. Proc Natl Acad Sci USA 96(26):15324–15329

Gott JM (2003) Expanding genome capacity via RNA editing. C R Biol 326(10–11):901–908

Gott JM, Emeson RB (2000) Functions and mechanisms of RNA editing. Annu Rev Genet 34:499–531

Grewe F, Viehoever P, Weisshaar B, Knoop V (2009) A trans-splicing group I intron and tRNA-hyperediting in the mitochondrial genome of the lycophyte *Isoetes engelmannii*. Nucleic Acids Res 37(15):5093–5104

Handa H (2003) The complete nucleotide sequence and RNA editing content of the mitochondrial genome of rapeseed (*Brassica napus* L.): comparative analysis of the mitochondrial genomes of rapeseed and *Arabidopsis thaliana*. Nucleic Acids Res 31(20):5907–5916

He G, Elling AA, Deng XW (2011) The epigenome and plant development. Annu Rev Plant Biol 62:411–435

Henderson IR, Jacobsen SE (2007) Epigenetic inheritance in plants. Nature 447:418–424

Henze K, Martin W (2003) Evolutionary biology: essence of mitochondria. Nature 426:127–128

Hoch B, Maier RM, Appel K, Igloi GL, Kossel H (1991) Editing of a chloroplast mRNA by creation of an initiation codon. Nature 353(6340):178–180

Lee CY, Agrawal DC, Wang CS, Yu SM, Chen JJ, Tsay HS (2008) T-DNA activation tagging as a tool to isolate *Salvia miltiorrhiza* transgenic lines for higher yields of tanshinones. Planta Med 74(7):780–786

Li M, Chen J, Peng Y, Wu Q, Xiao P (2008) Investigation of Danshen and related medicinal plants in China. J Ethnopharmacol 120(3):419–426

Li Y, Song L, Liu M, Hu Z, Wang Z (2009) Advancement in analysis of *Salviae miltiorrhizae* Radix et Rhizoma (Danshen). J Chromatogr A 1216(11):1941–1953

Ma Y, Yuan L, Wu B, Li X, Chen S, Lu S (2012) Genome-wide identification and characterization of novel genes involved in terpenoid biosynthesis in *Salvia miltiorrhiza*. J Exp Bot 63(7):2809–2823

Mallela A, Nishikura K (2012) A-to-I editing of protein coding and noncoding RNAs. Crit Rev Biochem Mol Biol 47(6):493–501

Martienssen RA, Colot V (2001) DNA methylation and epigenetic inheritance in plants and filamentous fungi. Science 293(5532):1070–1074

Martin JA, Wang Z (2011) Next-generation transcriptome assembly. Nat Rev Genet 12:671–682

Mower JP, Palmer JD (2006) Patterns of partial RNA editing in mitochondrial genes of *Beta vulgaris*. Mol Genet Genomics 276(3):285–293

Notsu Y, Masood S, Nishikawa T, Kubo N, Akiduki G, Nakazono M, Hirai A, Kadowaki K (2002) The complete sequence of the rice (*Oryza sativa* L.) mitochondrial genome: frequent DNA sequence acquisition and loss during the evolution of flowering plants. Mol Genet Genomics 268(4):434–445

Picardi E, Horner DS, Chiara M, Schiavon Valle G, Pesole G (2010) Large-scale detection and analysis of RNA editing in grape mtDNA by RNA deep-sequencing. Nucleic Acids Res 38(14):4755–4767

Stern DS, Higgs DC, Yang J (1997) Transcription and translation in chloroplasts. Trends Plant Sci 2(8):308–315

Sultan M, Dokel S, Amstislavskiy V, Wuttig D, Sultmann H, Lehrach H, Yaspo ML (2012) A simple strand-specific RNA-Seq library preparation protocol combining the Illumina TruSeq RNA and the dUTP methods. Biochem Biophys Res Commun 422 (4):643–646

Takenaka M, Verbitskiy D, van der Merwe JA, Zehrmann A, Brennicke A (2008) The process of RNA editing in plant mitochondria. Mitochondrion 8 (1):35–46

Takenaka M, Zehrmann A, Verbitskiy D, Hartel B, Brennicke A (2013) RNA editing in plants and its evolution. Annu Rev Genet 47:335–352

Trapnell C, Roberts A, Goff L, Pertea G, Kim D, Kelley DR, Pimentel H, Salzberg SL, Rinn JL, Pachter L (2012) Differential gene and transcript expression analysis of RNA-seq experiments with TopHat and Cufflinks. Nat Protoc 7(3):562–578

Verma D, Daniell H (2007) Chloroplast vector systems for biotechnology applications. Plant Physiol 145 (4):1129–1143

Wang B (2010) *Salvia miltiorrhiza*: chemical and pharmacological review of a medicinal plant. J Med Plants Res 4(25):2813–2820

Wang Z, Gerstein M, Snyder M (2009) RNA-Seq: a revolutionary tool for transcriptomics. Nat Rev Genet 10(1):57–63

Wu B, Chen H, Shao J, Zhang H, Wu K, Liu C (2017) Identification of symmetrical RNA editing events in the mitochondria of *Salvia miltiorrhiza* by strand-specific RNA sequencing. Sci Rep 7:42250

Xiao Y, Zhang L, Gao S, Saechao S, Di P, Chen J, Chen W (2011) The *c4h, tat, hppr* and *hppd* genes prompted engineering of rosmarinic acid biosynthetic pathway in *Salvia miltiorrhiza* hairy root cultures. PLoS ONE 6(12):e29713

Xu H, Song J, Luo H, Zhang Y, Li Q, Zhu Y, Xu J, Li Y, Song C, Wang B, Sun W, Shen G, Zhang X, Qian J, Ji A, Xu Z, Luo X, He L, Li C, Sun C, Yan H, Cui G,

Li X, Wei J, Liu J, Wang Y, Hayward A, Nelson D, Ning Z, Peters RJ, Qi X, Chen S (2016) Analysis of the genome sequence of the medicinal plant *Salvia miltiorrhiza*. Mol Plant 9(6):949–952

Yan Y, Wang Z (2007) Genetic transformation of the medicinal plant *Salvia miltiorrhiza* by *Agrobacterium tumefaciens*-mediated method. Plant Cell Tissue Organ Cult 88(2):175–184

Zghidi-Abouzid O, Merendino L, Buhr F, Malik Ghulam M, Lerbs-Mache S (2011) Characterization of plastid *psbT* sense and antisense RNAs. Nucleic Acids Res 39(13):5379–5387

Zhang G, Tian Y, Zhang J, Shu L, Yang S, Wang W, Sheng J, Dong Y, Chen W (2015) Hybrid de novo genome assembly of the Chinese herbal plant Danshen (*Salvia miltiorrhiza* Bunge). Gigascience 4:62

Zhong G, Li P, Zeng L, Guan J, Li D, Li S (2009) Chemical characteristics of *Salvia miltiorrhiza* (Danshen) collected from different locations in China. J Agric Food Chem 57(15):6879–6887

Zhou L, Zuo Z, Chow MS (2005) Danshen: an overview of its chemistry, pharmacology, pharmacokinetics, and clinical use. J Clin Pharmacol 45(12):1345–1359

Salvia miltiorrhiza Epigenetics

6

Xiaoxiao Qiu, Hong Zhou and Shanfa Lu

Abstract

Epigenetics refers to the heritable changes in phenotype or gene expression without changes in the underlying DNA sequence. It includes DNA methylation, histone modification, and noncoding RNA regulation. These epigenetic modifications play significant roles in plant growth and development, such as developmental regulation, stress response, and environmental adaption. This chapter reviews the studies on the regulation mechanisms and functions of microRNAs (miRNAs), long noncoding RNAs (lncRNAs), and DNA methylation in *Salvia miltiorrhiza*.

6.1 Introduction

'Epigenetics' was coined by Waddington and could be broadly defined as meiotically and mitotically inherited changes that can modulate gene expression without any changes in DNA sequences. It includes post-transcriptional regulation, reversible modification of DNA, and chromatin (Holliday 2006). *Salvia miltiorrhiza*, a well-known traditional Chinese medicine (TCM) with great medicinal and economic value, is becoming the model medicinal plant (Song et al. 2013). Its genome (Xu et al. 2016a; Zhang et al. 2015), transcriptomes from different tissues and treatments (Luo et al. 2014; Xu et al. 2015), DNA methylome, degradome, and small RNA datasets (Xu et al. 2014; Zhang et al. 2016) have been decoded successfully. These rich genomic resources have begun to provide insight into the epigenetic 'landscape' of this organism. Here, we review the emerging and prominent roles of miRNAs, long noncoding RNAs, and DNA methylation in epigenetic inheritance in *S. miltiorrhiza* (Fig. 6.1).

6.2 miRNAs in *S. miltiorrhiza*

Plant miRNAs are a class of small noncoding endogenous RNAs of about 21 nucleotides in length. They play important roles in a variety of developmental and physiological processes through regulating the expression of target genes at the transcriptional or post-transcriptional level (Voinnet 2009; Bartel 2004; Chen 2009). The biogenesis of miRNAs requires multiple steps in order to generate mature miRNAs from *MIR* genes. First, a *MIR* locus is transcribed to a

X. Qiu · H. Zhou · S. Lu (✉)
Institute of Medicinal Plant Development, Chinese Academy of Medical Sciences and Peking Union Medical College, No. 151 Malianwa North Road, Haidian District, Beijing 100193, China
e-mail: sflu@implad.ac.cn

X. Qiu
e-mail: 18813756675@163.com

H. Zhou
e-mail: zhouhong1013@126.com

S. Lu (ed.), *The Salvia miltiorrhiza Genome*, Compendium of Plant Genomes,
https://doi.org/10.1007/978-3-030-24716-4_6

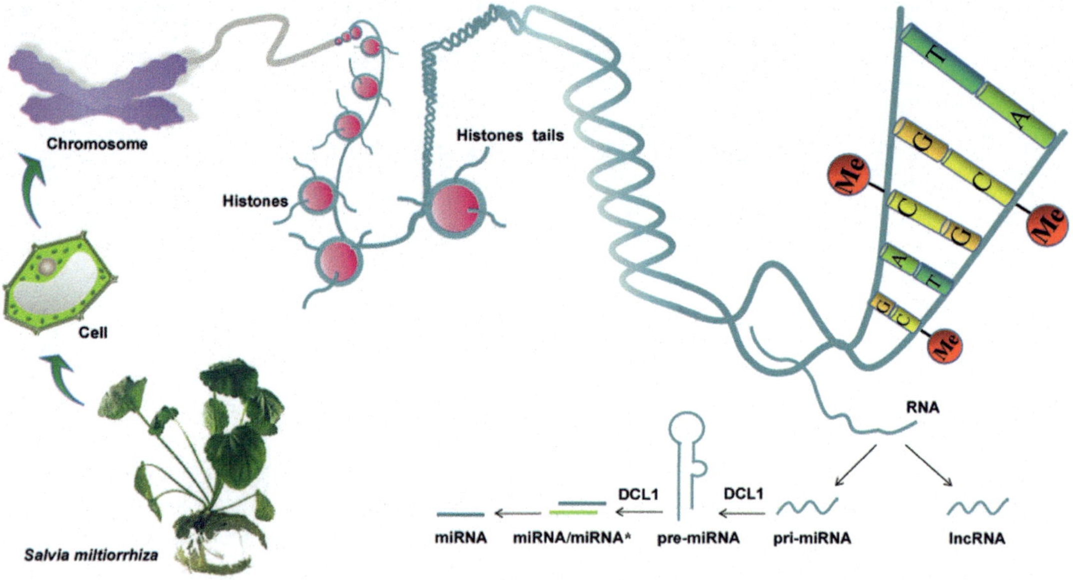

Fig. 6.1 Diagrammatic representation of epigenetic mechanisms in regulating the growth and development of *S. miltiorrhiza*

primary-*MIRNA* (*pri-MIR*), which is usually a long sequence of more than several hundred nucleotides and typically contains an imperfect hairpin structure. This step is catalyzed by Pol II enzymes. Second, the *pri-MIR* is processed to a stem-loop intermediate called *MIRNA* precursor (*pre-MIR*). This step is catalyzed by Dicer-like 1 enzyme (DCL1). Third, *pre-MIR* is cleaved into miRNA:miRNA* duplex, which comprises a mature miRNA on a side and a miRNA* on the other side, by DCL1 in the nucleus. Then, this duplex is translocated into the cytoplasm by HASTY. In the cytoplasm, the duplex is unwound to release a single-stranded mature miRNA by helicase. Finally, the mature miRNA is incorporated into a ribonucleoprotein complex, known as the RNA-induced silencing complex (RISC), to repress target gene expression (Voinnet 2009).

The first miRNA, termed lin-4, was first discovered in *Caenorhabditis elegans* by Lee et al. (1993). However, it was not until 2001 that *lin-4* was recognized as a founding member of a new and large class of small RNAs (Lee and Ambros 2001; Lau et al. 2001). Since then, thousands of miRNAs have been found in various organisms, including animals, plants, and viruses. Compared with animal miRNAs, plant miRNAs were identified in about 10 years later. In 2002, Reinhart et al. (2002) reported the first plant miRNAs in *Arabidopsis*, opening a new era in plant miRNA research. In the past decades, with the rapid development of high-throughput sequencing technology and bioinformatics, thousands of miRNAs have been identified in more than 100 plant species (Yi et al. 2015). To date, 10,414 miRNAs from 82 plant species have been deposited in the miRBase database (version 22, released in June 2018, http://www.mirbase.org/). However, there is only one *S. miltiorrhiza* miRNA, termed Smi-miR12112, has been included (Kozomara et al. 2019; Li et al. 2017).

Currently, several small RNA (sRNA) datasets have been generated from different tissue type and developmental stages of *S. miltiorrhiza* (Xu et al. 2014; Zhang et al. 2016). It greatly enriches the knowledge of miRNAs in *S. miltiorrhiza*. Xu et al. (2014) constructed four sRNA libraries from root, stem, leaf, and flower of *S. miltiorrhiza* and obtained approximately 40 million raw data. A total of 589 *pre-MIRs*, which can generate 452 known miRNAs, have been identified by mapping the sRNAs to the known plant precursors or miRNAs in miRbase. In addition, 40 novel miRNAs corresponding to 24 *pre-MIRs* were identified through blast against the specific genomes. Tissue-specific expression analysis showed that different miRNAs had different expression patterns. Among the miRNAs identified in *S. miltiorrhiza*, 62 were root-specific, 95 were stem-specific, 19 were leaf-specific, and 71 were flower-specific. The results are consistent with the opinion that the expression of miRNAs has spatiotemporal specificity. Tissue-specific miRNAs may have some crucial functions in specific developmental stages and tissues. In order to identify the miRNAs involved in plant response to replanting disease, Zhang et al. (2016) constructed two sRNA libraries for roots from first-year (FPR) and second-year (SPR) plants. A total of 14.73 million raw data were obtained. Analysis of these data revealed 110 known miRNAs belonging to 31 families and 7 novel miRNAs. Among them, 39 known and 2 novel miRNAs were expressed differentially in FPR and SPR. miR160, miR156, miR164, miR165, and miR396 may be involved in the response of *S. miltiorrhiza* to replanting disease by regulating genes encoding auxin response factors (*ARFs*), SQUAMOSA promoter binding protein-likes (*SPLs*), *NAC1*, class-III homeodomain leucine zippers (*HD-ZIP IIIs*) and *GRFs*, respectively. These ever-increasing sRNA databases shed light on the overall features of miRNAs in *S. miltiorrhiza* and laid the data foundation for further functional studies. It is worth mentioning that these reported *S. miltiorrhiza* miRNAs need to be further verified since their *pre-MIRs* lack genome-wide analysis and the hairpin structure of most precursors does not conform to existing standards (Meyers et al. 2008). Therefore, genome-wide characterization of small noncoding RNAs is necessary to be carried out in *S. miltiorrhiza*.

Although miRNAs are relatively small, they play important roles in plant growth and development by regulating gene expression at post-transcriptional level (Voinnet 2009; Chen 2009). The versatile functions of *S. miltiorrhiza* miRNAs are becoming clearer based on the analysis of degradome and transcriptome (Xu et al. 2014; Zhang et al. 2016). What is more, the discovery of gene families will be more conducive to accurately predict the target genes and functions of miRNAs. Currently, several transcription factors and protein gene families have been identified in *S. miltiorrhiza*, such as *SmSPLs* (Zhang et al. 2014), *SmARFs* (Xu et al. 2016b), *SmDCLs* (Shao et al. 2015), *SmAGOs* (Shao and Lu 2013), *SmMYBs* (Li and Lu 2014), *SmPPOs* (Li et al. 2017), and *SmDMLs* (Li et al. 2018b). Post-transcriptional regulation analysis showed that miRNAs could be involved in the development and secondary processes of *S. miltiorrhiza* by controlling the levels of these transcription factors and protein genes. The majority of functions of miRNAs in *S. miltiorrhiza* were investigated through bioinformatics analysis and a few were verified by 5′ RLM-RACE experiments. Functions of several *S. miltiorrhiza* miRNAs (Smi-miRNAs) are discussed below.

6.2.1 miRNAs and Development of *S. miltiorrhiza*

The miR156/157 family is highly conserved in plants (Axtell and Bowman 2008). It is involved in morphology and development of plants through direct cleavage of *SPL* transcripts. In *Arabidopsis*, 11 of the 17 identified *AtSPLs* are targeted by miR156/157 (Spanudakis and Jackson 2014). Among them, *AtSPL3*, *AtSPL4*, and *AtSPL5* redundantly control floral transition (Cardon et al. 1997; Jung et al. 2011), *AtSPL9* and *AtSPL15* regulate shoot maturation and flowering (Schwarz et al. 2008), and *AtSPL9* can induce the expression of miR172 (Wu et al. 2009). Both the miR156/157 and the miR172 families are major constituents of the aging pathway and act sequentially to regulate flowering time (Wu et al. 2009). In *S. miltiorrhiza*, 15 *SmSPLs* were identified from the genome-wide level. Among them, eight *SmSPLs* (Table 6.1) were predicted and experimentally validated to be regulated by miR156/157. The age-dependent decrease in miR156/157 levels also exists in

Table 6.1 The roles of miRNAs in *S. miltiorrhiza*

miRNA family	Target family	Confirmed targets	Function	References
miR156/157	*SPL*	*SmSPL2, SmSPL3, SmSPL5, SmSPL6, SmSPL8, SmSPL11, SmSPL14, SmSPL15*	Growth and development	Zhang et al. (2014)
miR159/319	*MYB*	*SmMYB96, SmMYB78, SmMYB80, SmMYB99*	Growth and development	Li and Lu (2014)
miR828	*MYB*	*SmMYB2, SmMYB36, SmMYB49*	Secondary metabolites	Li and Lu (2014)
miR858	*MYB*	*SmMYB2, SmMYB6, SmMYB12, SmMYB19, SmMYB22, SmMYB24, SmMYB28, SmMYB31, SmMYB33, SmMYB38, SmMYB40, SmMYB43, SmMYB46, SmMYB48, SmMYB52, SmMYB56, SmMYB62, SmMYB69, SmMYB75, SmMYB76, SmMYB79, SmMYB80, SmMYB81, SmMYB83, SmMYB84, SmMYB85, SmMYB87, SmMYB88, SmMYB89, SmMYB91, SmMYB92, SmMYB97, SmMYB101*	Secondary metabolites	Li and Lu (2014)
miR168	*AGO*	*SmAGO1*	miRNA function	Shao and Lu (2013)
miR403	*AGO*	*SmAGO2*	miRNA function	Shao and Lu (2013)
miR160	*ARF*	*SmARF1, SmARF13, SmARF14, SmARF15, SmARF16*	Signaling transduction	Xu et al. (2016b)
miR167	*ARF*	*SmARF8, SmARF9, SmARF10, SmARF19, SmARF24, SmARF25*	Signaling transduction	Xu et al. (2016b)
miR7972	*DML*	*SmDML1*	DNA methylation	Li et al. (2018b)
miR12112	*PPO*	*SmPPO1, SmPPO2, SmPPO3, SmPPO5, SmPPO6, SmPPO7, SmPPO8, SmPPO9, SmPPO10, SmPPO11, SmPPO12, SmPPO13, SmPPO14, SmPPO16, SmPPO17*	Secondary metabolites	Li et al. (2017)

S. miltiorrhiza. This phenomenon is accompanied by a concomitant increase in the expression of miR156/157-targeted *SmSPLs*. It suggests the functions of miR156/157-*SmSPL* regulatory pathway in *S. miltiorrhiza* development. With the maturation of plants, miR156/157 levels decline, resulting in a concomitant increase in *SPL* and miR172 expression in *Arabidopsis*. Negative correlation of the expression between miR156/157 and miR172 also exists in *S. miltiorrhiza*. Taken together, the *SmSPL*-, miR156/157-, and miR172-mediated regulation of developmental timing in *S. miltiorrhiza* were significant and intricate (Zhang et al. 2014).

6.2.2 miRNAs Involved in Signal Transduction

miRNAs play vital roles in hormone-mediated signal transduction through regulating key components of hormone signaling pathways and further regulating hormone homeostasis and related developmental processes (Mallory et al. 2005; Guo et al. 2005). Auxin, typified by indole-3-acetic acid (IAA), is an important hormone in plants. It can affect many aspects of plant growth and development (Santner and Estelle 2009). Auxin response factors (ARFs) can function as transcriptional activators or repressors to regulate the expression of auxin response genes by specifically binding to auxin response elements (AREs) (Hagen and Guilfoyle 2002). A total of 25 *ARF* genes have been identified in *S. miltiorrhiza* from the genome-wide level. Among them, 11 *SmARFs* have complementary sites with miRNAs. *SmARF1, SmARF13, SmARF14, SmARF15,* and *SmARF16* are potential targets of miR160, and *SmARF8, SmARF9, SmARF10, SmARF19, SmARF24,* and *SmARF25* are potential targets of miR167 (Xu et al. 2016b). In *Arabidopsis*, miR160 controls the expression patterns of *AtARF10, AtARF16,* and *AtARF17* to negatively regulate seed germination and post-germination activities (Liu et al. 2007, 2010). Phylogenetic tree analysis showed that these three *AtARFs* were closely related to miR160-targeted *SmARFs*. This finding implies that miR160 in *S. miltiorrhiza* may perform similar functions with Ath-miR160 through inhibiting the expression of *SmARF1, SmARF13, SmARF14, SmARF15,* and *SmARF16*. Additionally, miR167 was found to target *AtARF6* and *AtARF8* to regulate female and male reproduction or to promote jasmonic acid production and flower maturation in *Arabidopsis* (Nagpal et al. 2005). *AtARF6* and *AtARF8* were closely related to miR167-targeted *SmARFs* in the NJ tree (Xu et al. 2016b). It suggests that miR167 in *S. miltiorrhiza* might be involved in certain developmental processes by inhibiting the expression of *SmARF8*, 10, 19, 24, and 25 as described for *AtARF6* and *AtARF8*. Since *SmARF25* was identified as the best candidate regulator of flower development, Smi-miR167 targeting *SmARF25* for cleavage may also play vital roles in organ formation.

6.2.3 miRNAs Regulating miRNA Biogenesis and Function

The biogenesis and function of plant small RNAs involve a series of protein families, such as Dicer-likes (DCLs), RNA-dependent RNA polymerases (RDRs), and Argonautes (AGOs) (Voinnet 2009; Bartel 2004; Chen 2009). Among them, DCLs were the core components of small RNA biogenesis. A total of four *DCLs* were found in *Arabidopsis*, of which *AtDCL1* was one of the key enzymes in miRNA biogenesis, whereas *AtDCL2, AtDCL3,* and *AtDCL4* were mainly involved in the derivation of siRNAs (Bouché et al. 2006). Meanwhile, *AtDCL1* was negatively regulated by miR162 (Xie et al. 2003). There are five *DCL* genes in *S. miltiorrhiza*. They were named *SmDCL1, SmDCL2, SmDCL3, SmDCL4a,* and *SmDCL4b*, respectively. SmDCL1 clusters with its *Arabidopsis* counterpart, AtDCL1, in the NJ tree. However, the miR162-mediated feedback regulation of *DCL1* was absent in *S. miltiorrhiza* as the loss of miR162 target site in *SmDCL1* (Shao et al. 2015). There may be an alternative mechanism for maintaining *SmDCL1* at a proper level in *S. miltiorrhiza*, and its regulatory mechanism still needs further analysis.

It has been shown that there is a feedback regulation mechanism existing in the miRNA-*AGO* modular. miRNA-guided gene regulation requires a miRNA to form a complex, known as RISC. AGO proteins, which contain a single-stranded RNA binding site, are the central component of RISC (Yan et al. 2003; Vaucheret et al. 2004). In *Arabidopsis*, *AtAGO1* is the target of miR168 and *AtAGO2* is the target of miR403 (Vaucheret et al. 2004; Allen et al. 2005). Genome-wide analysis identified ten *SmAGO* genes in *S. miltiorrhiza*. Consistent with their counterparts in *Arabidopsis*, *SmAGO1* and *SmAGO2* can also be regulated by *S. miltiorrhiza* miR168a/b and miR403, respectively (Shao and Lu 2013). Since AtAGO1 is the core component of RISCs associated with the action of miRNAs, trans-acting siRNAs (ta-siRNAs), and transgene-derived siRNAs (Baumberger and Baulcombe 2005), the miR168-AGO1 modular may be crucial for the function of RISCs and miRNAs. AtAGO2 protein was found to be involved in antiviral defense by catalyzing viral RNA cleavage in *Arabidopsis* plants (Carbonell et al. 2012). The miR403-*AGO2* modular may also be involved in the antiviral defense in *S. miltiorrhiza*.

6.2.4 miRNAs Involved in Secondary Metabolites Biosynthesis in *S. miltiorrhiza*

Understanding the regulatory mechanisms of medicinal components biosynthesis and metabolism is vital for *S. miltiorrhiza* quality improvement. The main active ingredients of *S. miltiorrhiza* are hydrophilic phenolic acids and lipophilic tanshinones, both of them are secondary metabolites (Wang et al. 2007). Several studies have found that miRNAs play important roles in the synthesis of plant secondary metabolites. For instance, miR828 and miR858 promote anthocyanin and flavonol accumulation in grapes through inhibiting the expression of *VvMYB114* (Tirumalai et al. 2019). Recent studies showed that miR159/319-, miR828-, miR858-mediated post-transcriptional regulation of *MYB* transcription factor were also conserved in *S. miltiorrhiza* (Li and Lu 2014). A total of 110 *R2R3-MYBs* were identified in *S. miltiorrhiza*, of which 4, 3, and 33 were predicted targets of miR159/319, miR828, and miR858, respectively. In addition, *SmMYB2* has the complementary sites with both miR828 and miR858. *SmMYB80* is co-regulated by miR159/319 and miR858. All target sites of miR828 and miR858 locate in R3, miR828 and miR858 target to the less conserved 3′ end and more conserved 5′ end, respectively. Phylogenetic analysis showed that one of the miR828-regulated *SmMYBs*, *SmMYB36*, was clustered with *AtMYB75*, *AtMYB90*, *AtMYB113*, and *AtMYB114* in the NJ tree (Li and Lu 2014). Since these *AtMYBs* regulate anthocyanin biosynthesis (Gonzalez et al. 2008). It implies that *S. miltiorrhiza* miR828 may be also involved in anthocyanin biosynthesis through targeting *SmMYB36* for cleavage. Similarly, miR858-regulated *SmMYB*, *SmMYB97*, is clustered with *AtMYB11*, *AtMYB12*, and *AtMYB111* in the same subgroup. Since these three *AtMYBs* control flavonoid biosynthesis in *Arabidopsis* (Stracke et al. 2007; Mehrtens et al. 2005), miR858-*SmMYB97* is a potential regulator of flavonol biosynthesis in *S. miltiorrhiza*.

Furthermore, recent studies have shown that a novel miRNA, termed Smi-miR12112, regulates a subset of polyphenol oxidase genes (*PPOs*) in *S. miltiorrhiza*. All of the 15 miR12112-regulated *SmPPOs* contain a complementary sequence, which encodes KFDV, to miR12112. Among them, five *SmPPOs*, including *SmPPO3*, *SmPPO5*, *SmPPO9*, *SmPPO11*, and *SmPPO13*, were experimentally verified to be cleaved by Smi-miR12112 (Li et al. 2017). PPOs have been shown to be involved in the secondary metabolic pathway. For instance, PPOs in walnut are involved in the phenylpropanoid pathway and the tyramine pathway (Araji et al. 2014). PPOs are able to participate in the biosynthesis of water-soluble pigment betalains in *Portulaca grandiflora* and other *Caryophyllales* plants (Steiner et al. 1999; Strack et al. 2003; Gandía-Herrero et al. 2009). Thus, Smi-miR12112 may be involved in the biosynthesis of phenolic acids by regulating *SmPPOs*.

6.2.5 miRNAs Involved in DNA Methylation Change

A lineage-specific miRNA, termed Smi-miR7972, was found to be involved in the regulation of DEMETER-like DNA glycosylase 1 gene (Sm*DML1*) in *S. miltiorrhiza* (Li et al. 2018b). DMLs act as bifunctional glycosylase/AP-lyase and initiate the base excision repair-dependent DNA demethylation to regulate a wide range of biological processes in plants (Zhang and Zhu 2012). Smi-miR7972 could be involved in genome DNA methylation change through inhibiting the expression of *SmDML1*.

In conclusion, miRNAs in *S. miltiorrhiza* have versatile roles in plant growth and development. To date, hundreds of *S. miltiorrhiza* miRNAs have been identified by computational approaches, genetic screening, and EST analysis. However, the studies about their functions continue to lag behind the more mature areas of gene research. Only a small part of Smi-miRNAs-directed cleavage has been preliminarily verified by 5′ RLM-RACE experiment.

Further, analysis of Smi-miRNAs through genetic transformation will definitely shed lights on the regulatory mechanism and biological function of Smi-miRNAs in *S. miltiorrhiza*.

6.2.6 Evolution of miRNAs and Their Target Genes in *S. miltiorrhiza*

The origination and evolution of miRNAs exhibit functional adaptation and adaptation to changing environments in evolution (Cui et al. 2017; Gramzow and Theißen 2019). In *S. miltiorrhiza*, several conserved or known miRNA-target pairs have been lost and it reconstructed an alternative regulation of the target genes. For instance, miR162-mediated negative-feedback regulation of *DCL1*, which can maintain homeostasis of miRNA biogenesis in plants, was absent in *S. miltiorrhiza* (Shao et al. 2015). Additionally, miR1444 that targets polyphenol oxidase genes (*PPOs*) in *Salicaceae* plants (Wang et al. 2017) is missing in *S. miltiorrhiza*, but a novel *PPO*-targeting *MIRNA*, *Smi-MIR12112*, was originated through generation of short inverted repeats in target genes (Li et al. 2017; Lu 2019).

6.3 Long Noncoding RNAs in *S. miltiorrhiza*

Long noncoding RNAs (lncRNAs) are noncoding RNAs longer than 200 bp (or 40 bp in some references). They do not have recognizable potential to encode functional proteins in general. Based on their locations to nearby protein-coding genes, lncRNAs can be roughly classified into four types: long noncoding natural antisense transcripts (lincNATs), intron lncRNAs, promoter lncRNAs, and long intergenic ncRNAs (lincRNAs). LncRNAs can serve as recruiters, tethers, guides, decoys, and signals when they interact with other biomolecules. They regulate genes expression at the epigenetic, transcriptional, post-transcriptional, and translational levels (Zhang et al. 2017; Gomes et al. 2013) and are involved in a wide range of biological processes, such as genomic imprinting, chromatin remodeling, alternative splicing, transcriptional activation, etc. (Shafiq et al. 2016). So far, a larger number of lncRNAs have been identified from plants using different approaches. They are important regulators during plant growth and development, such as vernalization and flowering induction, pollen development, male sterility, lateral root development, auxin transport, development signaling, plant responses to diverse stresses and diseases, and so on (Tan et al. 2018).

Some lncRNAs do not have a poly(A) tail (Wang et al. 2014a), whereas the others are mRNA-like (mRNA-like noncoding RNAs, mlncRNAs). These mlncRNAs are transcribed by RNA polymerase II and polyadenylated, capped and spliced as conventional mRNAs, but have no protein-coding ability. mlncRNAs may generate from mlncRNA loci located between protein-coding genes. They can also be noncoding alternative transcripts of protein-coding genes, natural antisense transcripts, and pri-miRNAs (Rodriguez et al. 2004; Griffiths-Jones 2007). It has been shown that plant mlncRNAs are involved in cell differentiation, plant response to biotic and abiotic stresses, and nodulation. In addition, some mlncRNAs may serve as endogenous target mimics (eTMs) to inhibit the activity of miRNAs in plants (Wu et al. 2013). To date, thousands of mlncRNAs have been identified from various plant species, such as *A. thaliana* (MacIntosh et al. 2001), *Medicago truncatula* (Wen et al. 2007), *Panax ginseng* (Wang et al. 2014b), *Digitalis purpurea* (Wu et al. 2012), and *Zea mays* (Boerner and McGinnis 2012). In 2015, Li et al. reported the identification of a total of 5446 mlncRNAs from the transcriptome of *S. miltiorrhiza*. Among them, two are primary transcripts of miR156 that targets *SPLs* for cleavage, and 2030 can be grouped into 470 families with at least two members in a family (Zhang et al. 2014). Sequence conservation analysis showed that the majority of the identified *S. miltiorrhiza* mlncRNAs exhibited low homology with known plant lncRNAs (Li et al. 2015). It indicated that these mlncRNAs could be species- or lineage-specific. Quantitative real-time PCR analysis and RNA-seq data analysis showed that the majority of the 5446 mlncRNAs were differentially expressed in different tissues and were responsive to MeJA, yeast extract, and Ag^+ treatments (Li et al. 2015). These results suggest the significance of mlncRNAs in *S. miltiorrhiza* growth and development and in plant response to stress.

6.4 DNA Methylation in *S. miltiorrhiza*

DNA methylation is a highly conserved epigenetic mark that can be accurately replicated during both mitosis and meiosis in eukaryotes for maintaining cell phenotype. DNA methylation is a major chemical modification of DNA bases. It transfers a methyl (CH_3) group from S-adenyl methionine (SAM) to the fifth carbon of a cytosine residue to form 5mC. Compared with the level of DNA methylation in animals (about 3–8%), plants have higher level of DNA methylation (6–30%). DNA methylation regulates a wide range of biological processes, including embryonic development, transcription, chromatin structure, X-chromosome inactivation, genomic imprinting, and chromosome stability (Illingworth and Bird 2009; Li 2002; Bernstein et al. 2007).

In plants, DNA methylation has species, tissues, organs, and age specificity. It plays an important regulatory role in plant growth, development, and evolution. It influences gene expression, cell division, fertility, florescence, and morphology. It also influences plant imprinting, stress response, and heterosis. In plant genomes, the cytosine DNA methylation occurs in three DNA sequence contexts: CG (termed CpG sites), CHG and CHH (termed non-CpG sites, where H is A, C, or T). Both CpG and non-CpG sites can be methylated in promoters of silent genes and transposable elements (TEs) (To et al. 2015). However, only CpG sites can be methylated in exons and introns of gene body except for the TSS (Transcription Start Site)-proximal region (To et al. 2015).

DNA cytosine methylation in higher plants is maintained by DNA methyltransferases, DNA demethylases, histone-modifying enzymes, nuclear chromatin remodeling factors, and RNA interference mechanism (Bewick and Schmitz 2017; Cokus et al. 2008; Zhang et al. 2018; Law and Jacobsen 2010). There are two types of DNA methylation in plants. One is known as maintenance methylation, in which the parental methylation mode is passed to the offspring through semi-conservative replication. The other is known as de novo methylation. It does not rely on DNA replication and requires DNA methyltransferases differing from those in maintenance methylation.

Plant CG and CHG methylation are established through a semi-conservative mechanism that requires the DNA methyltransferases METHYLTRANSFERASE 1 (MET1) and CHROMOMETHYLASE 3 (CMT3), respectively (Finnegan and Kovac 2000; Lindroth et al. 2001). Asymmetric CHH methylation can be maintained by CMT2 and DOMAINS REARRANGED METHYLASEs (DRM1 and DRM2) through the RNA-directed DNA methylation (RdDM) pathway (Zhang and Zhu 2011; Zemach et al. 2013; Matzke and Mosher 2014; Stroud et al. 2014).

In the RdDM mechanism, RNA POLYMERASE IV (Pol IV) generates short single-stranded RNAs, which are subsequently copied into double-stranded RNAs by RNA-DEPENDENT RNA POLYMERASE 2 (RDR2) and then cleaved into 24-nt small interfering RNAs (24nt-siRNAs) by DICER-LIKE PROTEIN 3 (DCL3). These 24nt-siRNAs are loaded into ARGONAUTE (AGO) effector complexes containing AGO4, AGO6 and, AGO9. Next, Pol V generates longer noncoding transcripts, which serve as scaffolds for the recruitment of additional RdDM factors, such as 24nt-siRNA-loaded ARGONAUTE proteins, through the base-pairing mode. Finally, DRM2 is recruited to these complexes to direct de novo cytosine methylation at target loci and deposit

DNA methylation throughout the genome (Zhou et al. 2019; Zhai et al. 2015; Blevins et al. 2015; Li et al. 2006; Xie et al. 2004; Matzke et al. 2009; Wierzbicki et al. 2008, 2009; Zhang and Zhu 2011; Bohmdorfer et al. 2014).

Once the RdDM pathway established, distinct DNA methyltransferases of the maintenance pathways will take over to insure the faithful inheritance of DNA methylation patterns by DNA replication process. Moreover, DNA methylation status also needs demethylases to maintain. Active DNA demethylation requires ROS1 (Repressor of Silencing 1), DME (demeter DNA glycosylase), DML2 (Demeter-like 2), and DML3 (DME-like 3) to remove 5-mC directly from the DNA backbone, which creates an abasic site to be filled up with a non-methylated cytosine by the BER-dependent mechanism (Zhu 2009; Li et al. 2018c; Zhang et al. 2018).

To date, multiple techniques have been developed to assess DNA methylation. The most commonly used are the Methylation Sensitive Amplified Polymorphism (MSAP) PCR, Enzyme-Linked Immunosorbent Assay (ELISA), Methylated DNA Immunoprecipitation-sequencing (MeDIP-seq), and Whole-Genome Bisulfite Sequencing (WGBS) techniques (Shi et al. 2019). Among them, the WGBS technique makes it possible to construct single-base resolution maps of DNA methylation across entire genomes be possible (Cokus et al. 2008; Lister et al. 2008). This method was first applied to *A. thaliana*. Now, it has been widely applied to oil palm, maize, and various other plants (Ong-Abdullah et al. 2015; Han et al. 2018).

Recently, Li et al. (2018a) analyzed and identified eight putative cytosine-5 DNA methyltransferase genes (*SmC5-MTases*) from the genome assembly of *S. miltiorrhiza*. Conserved domain analysis and phylogenetic tree construction showed that SmC5-MTases could be divided into four subfamilies, including MET, CMT, DRM, and DNMT2. The sequences and conserved motifs of SmC5-MTases were similar to their counterparts from *Arabidopsis*. Among the eight *SmC5-MTases*, six, including *SmMET1*, *SmCMT1*, *SmCMT2a*, *SmCMT2b*, *SmCMT3*, and *SmDRM1* were salicylic acid-responsive. It suggests the involvement of *SmC5-MTases* in salicylic acid-dependent abiotic stress. The abundance of *SmC5-MTase* transcripts exhibited specific spatiotemporal patterns in *S. miltiorrhiza*. It indicates that DNA methylation is important in the growth and development of *S. miltiorrhiza*. In addition, *SmCMT2a* and *SmDRM1* showed differential expression in periderm, phloem, and xylem of roots, where the active ingredients accumulated. It indicates that DNA methylation may be involved in regulation of secondary metabolism.

In addition to *SmC5-MTases*, there are six putative DEMETER-like DNA glycosylase genes (*SmDMLs*) in *S. miltiorrhiza* (Li et al. 2018b). The number of *SmDML* genes is similar to the number of *DMLs* in other plant species, such as castor bean, tomato, *Arabidopsis*, and rice, which ranges from three to six. Phylogenetic analysis of 66 DMLs from 26 plant species showed that plant DMLs could be divided into three groups, including the ROS1 group, the DME group, and the DML3 group. Among the six SmDMLs, SmDML1 belonged to the ROS1 group, SmDML2, SmDML3, and SmDML4 were members of the DME group, and SmDML5 and SmDML6 were classified in the DML3 group (Li et al. 2018b). *SmDMLs* were differentially expressed in differential tissues, and *SmDML1*, *SmDML2*, and *SmDML4* were significantly down-regulated in leaves of *S. miltiorrhiza* plants treated with the general DNA methylation inhibitor, 5-aza-2′-deoxycytidine (5Aza-dC) (Li et al. 2018b). It confirms the involvement of *SmDMLs* in genome DNA methylation change.

Balance between DNA methylation and demethylation is important for maintaining normal growth and development during the lifespan of plants. Fine regulation of the genes involved in DNA methylation and demethylation could be vital for controlling the balance. Consistently, post-transcriptional analysis showed that *SmDML1* was regulated by a miRNA, termed Smi-miR7972 (Li et al. 2018b). This miRNA is distributed in some plants of the three orders of Lamiids, including Lamiales, Solanales, and Boraginales. Although the actual origin and evolution mechanism of *MIR7972* is unknown, the gain and loss of *MIR7972* could be important for plants of some lineages to survive in the stressful environments (Li et al. 2018b).

6.5 Conclusions

Epigenetic mechanisms are thought to act as a memory of a cell. The epigenetic modifications can be passed over the generations. Epigenetic mechanisms are also affected by environmental factors, such as biotic and abiotic stresses. It regulates gene expression during plant growth and development. Studies have shown that the growth and development of *S. miltiorrhiza* are dynamically regulated by miRNAs, lncRNAs, and DNA methylation. Moreover, the biosynthesis of secondary metabolites in *S. miltiorrhiza* is also modulated via the epigenetic mechanisms. Epigenetics is of great significance to control the quality and environmental adaptability of *S. miltiorrhiza* and other medicinal plants. Current results provide useful information for understanding the role of epigenetic regulation in medicinal plant development and bioactive compound biosynthesis.

Acknowledgements This work was supported by the National Natural Science Foundation of China (81773836) and the CAMS Innovation Fund for Medical Sciences (CIFMS) (2016-I2M-3-016).

References

Allen E, Xie Z, Gustafson AM, Carrington JC (2005) microRNA-directed phasing during trans-acting siRNA biogenesis in plants. Cell 121(2):207–221

Araji S, Grammer TA, Gertzen R, Anderson SD, Mikulic-Petkovsek M, Veberic R, Phu ML, Solar A, Leslie CA, Dandekar AM, Escobar MA (2014) Novel roles for the polyphenol oxidase enzyme in secondary metabolism and the regulation of cell death in walnut. Plant Physiol 164(3):1191–1203

Axtell MJ, Bowman JL (2008) Evolution of plant microRNAs and their targets. Trends Plant Sci 13:343–349

Bartel DP (2004) MicroRNAs: genomics, biogenesis, mechanism, and function. Cell 116(2):281–297

Baumberger N, Baulcombe DC (2005) *Arabidopsis* ARGONAUTE1 is an RNA slicer that selectively recruits microRNAs and short interfering RNAs. Proc Natl Acad Sci U S A 102(33):11928–11933

Bernstein BE, Meissner A, Lander ES (2007) The mammalian epigenome. Cell 128:669–681

Bewick AJ, Schmitz RJ (2017) Gene body DNA methylation in plants. Curr Opin Plant Biol 36:103–110

Blevins T, Podicheti R, Mishra V, Marasco M, Wang J, Rusch D, Tang H, Pikaard CS (2015) Identification of Pol IV and RDR2-dependent precursors of 24 nt siRNAs guiding *de novo* DNA methylation in *Arabidopsis*. Elife 4:e09591

Boerner S, McGinnis KM (2012) Computational identification and functional predictions of long noncoding RNA in *Zea mays*. PLoS ONE 7:e43047

Bohmdorfer G, Rowley MJ, Kuciński J, Zhu Y, Amies I, Wierzbicki AT (2014) RNA-directed DNA methylation requires stepwise binding of silencing factors to long non-coding RNA. Plant J 79:181–191

Bouché N, Lauressergues D, Gasciolli V, Vaucheret H (2006) An antagonistic function for *Arabidopsis* *DCL2* in development and a new function for *DCL4* in generating viral siRNAs. EMBO J 25(14):3347–3356

Carbonell A, Fahlgren N, Garcia-Ruiz H, Gilbert KB, Montgomery TA, Nguyen T, Cuperus JT, Carrington JC (2012) Functional analysis of three *Arabidopsis* ARGONAUTES using slicer-defective mutants. Plant Cell 24(9):3613–3629

Cardon G, Höhmann S, Nettesheim K, Saedler H, Huijser P (1997) Functional analysis of the *Arabidopsis thaliana* SBP-box gene *SPL3*: a novel gene involved in the floral transition. Plant J 12:367–377

Chen X (2009) Small RNAs and their roles in plant development. Annu Rev Cell Dev Biol 25:21–44

Cokus SJ, Feng S, Zhang X, Chen Z, Merriman B, Haudenschild CD, Pradhan S, Nelson SF, Pellegrini M, Jacobsen SE (2008) Shotgun bisulfite sequencing of the *Arabidopsis* genome reveals DNA methylation patterning. Nature 452(7184):215–219

Cui J, You C, Chen X (2017) The evolution of microRNAs in plants. Curr Opin Plant Biol 35:61–67

Finnegan EJ, Kovac KA (2000) Plant DNA methyltransferases. Plant Mol Biol 43:189–201

Gandía-Herrero F, Jiménez-Atiénzar M, Cabanes J, Escribano J, García-Carmona F (2009) Fluorescence detection of tyrosinase activity on dopamine-betaxanthin purified from *Portulaca oleracea* (common purslane) flowers. J Agric Food Chem 57 (6):2523–2528

Gomes AQ, Nolasco S, Soares H (2013) Non-coding RNAs: multi-tasking molecules in the cell. Int J Mol Sci 14(8):16010–16039

Gonzalez A, Zhao M, Leavitt JM, Lloyd AM (2008) Regulation of the anthocyanin biosynthetic pathway by the *TTG1/bHLH/Myb* transcriptional complex in *Arabidopsis* seedlings. Plant J 53(5):814–827

Gramzow L, Theißen G (2019) Plant miRNA conservation and evolution. Methods Mol Biol 1932:41–50
Griffiths-Jones S (2007) Annotating noncoding RNA genes. Annu Rev Genomics Hum Genet 8:279–298
Guo HS, Xie Q, Fei JF, Chua NH (2005) MicroRNA directs mRNA cleavage of the transcription factor *NAC1* to downregulate auxin signals for *Arabidopsis* lateral root development. Plant Cell 17:1376–1386
Hagen G, Guilfoyle T (2002) Auxin-responsive gene expression: genes, promoters and regulatory factors. Plant Mol Biol 49:373–385
Han Z, Crisp PA, Stelpflug S, Kaeppler SM, Li Q, Springer NM (2018) Heritable epigenomic changes to the maize methylome resulting from tissue culture. Genetics 209(4):983–995
Holliday R (2006) Epigenetics: a historical overview. Epigenetics 1(2):76–80
Illingworth RS, Bird AP (2009) CpG islands—'a rough guide'. FEBS Lett 583:1713–1720
Jung JH, Seo PJ, Kang SK, Park CM (2011) miR172 signals are incorporated into the miR156 signaling pathway at the *SPL3/4/5* genes in *Arabidopsis* developmental transitions. Plant Mol Biol 76:35–45
Kozomara A, Birgaoanu M, Griffiths-Jones S (2019) miRBase: from microRNA sequences to function. Nucleic Acids Res 47:D155–D162
Lau NC, Lim LP, Weinstein EG, Bartel DP (2001) An abundant class of tiny RNAs with probable regulatory roles in *Caenorhabditis elegans*. Science 294 (5543):858–862
Law JA, Jacobsen SE (2010) Establishing, maintaining and modifying DNA methylation patterns in plants and animals. Nat Rev Genet 11(3):204–220
Lee RC, Ambros V (2001) An extensive class of small RNAs in *Caenorhabditis elegans*. Science 294 (5543):862–864
Lee RC, Feinbaum RL, Ambros V (1993) The *C. elegans* heterochronic gene *lin-4* encodes small RNAs with antisense complementarity to *lin-14*. Cell 75(5):843–854
Li E (2002) Chromatin modification and epigenetic reprogramming in mammalian development. Nat Rev Genet 3:662–673
Li C, Lu S (2014) Genome-wide characterization and comparative analysis of *R2R3-MYB* transcription factors shows the complexity of *MYB*-associated regulatory networks in *Salvia miltiorrhiza*. BMC Genom 15:277
Li CF, Pontes O, ElShami M, Henderson IR, Bernatavichute YV, Chan SW, Lagrange T, Pikaard CS, Jacobsen SE (2006) An ARGONAUTE4-containing nuclear processing center colocalized with Cajal bodies in *Arabidopsis thaliana*. Cell 126:93–106
Li D, Shao F, Lu S (2015) Identification and characterization of mRNA-like noncoding RNAs in *Salvia miltiorrhiza*. Planta 241:1131–1143
Li C, Li D, Li J, Shao F, Lu S (2017) Characterization of the polyphenol oxidase gene family reveals a novel microRNA involved in posttranscriptional regulation of *PPOs* in *Salvia miltiorrhiza*. Sci Rep 7:44622
Li J, Li C, Lu S (2018a) Identification and characterization of the cytosine-5 DNA methyltransferase gene family in *Salvia miltiorrhiza*. PeerJ 6:e4461
Li J, Li C, Lu S (2018b) Systematic analysis of *DEMETER*-like DNA glycosylase genes shows lineage-specific Smi-miR7972 involved in *SmDML1* regulation in *Salvia miltiorrhiza*. Sci Rep 8:7143
Li Y, Kuma S, Qian W (2018c) Active DNA demethylation: mechanism and role in plant development. Plant Cell Rep 37:77–85
Lindroth AM, Cao X, Jackson JP, Zilberman D, McCallum CM, Henikoff S, Jacobsen SE (2001) Requirement of *CHROMOMETHYLASE3* for maintenance of CpXpG methylation. Science 292:2077–2080
Lister R, O'Malley RC, Tonti-Filippini J, Gregory BD, Berry CC, Millar AH, Ecker JR (2008) Highly integrated single-base resolution maps of the epigenome in *Arabidopsis*. Cell 133:523–536
Liu PP, Montgomery TA, Fahlgren N, Kasschau KD, Nonogaki H, Carrington JC (2007) Repression of AUXIN RESPONSE FACTOR10 by microRNA160 is critical for seed germination and post-germination stages. Plant J 52(1):133–146
Liu X, Huang J, Wang Y, Khanna K, Xie Z, Owen HA, Zhao D (2010) The role of floral organs in carpels, an *Arabidopsis* loss-of-function mutation in *MicroRNA160a*, in organogenesis and the mechanism regulating its expression. Plant J 62(3):416–428
Lu S (2019) *De novo* origination of *MIRNAs* through generation of short inverted repeats in target genes. RNA Biol 16(6):846–859
Luo H, Zhu Y, Song J, Xu L, Sun C, Zhang X, Xu Y, He L, Sun W, Xu H, Wang B, Li X, Li C, Liu J, Chen S (2014) Transcriptional data mining of *Salvia miltiorrhiza* in response to methyl jasmonate to examine the mechanism of bioactive compound biosynthesis and regulation. Physiol Plant 152 (2):241–255
MacIntosh GC, Wilkerson C, Green PJ (2001) Identification and analysis of *Arabidopsis* expressed sequence tags characteristic of non-coding RNAs. Plant Physiol 127:765–776
Mallory AC, Bartel DP, Bartel B (2005) MicroRNA-directed regulation of *Arabidopsis* AUXIN RESPONSE FACTOR17 is essential for proper development and modulates expression of early auxin response genes. Plant Cell 17:1360–1375
Matzke MA, Mosher RA (2014) RNA-directed DNA methylation: an epigenetic pathway of increasing complexity. Nat Rev Genet 15:394–408
Matzke M, Kanno T, Daxinger L, Huettel B, Matzke AJ (2009) RNA-mediated chromatin-based silencing in plants. Curr Opin Cell Biol 21:367–376
Mehrtens F, Kranz H, Bednarek P, Weissharr B (2005) The *Arabidopsis* transcription factor *MYB12* is a flavonol-specific regulator of phenylpropanoid biosynthesis. Plant Physiol 138(2):1083–1096
Meyers BC, Axtell MJ, Bartel B, Bartel DP, Baulcombe D, Bowman JL, Cao X, Carrington JC, Chen X, Green PJ, Griffiths-Jones S, Jacobsen SE,

Mallory AC, Martienssen RA, Poethig RS, Qi Y, Vaucheret H, Voinnet O, Watanabe Y, Weigel D, Zhu JK (2008) Criteria for annotation of plant microRNAs. Plant Cell 20(12):3186–3190

Nagpal P, Ellis CM, Weber H, Ploense SE, Barkawi LS, Guilfoyle TJ, Hagen G, Alonso JM, Cohen JD, Farmer EE, Ecker JR, Reed JW (2005) Auxin response factors *ARF6* and *ARF8* promote jasmonic acid production and flower maturation. Development 132(18):4107–4118

Ong-Abdullah M, Ordway JM, Jiang N, Ooi SE, Kok SY, Sarpan N, Azimi N, Hashim AT, Ishak Z, Rosli SK, Malike FA, Bakar NA, Marjuni M, Abdullah N, Yaakub Z, Amiruddin MD, Nookiah R, Singh R, Low ET, Chan KL, Azizi N, Smith SW, Bacher B, Budiman MA, Van Brunt A, Wischmeyer C, Beil M, Hogan M, Lakey N, Lim CC, Arulandoo X, Wong CK, Choo CN, Wong WC, Kwan YY, Alwee SS, Sambanthamurthi R, Martienssen RA (2015) Loss of *Karma* transposon methylation underlies the mantled somaclonal variant of oil palm. Nature 525(7570):533–537

Reinhart BJ, Weinstein EG, Rhoades MW (2002) MicroRNAs in plants. Genes Dev 16(13):1616–1626

Rodriguez A, Griffiths-Jones S, Ashurst JL, Bradley A (2004) Identification of mammalian microRNA host genes and transcription units. Genome Res 14:1902–1910

Santner A, Estelle M (2009) Recent advances and emerging trends in plant hormone signaling. Nature 459:1071–1078

Schwarz S, Grande AV, Bujdoso N, Saedler H, Huijser P (2008) The microRNA regulated SBP-box genes *SPL9* and *SPL15* control shoot maturation in *Arabidopsis*. Plant Mol Biol 67:183–195

Shafiq S, Li J, Sun Q (2016) Functions of plants long non-coding RNAs. Biochim Biophys Acta 1859 (1):155–162

Shao F, Lu S (2013) Genome-wide identification, molecular cloning, expression profiling and posttranscriptional regulation analysis of the *Argonaute* gene family in *Salvia miltiorrhiza*, an emerging model medicinal plant. BMC Genom 14:512

Shao F, Qiu D, Lu S (2015) Comparative analysis of the *Dicer*-like gene family reveals loss of miR162 target site in *SmDCL1* from *Salvia miltiorrhiza*. Sci Rep 5:9891

Shi P, Wang Y, Jin L, Zhang D, Zhao Z, Cao H, Lei X (2019) Research progress on DNA methylation during plant tissue culture. Chin J Trop Crops 40(1):199–207

Song JY, Luo HM, Li CF, Sun C, Xu J, Chen SL (2013) *Salvia miltiorrhiza* as medicinal model plant. Acta Pharm Sin 48(7):1099–1106

Spanudakis E, Jackson S (2014) The role of microRNAs in the control of flowering time. J Exp Bot 65(2):365–380

Steiner U, Schliemann W, Böhm H, Strack D (1999) Tyrosinase involved in betalain biosynthesis of higher plants. Planta 208:114–124

Strack D, Vogt T, Schliemann W (2003) Recent advances in betalain research. Phytochemistry 62:247–269

Stracke R, Ishihara H, Huep G, Barsch A, Mehrtens F, Niehaus K, Weisshaar B (2007) Differential regulation of closely related *R2R3-MYB* transcription factors controls flavonol accumulation in different parts of the *Arabidopsis thaliana* seedling. Plant J 50(4):660–677

Stroud H, Do T, Du J, Zhong X, Feng S, Johnson L, Johnson L, Patel DJ, Jacobsen SE (2014) Non-CG methylation patterns shape the epigenetic landscape in *Arabidopsis*. Nat Struct Mol Biol 21:64–72

Tan Y, Wang D, Gao X, Liu J (2018) Research advance on plant long noncoding RNAs. Biotechnol Bull 34 (10):1–10

Tirumalai V, Swetha C, Nair A, Pandit A, Shivaprasad PV (2019) miR828 and miR858 regulate *VvMYB114* to promote anthocyanin and flavonol accumulation in grapes. J Exp Bot. https://doi.org/10.1093/jxb/erz264

To TK, Saze H, Kakutani T (2015) DNA methylation within transcribed regions. Plant Physiol 168(4):1219–1225

Vaucheret H, Vazquez F, Crété P, Bartel DP (2004) The action of ARGONAUTE1 in the miRNA pathway and its regulation by the miRNA pathway are crucial for plant development. Genes Dev 18(10):1187–1197

Voinnet O (2009) Origin, biogenesis, and activity of plant microRNAs. Cell 136(4):669–687

Wang X, Morris-Natschke SL, Lee KH (2007) New developments in the chemistry and biology of the bioactive constituents of Tanshen. Med Res Rev 27 (1):133–148

Wang Y, Wang X, Deng W, Fan X, Liu TT, He G, Chen R, Terzaghi W, Zhu D, Deng XW (2014a) Genomic features and regulatory roles of intermediate-sized non-coding RNAs in *Arabidopsis*. Mol Plant 7:514–527

Wang M, Wu B, Chen C, Lu S (2014b) Identification of mRNA-like non-coding RNAs and validation of a mighty one named *MAR* in *Panax ginseng*. J Integr Plant Biol 57(3):256–270

Wang M, Li C, Lu S (2017) Origin and evolution of *MIR1444* genes in *Salicaceae*. Sci Rep 7:39740

Wen J, Parker BJ, Weiller GF (2007) *In silico* identification and characterization of mRNA-like noncoding transcripts in *Medicago truncatula*. Silico Biol 7:485–505

Wierzbicki AT, Haag JR, Pikaard CS (2008) Noncoding transcription by RNA polymerase Pol IVb/Pol V mediates transcriptional silencing of overlapping and adjacent genes. Cell 135:635–648

Wierzbicki AT, Ream TS, Haag JR, Pikaard CS (2009) RNA polymerase V transcription guides ARGONAUTE4 to chromatin. Nat Genet 41:630–634

Wu G, Park Y, Conway SR, Wang JW, Weigel D, Poethig RS (2009) The sequential action of miR156 and miR172 regulates developmental timing in *Arabidopsis*. Cell 138:750–759

Wu B, Li Y, Yan H, Ma Y, Luo H, Yuan L, Chen S, Lu S (2012) Comprehensive transcriptome analysis reveals novel genes involved in cardiac glycoside biosynthesis and mlncRNAs associated with secondary metabolism and stress response in *Digitalis purpurea*. BMC Genom 13:15

Wu HJ, Wang ZM, Wang M, Wang XJ (2013) Widespread long noncoding RNAs as endogenous target mimics for microRNAs in plants. Plant Physiol 161:1875–1884

Xie Z, Kasschau KD, Carrington JC (2003) Negative feedback regulation of Dicer-Like1 in *Arabidopsis* by microRNA-guided mRNA degradation. Curr Biol 13:784–789

Xie Z, Johansen LK, Gustafson AM, Kasschau KD, Lellis AD, Zilberman D, Jacobsen SE, Carrington JC (2004) Genetic and functional diversification of small RNA pathways in plants. PLoS Biol 2(5):E104

Xu X, Jiang Q, Ma X, Ying Q, Shen B, Qian Y, Song H, Wang H (2014) Deep sequencing identifies tissue-specific microRNAs and their target genes involving in the biosynthesis of tanshinones in *Salvia miltiorrhiza*. PLoS ONE 9(11):e111679

Xu Z, Peters RJ, Weirather J, Luo H, Liao B, Zhang X, Zhu Y, Ji A, Zhang B, Hu S, Au KF, Song J, Chen S (2015) Full-length transcriptome sequences and splice variants obtained by a combination of sequencing platforms applied to different root tissues of *Salvia miltiorrhiza* and tanshinone biosynthesis. Plant J 82 (6):951–961

Xu H, Song J, Luo H, Zhang Y, Li Q, Zhu Y, Xu J, Li Y, Song C, Wang B, Sun W, Shen G, Zhang X, Qian J, Ji A, Xu Z, Luo X, He L, Li C, Sun C, Yan H, Cui G, Li X, Li X, Wei J, Liu J, Wang Y, Hayward A, Nelson D, Ning Z, Peters RJ, Qi X, Chen S (2016a) Analysis of the genome sequence of the medicinal plant *Salvia miltiorrhiza*. Mol Plant 9(6):949–952

Xu Z, Ji A, Song J, Chen S (2016b) Genome-wide analysis of auxin response factor gene family members in medicinal model plant *Salvia miltiorrhiza*. Biol Open 5(6):848–857

Yan KS, Yan S, Farooq A, Han A, Zeng L, Zhou MM (2003) Structure and conserved RNA binding of the PAZ domain. Nature 426:468–474

Yi X, Zhang Z, Ling Y, Xu W, Su Z (2015) PNRD: a plant non-coding RNA database. Nucleic Acids Res 43(Database issue):D982–D989

Zemach A, Kim MY, Hsieh PH, Coleman-Derr D, Eshed-Williams L, Thao K, Harmer SL, Zilberman D (2013) The *Arabidopsis* nucleosome remodeler DDM1 allows DNA methyltransferases to access H1-containing heterochromatin. Cell 153:193–205

Zhai J, Bischof S, Wang H, Feng S, Lee TF, Teng C, Chen X, Park SY, Liu L, Gallego-Bartolome J, Liu W, Henderson IR, Meyers BC, Ausin I, Jacobsen SE (2015) A one precursor one siRNA model for Pol IV-dependent siRNA biogenesis. Cell 163:45–55

Zhang H, Zhu JK (2011) RNA-directed DNA methylation. Curr Opin Plant Biol 14:142–147

Zhang H, Zhu JK (2012) Active DNA demethylation in plants and animals. Cold Spring Harb Symp Quant Biol 77:161–173

Zhang L, Wu B, Zhao D, Li C, Shao F, Lu S (2014) Genome-wide analysis and molecular dissection of the *SPL* gene family in *Salvia miltiorrhiza*. J Integr Plant Biol 56:38–50

Zhang G, Tian Y, Zhang J, Shu L, Yang S, Wang W, Sheng J, Dong Y, Chen W (2015) Hybrid *de novo* genome assembly of the Chinese herbal plant danshen (*Salvia miltiorrhiza* Bunge). Gigascience 4:62

Zhang H, Jin W, Zhu X, Liu L, He Z, Yang S, Liang Z, Yan X, He Y, Liu Y (2016) Identification and characterization of *Salvia miltiorrhizain* miRNAs in response to replanting disease. PLoS ONE 11(8): e0159905

Zhang Y, Tao Y, Liao Q (2017) Long noncoding RNA: a crosslink in biological regulatory network. Brief Bioinform 19(5):930–945

Zhang H, Lang Z, Zhu JK (2018) Dynamics and function of DNA methylation in plants. Nat Rev Mol Cell Biol 19:489–506

Zhou M, Palanca AMS, Law JA (2019) Locus-specific control of the *de novo* DNA methylation pathway in *Arabidopsis* by the CLASSY family. Nat Genet 50 (6):865–873

Zhu JK (2009) Active DNA demethylation mediated by DNA glycosylases. Annu Rev Genet 43:143–166

7 Transcriptome Analysis of *Salvia miltiorrhiza*

Hongmei Luo

Abstract
Salvia miltiorrhiza is the source for the production of tanshinones and phenolic acids, which possess pharmacological properties for the treatment of cardiovascular and cerebrovascular diseases and hyperlipidemia. However, the biosynthetic mechanism of these bioactive secondary metabolites remains unclear. Transcriptome analysis is a promising tool to illustrate the biosynthesis, growth, and development of these bioactive compounds and the genetic diversity of *S. miltiorrhiza*. The next-generation sequencing (NGS) technologies, such as the second-generation sequencing (SGS) technologies (e.g., Illumina) and the third-generation sequencing technologies (e.g., PacBio), are suitable and widely used for transcriptome analysis of *S. miltiorrhiza*. NGS enables the identification of gene expression profiling and facilitates reliable discoveries of genes related to secondary metabolite biosynthetic pathway. At present, hybrid sequencing strategies integrating the strengths of SGS and PacBio sequencing have obtained considerable transcriptome information of medicinal plants. NGS provides useful information for the direct detection of genetic markers and alternative splicing events related to the biosynthesis of secondary metabolites that facilitate the rapid breeding of medicinal plants.

7.1 Introduction

Salvia miltiorrhiza Bunge, a well-known medicinal plant, belongs to the Lamiaceae family. Its dry roots or rhizomes (called "danshen" or "tanshen" in China) are the sources for the production of valuable compounds (Zhong et al. 2009), which are used to treat cardiovascular and cerebrovascular diseases and hyperlipidemia (Zhou et al. 2005; Wang 2010). These active compounds include lipid-soluble tanshinones (e.g., tanshinone I, tanshinone IIA, tanshinone IIB, and cryptotanshinone) and water-soluble phenolic acids (e.g., rosmarinic acid and salvianolic acid). These compounds possess antitumor, antioxidant, anti-inflammatory, antineoplastic, and antimicrobial properties, in addition to their pharmacological activity (Zhou et al. 2005). However, the biosynthetic pathway and regulatory mechanism of these compounds in *S. miltiorrhiza* remain unclear.

Transcriptome analysis is a promising and rapid approach to illustrate biological processes. In particular, whole transcriptome analysis provides comprehensive information on determining

H. Luo (✉)
Institute of Medicinal Plant Development, Chinese Academy of Medical Sciences & Peking Union Medical College, No. 151 Malianwa North Road, Haidian District, Beijing 100193, China
e-mail: hmluo@implad.ac.cn

S. Lu (ed.), *The Salvia miltiorrhiza Genome*, Compendium of Plant Genomes,
https://doi.org/10.1007/978-3-030-24716-4_7

genome structure and function by identifying genetic networks underlying unique cellular, physiological, biochemical, and biological systems and establishing molecular biomarkers that respond to diseases, pathogens, and environmental challenges for plants (Jiang et al. 2015). For medicinal plants, transcriptome analysis can provide insights into the biosynthesis, growth and development, and genetic diversity of bioactive compounds. This chapter systematically discusses the characteristics and methods of transcriptome research and its application in functional genomics research in *S. miltiorrhiza*.

7.2 Transcriptome Characteristics

Transcriptome generally refers to the complete ribonucleic acid (RNA) transcribed from the genome of a cell, tissue, or an organism, which is growing/developing at specific stages or under special conditions, including the complete set of protein-coding messenger RNA (mRNA) and various noncoding RNA (ncRNA) (Morozova et al. 2009). Transcriptome analysis presents the gene expression profiling at the level of all RNAs isolated from a cell/tissue genome and provides the expression characteristics of all genes and the functional information of corresponding proteins encoded by these genes.

The characteristics of transcriptome possess specificities of species, tissues, and cells and the developmental stages and multiple physiological effects on the development of materials used in transcriptome analysis. Transcriptome is the dynamic relationship between the genome and the external environment of a specific organ, tissue, or physiological stage of a particular development. External stimuli or environmental changes can lead to changes in gene expression levels of organisms. Transcriptome sequencing can quantitatively detect all transcripts in specific tissues/organs or cell types of organisms under specific environmental conditions. Furthermore, a gene can form different transcripts due to alternative splicing. Variations in the structures of transcripts lead to the diversity of post-transcriptional gene structure and gene expression. One of the precursors of some genes produces different splicing isomers through different splicing modes. Variable splicing enables a gene to produce multiple RNA transcripts that are translated into different proteins. Transcriptome sequencing has significant advantages in identifying structures and variations in transcripts, gene expression level, functions of ncRNA, and discovery of low abundance and new transcripts. Comparative transcriptome can be used to analyze the differential expression levels of genes in different tissues/organs or the same tissues/organs under different physiological conditions or external conditions, to enable the discovery of genes related to specific physiological functions, and to speculate the functions of unknown genes.

7.3 Transcriptome Sequencing Platforms

7.3.1 Expressed Sequence Tag (EST) and DNA Microarray

Previous approaches for detecting the transcriptomes of tissues or organisms are ESTs and DNA microarray analysis. EST analysis provides an alternative method for gene discovery because ESTs represent the parts of species transcriptome. The mining of ESTs from *Artemisia annua*, *Taxus chinensis,* and *Catharanthus roseus* has revealed the presence of transcripts involved in the biosynthesis of artemisinin, paclitaxel, vinblastine, and vindoline (Jennewein et al. 2004; Murata et al. 2008; Teoh et al. 2006). Despite the power of ESTs to detect the expression of thousands of genes simultaneously, cloning-based amplification for clone selection and Sanger sequencing for EST identification are labor-intensive and expensive processes (Hall 2007). Similarly, DNA microarray can simultaneously detect the expression levels of thousands of genes and compare the expression ratio of the same transcript under different conditions. Microarray assay can be used to estimate the abundance of a particular mRNA species by detecting the hybridization intensity of a cDNA and its complementary oligonucleotide probes (Schena et al. 1995).

The abundance of mRNA in different types of cells or tissues can be quantified with tag sequencing methods, such as serial analysis of gene expression (SAGE) and massively parallel signature sequencing (Sengoelge et al. 2014; Sánchez-León et al. 2012; Huang et al. 2007). With these methods, the 3′ or 5′ end of mRNA is sequenced for cDNA fragments, and the sequencing times of these fragments are counted to determine the transcriptome expression profiles (Velculescu et al. 1995; Brenner et al. 2000). LongSAGE and DeepSAGE (Nielsen et al. 2006) and 5′-robust analysis of 5′ transcript ends (Gowda et al. 2006) are combined with next-generation sequencing (NGS) technologies.

7.3.2 NGS Technologies

With the development of sequencing technologies, NGS technologies have revolutionized the transcriptome research through deep sequencing of transcriptomes from virtually any organism or cell type to investigate transcript profiling in various tissues with different physiological and pathological conditions (Morozova et al. 2009). NGS platforms, such as 454/Roche, Solexa/Illumina, and SOLiD/ABI, can produce large amounts of short DNA sequence reads from transcriptome sequencing in previous studies. However, 454 (Margulies et al. 2005) and SOLiD (Shendure et al. 2005) sequencing technologies have been eliminated, whereas the Illumina platform remains dominant for sequencing.

An Illumina sequencer includes a genome analyzer, HiSeq 2500, HiSeq 2000, HiSeq 1000, HiScanSQ, MiSeq, MiniSeq/NextSeq, and NovaSeq instruments. The Illumina HiSeq platform possesses library automation, provides low-input DNA, and captures hybridization multiplexing, and analysis pipeline improvements include the integration of variant calling and annotation tools. MiSeq integrates cluster generation and paired-end sequencing and reads into a fully automated "hands-off" workflow. The MiniSeq system supports various targeted DNA and RNA applications to investigate single genes or pathways. The new NextSeq 500 system can sequence exons, transcripts, and genomes. NovaSeq 6000 can run four different types of patterned flowcells and are suitable for exon and transcriptome sequencing.

RNA sequencing (RNA-Seq) is a common Illumina sequencing method and widely used for the transcriptome sequencing and analysis of transcript profiling, the identification of differentially expressed genes (DEGs) and alternative splicing (AS) events, and the discovery of novel transcripts (Maher et al. 2009; Mortazavi et al. 2008; Wang et al. 2009). RNA-Seq can provide genome-wide transcriptome profiling and reveal the precise location of transcription boundaries for genes. This method is an efficient tool for elucidating the important candidates involved in complex biosynthetic pathways irrespective of genome complexity even in cases of nonmodel plant species (Unamba et al. 2015). With RNA-Seq, information about the RNA content and transcriptional status of an experimental system can be depicted and quantified, and the quality of the sequenced transcripts can be validated in these experiments.

7.3.3 Third-Generation Sequencers

Third-generation sequencers (e.g., Heliscope, PacBio, and nanopore) have been developed with high-throughput, low costs, and long read-length direction for single-molecule sequencing. Single-molecule sequencing improves the sample throughput and sequencing speed. The first commercially released single-molecule sequencing instrument is Heliscope single-molecule sequencer (Helicos Biosciences), which was developed by Helicos Inc. in the USA (Harris et al. 2008; Lipson et al. 2009; Tessler et al. 2009). However, the development of the Helicos platform had been terminated since 2012 because of its low single-pass accuracy (81–83%), low throughput, low sequencing efficiency, and high sequencing cost.

Based on a novel single-molecular real-time (SMRT) technology, the PacBio *RS* system shows potential for DNA sequencing with long read lengths of several kilobases (Schadt et al. 2010). This sequencing system has been successfully applied to a large number of transcriptome sequencing because of its long read lengths and rapid turnaround times (Eid et al. 2009; Flusberg et al. 2010). PacBio has launched three sequencers based on SMRT technology, including PacBio RS, PacBio RS II, and PacBio Sequel system, in 2011. The advantage of PacBio sequencing is its long read length. Despite the remarkable advantages of SMRT sequencing, this technology is challenged with raw read error rates (approximately 11–15%) and modest throughput of a run.

In February 2012, a series of nanoporous DNA sequencing platforms (e.g., MinION, Promethion, and GridION systems) was released. Oxford Nanopore Technology is based on the implementation of "strand" nanopore sequencing approach. Nanopore sequencing is characterized by long-reading, real-time, single-molecule, and cost-effective sequencing. This technology is challenged in which sequencing errors may not be randomly distributed. Hybrid sequencing strategies have integrated the strengths and weaknesses of RNA-Seq and PacBio sequencing. These approaches are frequently used for full-length transcriptome sequences to understand the complete expression of gene isoforms.

7.4 Transcriptome Analysis of *S. miltiorrhiza*

S. miltiorrhiza is an important medicinal model plant. The transcriptome data of this medicinal plant through different sequencing platforms to sequence different tissues/organs under different treatments contribute to the identification of transcription profiling. Transcription profiling is widely used to compare gene sequences and functions and to identify gene expression. Shao et al. (2016) constructed the Danshen Transcriptional Resource Database (DsTRD) by using 76,531 transcribed sequences assembled from 12 RNA-Seq transcriptomes. This database facilitates browsing and searching for sequences and functional annotations of *S. miltiorrhiza*. The transcriptome of *S. miltiorrhiza* enables the discovery of key genes for the biosynthesis of bioactive ingredients not only to elucidate the secondary metabolic pathways and their regulatory mechanisms but also to detect genes or genetic markers related to growth, development, and disease/stress resistance (Table 7.1).

7.4.1 Gene Expression Profiling Analysis of *S. miltiorrhiza*

The abundance of mRNA in different cell types or under different conditions quantified through transcriptome sequencing represents the gene expression profiling of experimental materials. EST generation, cDNA microarray analysis, and RNA-Seq are widely used for the gene expression profiling analysis of medicinal plants. Yan et al. (2010) constructed a cDNA library of *S. miltiorrhiza* whole plantlet and sequenced 10,228 ESTs with the generation of 4225 unigenes. This set of ESTs represents a proportion of gene expression profiling for *S. miltiorrhiza* plantlet and provides insights into the gene complement of *S. miltiorrhiza*. Cui et al. (2011) identified the changes in the gene expression profile of 114 differentially expressed cDNAs at different stages of hairy root development in *S. miltiorrhiza* through cDNA microarray analysis. Among the DEGs, copalyl diphosphate synthase (CPS) gene, which encodes the key enzyme involved in catalyzing the conversion of diterpenoid precursor (*E,E,E*)-geranylgeranyl diphosphate (GGPP) to form copalyl diphosphate, and eight candidate P450 genes are differentially expressed in the hairy roots of *S. miltiorrhiza*. Hua et al. (2011) investigated wide gene expression profiles associated with the entire growing cycle of *S. miltiorrhiza* through Illumina deep sequencing. The experimental materials used for transcriptome analysis include seeds, 2-day seedlings, and 2-week-old and 2-month-old plantlets (counted from germination) that

Table 7.1 Transcriptome analysis of *S. miltiorrhiza*

Samples	Sequencing platform	Functional identification	References
Tissue culture seedling	Sanger	Phenolic acid and diterpene biosynthesis; SSRs	Yan et al. (2010)
Root	454 GS FLX	Phenolic acid and diterpene biosynthesis; CYP450s; TFs	Li et al. (2010)
Seedling/vegetative/reproductive stages, flower/stem/leave/root	Solexa	Phenolic acid and diterpene biosynthesis; TFs	Hua et al. (2011)
Root, leaf tissues	454 GS-FLX	Diterpene biosynthesis; CYP450s; TFs	Yang et al. (2013)
Leaf, MeJA treatment	Illumina Hiseq 2000	Phenolic acid and diterpene biosynthesis; CYP450s; TFs	Luo et al. (2014)
Hairy roots (Ag^+) 0/12/24/36/48 h	454 GS FLX	Diterpene biosynthesis; CYP450s; TFs	Gao et al. (2014)
Leaf, root, flower	Illumina	Identification and systematic classification of CYP450s	Chen et al. (2014)
Periderm/phloem/xylem	PacBio RS; Illumina Hiseq 2500	Full-length transcriptome; alternative splicing; diterpene biosynthesis; CYP450s; SDRs; 2ODDs	Xu et al. (2015)
Hairy roots (MeJA/YE)	Hiseq 2000	Tanshinone and phenolic acid biosynthesis; CYP450s	Zhou et al. (2017)
Cell cultures (SA)	Hiseq 2500	SA signaling components	Zhang et al. (2016)
Mid-segments roots	Hiseq 2000	Phenylpropanoid and lignin biosynthesis	Song et al. (2017)
Root, stem, leaf, and flower	Hiseq 2500	miRNAs	Xu et al. (2014)
Plants in sterile tubes	Hiseq 2500	AtDREB1A-mediated drought tolerance; transcription factors	Wei et al. (2018)
Plants with MeJA treatment/control	Solexa	JAZ repressor	Ge et al. (2015)

represent seedling stage (SS) and vegetative stage (VS), respectively, and flowers, stems, leaves, and roots that represent the reproductive stage (RS). A total of 56,774 unigenes, including genes associated with secondary metabolite pathways and transcription factor (TF) genes, are covered in this *S. miltiorrhiza* transcriptome. Yang et al. (2013) detected the transcriptome profiling of *S. miltiorrhiza* root and leaf tissues by using a 454 GS-FLX sequencing platform to generate 550,546 and 525,292 reads. A total of 2863 highly expressed unigenes, such as *SmCPS*, kaurene synthase-like (*SmKSL*), and *CYP76AH1*, have been identified on the basis of the comparative transcriptome analyses of the two tissues in the roots.

7.4.2 Discovery of Genes Related to Tanshinone and Salvianolic Acid Biosynthesis from Transcriptome

The identification of the biosynthetic pathways of bioactive compounds for tanshinone and salvianolic acid, which are complex and composed of multiple steps of molecular regulation, is the focus of transcriptome research on *S. miltiorrhiza*. EST approach provides a rapid method to identify gene-encoding key enzymes and regulatory factors involved in secondary biosynthetic pathways (Ohlrogge and Benning 2000). Yan et al. (2010) sequenced the EST of *S. miltiorrhiza* tissue culture seedlings and identified 26 genes related to salvianolic acid synthesis and ten genes associated with diterpenoid synthesis for the first time. NGS technologies enable the large-scale analysis of transcriptomes from *S. miltiorrhiza* to identify key genes associated with the biosynthesis of tanshinone and salvianolic acid. Hua et al. (2011) used an Illumina platform to sequence the samples of mixed materials covering the entire life stage of *S. miltiorrhiza*. A total of 56,774 transcripts are assembled and annotated from these transcriptome data. Of these transcripts, 1539 are enriched in secondary metabolic pathways (e.g., phenylpropane synthesis and terpenoid synthesis pathways). To determine the mechanism of root-specific accumulation and synthesis of tanshinone, Yang et al. (2013) sequenced the transcriptome of the root and leaf tissues of *S. miltiorrhiza* and discovered that the key genes (e.g., *SmCPS*, *SmKSL*, and *CYP76AH1*) involved in tanshinone biosynthesis are highly expressed in roots. Li et al. (2010) constructed 454 ESTs from the root of *S. miltiorrhiza* and obtained 27 genes related to tanshinone biosynthesis and 29 genes related to phenolic acid biosynthesis.

Genes involved in a natural product biosynthetic pathway are usually co-expressed in medicinal plants. Co-expression analysis contributes to the effective screening of genes involved in a specific pathway and limits the number of candidate genes based on transcriptome data generated from multiple samples. Guo et al. (2013) investigated the silver-induced hairy root transcriptome of *S. miltiorrhiza* and discovered that six cytochrome P450 genes co-expressed with *SmCPS* and *SmKSL*, the genes encode two identified enzymes involved in biosynthesized tanshinone skeleton. Among these candidate cytochrome P450s (CYP450s), *CYP76AH1* was confirmed to be involved in tanshinone biosynthetic pathway. Xu et al. (2015) observed the synthesis of tanshinones and salvianolic acid B and their accumulation in different tissues of *S. miltiorrhiza* root and screened 15 CYP450s, one 2-oxoglutarate-dependent dioxygenase, and five short-chain alcohol dehydrogenase genes involved in tanshinone biosynthesis. Two CYP450s are likely involved in the synthesis pathway of rosmarinic acid in *S. miltiorrhiza*.

7.4.3 Comparative Transcriptome Study of *S. miltiorrhiza*

MeJA induction can significantly increase the yield of tanshinone and salvianolic acid compounds. Luo et al. (2014) used an Illumina HiSeq 2000 sequencing platform to sequence the transcriptome of *S. miltiorrhiza* leaves (control) and those of *S. miltiorrhiza* leaves treated with MeJA for 12 h. A total of 5287 transcripts respond to MeJA induction. It includes the well-known *S. miltiorrhiza* genes related to the synthesis pathway of ketones and salvianolic acids. In addition, three CYP450s related to tanshinone synthesis and four CYP450s associated with salvianolic acid synthesis are predicted on the basis of the co-expression analysis of *SmCPS1*, *SmKSL1*, and rosmarinic acid synthase (*RAS*) genes. Ge et al. (2015) investigated the expression of JAZ inhibitory protein in MeJA signaling pathway after 6 h of MeJA induction. Yeast elicitors can significantly induce the accumulation of tanshinone in the hairy roots of *S. miltiorrhiza*. Gao et al. (2014) sequenced the transcriptomes of four hairy roots of blank control and 100 μg/mL yeast elicitors for 12, 24, and 36 h after induction. They monitored the blank control and treated materials by using an Agilent 1290 Infinity HPLC system. According to gene co-expression analysis, 70 TFs and eight CYP450s are predicted to be involved in tanshinone biosynthesis. Zhou et al. (2017) conducted a comprehensive transcriptome profiling analysis of *S. miltiorrhiza* hairy roots treated with either MeJA or yeast extract (YE) to identify the downstream and regulatory genes involved in tanshinone and phenolic acid biosyntheses. Candidate downstream genes, including several CYP450s, are screened from the RNA-Seq dataset based on the co-expression pattern analysis with key biosynthetic genes. Zhang et al. (2016) demonstrated the genes involved in salicylic acid (SA) signaling mechanisms by using

RNA-Seq to evaluate the transcriptional profiles in *S. miltiorrhiza* cell cultures. The upregulated genes in response to SA induction are mainly related to stimulus-response and multiorganism process. TFs, such as WRKY, bHLH, and GRAS, and genes involved in hormone signal transduction are differentially expressed under SA treatment. The genes associated with defense signaling, such as antioxidant system genes, cytochrome P450s, and ATP binding cassette transporters, are significantly upregulated in *S. miltiorrhiza*.

Xu et al. (2015) compared the content of tanshinones in different tissues of *S. miltiorrhiza* root and found that the distribution of these compounds is mostly abundant in the periderm, followed by the phloem and xylem of the roots. The expression patterns of *SmDXS2*, *SmDXR*, *SmHDS*, *SmHDR1*, *SmHDR3*, *SmIPI1*, *SmGGPPS1*, *SmCPS1*, *SmCPS5*, *SmKSL1*, and *SmKSL7* are identical to the distribution of tanshinones in the roots, indicating that the periderm of *S. miltiorrhiza* accumulates large amounts of tanshinone and actively synthesizes these compounds in root tissues. In addition, *SmPAL1*, *SmPAL3*, *SmC4H1*, *Sm4CL3*, *Sm4CL-like 1*, *Sm4CL-like 4*, *SmTAT1*, *SmHPPR3*, *SmRAS*, and *SmCYP98A78* related to rosmarinic acid synthesis are expressed more abundantly in the phloem and xylem of *S. miltiorrhiza* than in the periderm. This observation is consistent with the distribution of salvianolic acid B in different root tissues. The content of salvianolic acid B in the phloem and xylem is five times higher than that in the periderm. Song et al. (2017) performed de novo RNA-Seq and comparative transcriptome analysis in two *S. miltiorrhiza* genotypes (BH18 and ZH23) with different phenolic acid concentrations. Among the 670 DEGs identified between BH18 and ZH23, 250 DEGs, including genes involved in the phenylpropanoid biosynthesis pathway, are upregulated in ZH23. Nine genes involved in the lignin biosynthesis pathway are upregulated in BH18, and they are abundant with lignin content. Most of the genes involved in the core common upstream phenylpropanoid biosynthesis pathway are higher in ZH23 than in BH18. Lignin biosynthesis is a putative partially competing pathway with phenolic acid biosynthesis in *S. miltiorrhiza*.

7.4.4 Identification of Candidate TFs Involved in Bioactive Compound Biosynthesis and Stress Response from *S. miltiorrhiza* Transcriptome Data

TFs are protein molecules composed of a special structure and involved in regulating gene expression, which are also known as *trans*-acting factors. Two kinds of TFs are found in plants: the first is called nonspecific TF, which cannot selectively regulate gene transcription and expression, and the second is called specific TF, which can selectively regulate the transcription and expression of some genes.

Li et al. (2010) completed the transcriptome sequencing of *S. miltiorrhiza* root and discovered 577 TF genes. Zhang et al. (2015a) reported the whole genome and transcriptome information of *S. miltiorrhiza*, identified the bHLH TF gene family in *S. miltiorrhiza*, and provided basic data to study the synthesis and regulation of effective components of *S. miltiorrhiza*. A total of 127 bHLH TF genes are also identified in the genome of *S. miltiorrhiza* and classified into 25 subfamilies. The gene-specific expression patterns and upregulated expression patterns in response to MeJA treatment of these bHLH genes are analyzed on the basis of the RNA-Seq data of 19 sequencing libraries. Consequently, seven bHLH genes are predicted to be involved in the regulation of tanshinone biosynthesis. Ji et al. (2016) identified the AP2/ethylene responsive factor (ERF) gene family in the genome-wide and obtained the candidate TF genes involved in active constituent biosynthesis in *S. miltiorrhiza*. A total of 170 AP2/ERF genes are identified in *S. miltiorrhiza* genome and divided into five subfamilies, namely, AP2 (25 genes), DREB (61 genes), ERF (79 genes), RAV (four genes), and soloist (one gene). Four genes related to the biosynthesis of bioactive constituents of *S. miltiorrhiza* are selected on the basis of the distribution of bioactive constituents and the co-expression profiling of AP2/ERF genes in different organs and root tissues. Zhang et al. (2018) conducted a genome-wide survey of the basic leucine zipper (bZIP) gene family and identified 70 SmbZIP TFs in *S. miltiorrhiza*. The phylogeny, gene structure, conserved motifs, and the AS events of these SmbZIPs are observed in this study. The two candidate SmbZIP (*SmbZIP7* and *SmbZIP20*) genes that may be responsible for the regulation of tanshinone biosynthesis are selected on the basis of the transcriptomic data with the highest expression levels in the root and periderm of *S. miltiorrhiza*. Wei et al. (2018) compared the global transcriptional profiles of wild-type and *AtDREB1A*-expressing transgenic plants of *S. miltiorrhiza* by using RNA-Seq to determine the molecular mechanisms underpinning the increased drought tolerance in transgenic plants. Among the 3904 DEGs, various TFs significantly increase after drought stress, especially AP2/ERF, bZIP, and MYB protein families.

7.4.5 Full-Length Transcriptome Study of *S. miltiorrhiza*

Full-length transcriptome sequencing based on PacBio SMRT sequencing enables the acquisition of transcriptomic information of almost all transcripts in a specific tissue or organ of any species. Superlong sequences obtained through PacBio sequencing can span the complete sequence of transcripts from 5′ to 3′ polyA tail, which contributes to identify accurate gene isomers, variable AS, fusion genes, and gene full-length sequence because of the advantage of PacBio sequencing (Gordon et al. 2015). AS is a post-transcriptional mechanism that plays essential roles in the regulation of gene expression and increases the diversity of transcriptome and proteome (Barash et al. 2010). AS enables an RNA precursor to produce different splicing isomers under different splicing modes. The accuracy of AS characterization predicted through SMRT sequencing technology can be increased compared with that of short transcript assembled from RNA-Seq reads (Dong et al. 2015). However, PacBio sequencing exhibits substantial sequencing errors ($\sim 10\%$), which can be corrected by RNA-Seq data (Koren et al. 2012). Therefore, hybrid sequencing strategies use RNA-Seq short reads to correct the error inherent in long single-molecule sequences have been developed and widely used in full-length transcriptome analysis (Koren et al. 2012). Hybrid sequencing approaches enable the detection of sensitive isoforms for transcripts based on reference genome sequences (Au et al. 2013).

Xu et al. (2015) conducted a full-length transcriptome study of *S. miltiorrhiza* roots by using PacBio combined with Illumina sequencing technology and identified some gene isomers and AS phenomena of the transcripts in *S. miltiorrhiza* roots. The variable splicing of *S. miltiorrhiza* transcripts, especially the genes responsible for the synthesis of active ingredients, has been comprehensively identified in this study. Among the 636,805 high-quality transcripts, 71% of full-length transcripts in *S. miltiorrhiza* genome are covered. In addition, 4035 gene isomers are identified on IDP software, and 16,241 gene isomers are predicted. The variable splicing of gene loci and AS types is complex and diverse.

7.4.6 Gene Annotation for *S. miltiorrhiza* Genome by Using Transcriptome Data

Genome annotation is a high-throughput annotation of the biological functions of all genes in the genome by using bioinformatics methods and tools, such as gene recognition and gene function annotation. The core of gene recognition is to determine the exact location of all genes in the whole genome sequence. The transcriptome data of ESTs and RNA-Seq enable the prediction of new genes and protein-coding gene in a genome. Protein-coding gene annotation, including the identification of promoters and operon, determination of functional gene categories, characterization of other features in a genome and genome-scale gene mapping, elucidates the essential information of genomes. Genome annotation includes the discovery of novel noncoding RNA genes, the identification of regulatory elements that monitor temporal or spatial gene expression and splice variants, and the illustration of the transcription start and polyadenylation sites and exon–intron structures. The data generated from transcriptome studies can provide protein-coding gene annotation information. In particular, ESTsor full-length complementary DNA sequences are aligned and mapped to reference genome sequences, illustrating the presence of exons, introns, exon junctions, and transcription boundaries for genes and providing an accurate and effective annotation of protein-coding genes. The annotation of protein-coding genes of a genome can illustrate the dynamic genetic architectures, reveal critical developmental pathways, and enable the understanding of evolution.

The genome of *S. miltiorrhiza* has been sequenced and assembled (Xu et al. 2016; Zhang et al. 2015b). A total of 30,478 are predicted to be putative genes in the *S. miltiorrhiza* genome, with an average transcript length of 2825 base pairs (bp) by using Ab initio gene prediction and homology-based annotation (Xu et al. 2016). Most of the genes (27,793; 91.19%) have homologs in the nr database (with *E* value less than 10^{-5}). Zhang et al. (2015b) used three methods, namely, EST and transcriptome, de novo, and homology-based predictions, to perform gene annotation for *S. miltiorrhiza* genome. RNA-Seq datasets generated from the leaf, root, and flower tissues of *S. miltiorrhiza* were obtained from the National Center for Biotechnology Information database (SRX388784, SRX371961, SRX370399, and 10,494 ESTs) and subsequently used for the de novo assembly of the transcriptome. First, all RNA reads are aligned to the *S. miltiorrhiza* genome by using TopHat (Trapnell et al. 2012). Second, the transcripts are assembled using Cufflinks (Trapnell et al. 2012). The open-reading frames of reliable transcripts are predicted with hidden Markov model-based training parameters. The transcriptome data generated from different assembled tissues result in 46–68 Mb. Finally, a total of 40,700 transcripts with an average length of 2606 bp are obtained in this study. In addition, 10,494 ESTs are blasted against the assembled *S. miltiorrhiza* genome, and 3974 transcripts are identified with an average length of 1596 bp (Zhang et al. 2015b).

7.4.7 Noncoding RNA and Novel Small RNA Discovery from Transcriptome

Small ncRNAs, which are types of RNA molecules with a length of 18–30 nucleotides and crucial regulators of development and cell fate determination, are not translated into protein products. ncRNAs include transfer RNA, ribosomal RNA, microRNA (miRNA), and small interfering RNA, where the latter two types are involved in the post-transcriptional regulation of gene expression in various organisms (Filipowicz et al. 2008; Gowda et al. 2007). The high-throughput miRNA profiling by NGS technologies provides several advantages for ncRNA sequencing studies, such as decreased procedural complexity and cost, increased high-throughput and coverage depth, and detection of variants in mature miRNA length, miRNA editing events, and miRNA–target RNA pairs (German et al. 2008; Reid et al. 2008).

Xu et al. (2014) identified 452 known miRNAs corresponding to 589 precursor miRNAs (pre-miRNAs) and 40 novel miRNAs corresponding to 24 pre-miRNAs in different tissues of *S. miltiorrhiza* through transcriptome sequencing. Sixty-nine targets are verified to be potentially cleaved by 25 miRNAs through degradome analysis, especially acetyl-CoA C-acetyltransferase, which is cleaved by miR5072 and involved in the biosynthesis of tanshinones. mRNA-like noncoding RNAs (mlncRNAs), which are ncRNAs transcribed by RNA polymerase II and are polyadenylated, capped, and spliced, play important roles in plant development and defense responses. Li et al. (2015) identified and characterized 5446 mlncRNAs from *S. miltiorrhiza* through transcriptome analysis. The majority of identified mlncRNAs are involved in stress response in *S. miltiorrhiza*. Among the 5446 mlncRNAs, 3044 mlncRNAs are expressed in the hairy roots of *S. miltiorrhiza*. The expression of 1904 out of 3044 mlncRNAs is altered through YE and Ag^+ treatments.

7.4.8 Genetic Marker Identification from Transcriptome Data

Genetic markers are generated from variations in the nucleotide sequences of individual species. The two typical classes of genetic markers are simple sequence repeats (SSRs) and single nucleotide polymorphisms (SNPs). SSR, also known as microsatellite, is a type of second-generation DNA molecular marker composed of 1–5 bp with repeat several times to reach the length of approximately 100 bp. EST-SSR is a molecular marker presented in EST sequences that are conservative in a gene-coding region. EST-SSR is widely used in genetic mapping and genetic diversity evaluation, germplasm identification, phylogenetic, and evolutionary studies. SNPs, also termed as third-generation markers, have been widely used in genetic studies on many plants (Gupta et al. 2001). Transcriptome data are a large resource for the identification of SSRs and SNPs (Tang et al. 2008). SSRs and SNPs from expressed sequences that are more conserved than nongenic sequences are potentially tightly linked with functional genes that may control certain important agronomic characters (Bozhko et al. 2003; Graham et al. 2010). EST-SSRs and SNPs are identified from transcriptome data generated from NGS sequencing platforms in *S. miltiorrhiza*. These SSRs are powerful resources for subsequent studies on taxonomy, molecular breeding, genetics, genomics, and secondary metabolism in medicinal plants.

Yan et al. (2010) identified 122 microsatellites from 112 unigene sequences. The major types of these microsatellites are dinucleotide and trinucleotide. Among the 112 EST sequences, 78 have sufficient flanking sequences for the design of 82 primers, ensuring their suitability for genetic linkage mapping. Eighty-two of 112 microsatellite-containing ESTs are annotated to be homologous to stress-related proteins, such as bZIP TF family proteins, universal stress family proteins, and dehydration proteins. The microsatellite information obtained in this study can be used for genetic linkage mapping in *S. miltiorrhiza*.

7.5 Transcriptome Combining Other Omic Analyses of *S. miltiorrhiza*

The active ingredients of *S. miltiorrhiza* determine its quality and quantity and are affected by harvesting seasons, harvesting time, and harvesting methods. Therefore, *S. miltiorrhiza* transcriptome analysis is related to the rationality of medicinal parts and harvesting time. The active components of *S. miltiorrhiza* are mostly secondary metabolites, and their synthesis and accumulation are affected by developmental stages and environmental conditions in different tissues/organs. The biosynthesis pathways of active components can be systematically investigated, and the molecular mechanism of biosynthesis and regulation of secondary metabolites can be effectively revealed by constructing transcriptome libraries of different stages, organs, and origins of *S. miltiorrhiza*.

Ge et al. (2015) evaluated the metabolomic and transcriptomic signatures to determine the functions of genes involved in the jasmonate (JA)-mediated accumulation of secondary metabolites in *S. miltiorrhiza*. Among MeJA-induced genes, SmJAZ8 is identified as a transcriptional regulator in the JA signal pathway of *S. miltiorrhiza* and has a novel function in the constitutive accumulation of secondary metabolites. Gao et al. (2014) used a combined metabolomics and transcriptomics method to investigate the inducible biosynthesis of bioactive diterpenoid tanshinones from *S. miltiorrhiza*. Tanshinone production is a dominant component of metabolites and increases at later time points in the elicited hairy root cultures. Consistent with metabolomic analysis results, transcript levels of biosynthetic genes, including 70 TFs and 8

CYP450 genes, which are related to tanshinone biosynthesis and regulation increase.

7.6 Concluding Remarks

The transcriptome of *S. miltiorrhiza* is extremely complex. It shows spatial- and temporal-specific patterns in terms of tissues and various stresses that are usually related to bioactive compounds. The growth and development of *S. miltiorrhiza* are related to the synthesis and accumulation of active secondary metabolites in vivo. Thus, growth and development patterns are observed in different *S. miltiorrhiza*. Transcriptome studies can determine the growth and development mechanism of *S. miltiorrhiza*. Studies on the transcriptome of *S. miltiorrhiza* have provided valuable insights into gene differential expression, gene splicing, and AS through NGS. The characterization of the transcriptome of *S. miltiorrhiza* has revealed many active genes with spatiotemporal regulation and evolutionary significance. Investigating the response mechanism of *S. miltiorrhiza* to adverse events helps understand their growth and development under adverse conditions and provides an important guiding role in cultivating good varieties with resistance to adverse environmental characteristics. Transcriptome research of *S. miltiorrhiza* is expected to provide a new source for natural drugs by analyzing the synthetic pathways of natural products. High-throughput transcriptome sequencing provides rapid, efficient, and comprehensive interpretation on the genetic information of *S. miltiorrhiza*, thereby promoting the development of secondary metabolic engineering and synthetic biology of *S. miltiorrhiza* and laying a foundation for functional gene mining, biosynthesis and regulation of active components of *S. miltiorrhiza*, and evaluation of excellent germplasm resources.

Acknowledgements This work was supported by the National Natural Science Foundation (31570302) and Chinese Academy of Medical Sciences (CAMS) Innovation Fund for Medical Sciences (CIFMS, 2016-I2M-3-016).

References

Au KF, Sebastiano V, Afshar PT, Durruthy JD, Lee L, Williams BA, van Bakel H, Schadt EE, Reijo-Pera RA, Underwood JG, Wong WH (2013) Characterization of the human ESC transcriptome by hybrid sequencing. Proc Natl Acad Sci U S A 110:4821–4830

Barash Y, Calarco JA, Gao W, Pan Q, Wang X, Shai O, Blencowe BJ, Frey BJ (2010) Deciphering the splicing code. Nature 465:53–59

Bozhko M, Riegel R, Schubert R, Muller-Starck G (2003) A cyclophilin gene marker confirming geographical differentiation of Norway spruce populations and indicating viability response on excess soil-born salinity. Mol Ecol 12:3147–3155

Brenner S, Johnson M, Bridgham J, Golda G, Lloyd DH, Johnson D, Luo S, McCurdy S, Foy M, Ewan M, Roth R, George D, Eletr S, Albrecht G, Vermaas E, Williams SR, Moon K, Burcham T, Pallas M, DuBridge RB, Kirchner J, Fearon K, Mao J, Corcoran K (2000) Gene expression analysis by massively parallel signature sequencing (MPSS) on microbead arrays. Nat Biotechnol 18:630–634

Chen HM, Wu B, Nelson DR, Wu K, Liu C (2014) Computational identification and systematic classification of novel cytochrome P450 genes in *Salvia miltiorrhiza*. PLoS ONE 9(12):e115149

Cui G, Huang L, Tang X, Zhao J (2011) Candidate genes involved in tanshinone biosynthesis in hairy roots of *Salvia miltiorrhiza* revealed by cDNA microarray. Mol Biol Rep 38:2471–2478

Dong L, Liu H, Zhang J, Yang S, Kong G, Chu JS, Chen N, Wang D (2015) Single-molecule real-time transcript sequencing facilitates common wheat genome annotation and grain transcriptome research. BMC Genom 16:1039

Eid J, Fehr A, Gray J, Luong K, Lyle J, Otto G, Peluso P, Rank D, Baybayan P, Bettman B, Bibillo A, Bjornson K, Chaudhuri B, Christians F, Cicero R, Clark S, Dalal R, Dewinter A, Dixon J, Foquet M, Gaertner A, Hardenbol P, Heiner C, Hester K, Holden D, Kearns G, Kong X, Kuse R, Lacroix Y, Lin S, Lundquist P, Ma C, Marks P, Maxham M, Murphy D, Park I, Pham T, Phillips M, Roy J, Sebra R, Shen G, Sorenson J, Tomaney A, Travers K, Trulson M, Vieceli J, Wegener J, Wu D, Yang A, Zaccarin D, Zhao P, Zhong F, Korlach J, Turner S (2009) Real-time DNA sequencing from single polymerase molecules. Science 323:133–138

Filipowicz W, Bhattacharyya SN, Sonenberg N (2008) Mechanisms of post-transcriptional regulation by microRNAs: are the answers in sight? Nat Rev Genet 9:102–114

Flusberg BA, Webster DR, Lee JH, Travers KJ, Olivares EC, Clark TA, Korlach J, Turner SW (2010) Direct detection of DNA methylation during single-molecule, real-time sequencing. Nat Methods 7:461–465

Gao W, Sun HX, Xiao H, Cui G, Hillwig ML, Jackson A, Wang X, Shen Y, Zhao N, Zhang L, Wang XJ, Peters RJ, Huang L (2014) Combining metabolomics and transcriptomics to characterize tanshinone biosynthesis in *Salvia miltiorrhiza*. BMC Genom 115:73

Ge Q, Zhang Y, Hua WP, Wu YC, Jin XX, Song SH, Wang ZZ (2015) Combination of transcriptomic and metabolomic analyses reveals a JAZ repressor in the jasmonate signaling pathway of *Salvia miltiorrhiza*. Sci Rep 5:14048

German MA, Pillay M, Jeong DH, Hetawal A, Luo S, Janardhanan P, Kannan V, Rymarquis LA, Nobuta K, German R, De Paoli E, Lu C, Schroth G, Meyers BC, Green PJ (2008) Global identification of microRNA-target RNA pairs by parallel analysis of RNA ends. Nat Biotechnol 26:941–946

Gordon SP, Tseng E, Salamov A, Zhang J, Meng X, Zhao Z, Kang D, Underwood J, Grigoriev IV, Figueroa M, Schilling JS, Chen F, Wang Z (2015) Widespread polycistronic transcripts in fungi revealed by single-molecule mRNA sequencing. PLoS ONE 10:e0132628

Gowda M, Li H, Alessi J, Chen F, Pratt R, Wang GL (2006) Robust analysis of 5'-transcript ends (5'-RATE): a novel technique for transcriptome analysis and genome annotation. Nucleic Acids Res 34:e126

Gowda M, Li H, Wang GL (2007) Robust analysis of 5'-transcript ends: a high-throughput protocol for characterization of sequence diversity of transcription start sites. Nat Protoc 2:1622–1632

Graham IA, Besser K, Blumer S, Branigan CA, Czechowski T, Elias L, Guterman I, Harvey D, Isaac PG, Khan AM, Larson TR, Li Y, Pawson T, Penfield T, Rae AM, Rathbone DA, Reid S, Ross J, Smallwood MF, Segura V, Townsend T, Vyas D, Winzer T, Bowles D (2010) The genetic map of *Artemisia annua* L. identifies loci affecting yield of the antimalarial drug artemisinin. Science 327:328–331

Guo J, Zhou YJ, Hillwig ML, Shen Y, Yang L, Wang Y, Zhang X, Liu W, Peters RJ, Chen X, Zhao ZK, Huang L (2013) CYP76AH1 catalyzes turnover of miltiradiene in tanshinones biosynthesis and enables heterologous production of ferruginol in yeasts. Proc Natl Acad Sci U S A 110:12108–12113

Gupta PK, Roy JK, Prasad M (2001) Single nucleotide polymorphisms: a new paradigm for molecular marker technology and DNA polymorphism detection with emphasis on their use in plants. Curr Sci 80:524–536

Hall N (2007) Advanced sequencing technologies and their wider impact in microbiology. J Exp Biol 210:1518–1525

Harris TD, Buzby PR, Babcock H, Beer E, Bowers J, Braslavsky I, Causey M, Colonell J, Dimeo J, Efcavitch JW, Giladi E, Gill J, Healy J, Jarosz M, Lapen D, Moulton K, Quake SR, Steinmann K, Thayer E, Tyurina A, Ward R, Weiss H, Xie Z (2008) Single-molecule DNA sequencing of a viral genome. Science 320:106–109

Hua W, Zhang Y, Song J, Zhao L, Wang Z (2011) *De novo* transcriptome sequencing in *Salvia miltiorrhiza* to identify genes involved in the biosynthesis of active ingredients. Genomics 98(4):272–279

Huang J, Hao P, Zhang YL, Deng FX, Deng Q, Hong Y, Wang XW, Wang Y, Li TT, Zhang XG, Li YX, Yang PY, Wang HY, Han ZG (2007) Discovering multiple transcripts of human hepatocytes using massively parallel signature sequencing (MPSS). BMC Genom 8:207

Jennewein S, Wildung MR, Chau M, Walker K, Croteau R (2004) Random sequencing of an induced *Taxus* cell cDNA library for identification of clones involved in Taxol biosynthesis. Proc Natl Acad Sci U S A 101:9149–9154

Jiang Z, Zhou X, Li R, Michal JJ, Zhang S, Dodson MV, Zhang Z, Harland RM (2015) Whole transcriptome analysis with sequencing: methods, challenges and potential solutions. Cell Mol Life Sci 72(18):3425–3439

Ji A, Luo H, Xu Z, Zhang X, Zhu Y, Liao B, Yao H, Song J, Chen S (2016) Genome-wide identification of the AP2/ERF gene family involved in active constituent biosynthesis in *Salvia miltiorrhiza*. Plant Genome 9(2). https://doi.org/10.3835/plantgenome2015.08.0077

Koren S, Schatz MC, Walenz BP, Martin J, Howard JT, Ganapathy G, Wang Z, Rasko DA, McCombie WR, Jarvis ED, Phillippy Adam M (2012) Hybrid error correction and de novo assembly of single-molecule sequencing reads. Nat Biotechnol 30:693–700

Li Y, Luo HM, Sun C, Song JY, Sun YZ, Wu Q, Wang N, Yao H, Steinmetz A, Chen SL (2010) EST analysis reveals putative genes involved in glycyrrhizin biosynthesis. BMC Genom 11:268

Li D, Shao F, Lu S (2015) Identification and characterization of mRNA-like noncoding RNAs in *Salvia miltiorrhiza*. Planta 241:1131–1143

Lipson D, Raz T, Kieu A, Jones DR, Giladi E, Thayer E, Thompson JF, Letovsky S, Milos P, Causey M (2009) Quantification of the yeast transcriptome by single-molecule sequencing. Nat Biotechnol 27:652–658

Luo H, Zhu Y, Song J, Xu L, Sun C, Zhang X, Xu Y, He L, Sun W, Xu H, Wang B, Li X, Li C, Liu J, Chen S (2014) Transcriptional data mining of *Salvia miltiorrhiza* in response to methyl jasmonate to examine the mechanism of bioactive compound biosynthesis and regulation. Physiol Plant 152 (2):241–255

Maher CA, Kumar-Sinha C, Cao X, Kalyana-Sundaram S, Han B, Jing X, Sam L, Barrette T, Palanisamy N, Chinnaiyan AM (2009) Transcriptome sequencing to detect gene fusions in cancer. Nature 458:97–101

Margulies M, Egholm M, Altman WE, Attiya S, Bader JS, Bemben LA, Berka J, Braverman MS, Chen YJ, Chen Z, Dewell SB, Du L, Fierro JM, Gomes XV, Godwin BC, He W, Helgesen S, Ho CH, Irzyk GP, Jando SC, Alenquer ML, Jarvie TP, Jirage KB, Kim JB, Knight JR, Lanza JR, Leamon JH, Lefkowitz SM, Lei M, Li J, Lohman KL, Lu H, Makhijani VB, McDade KE, McKenna MP, Myers EW, Nickerson E, Nobile JR, Plant R,

Puc BP, Ronan MT, Roth GT, Sarkis GJ, Simons JF, Simpson JW, Srinivasan M, Tartaro KR, Tomasz A, Vogt KA, Volkmer GA, Wang SH, Wang Y, Weiner MP, Yu P, Begley RF, Rothberg JM (2005) Genome sequencing in microfabricated high-density picolitre reactors. Nature 437:376–380

Morozova O, Hirst M, Marra MA (2009) Applications of new sequencing technologies for transcriptome analysis. Annu Rev Genom Hum Genet 10:135–151

Mortazavi A, Williams BA, McCue K, Schaeffer L, Wold B (2008) Mapping and quantifying mammalian transcriptomes by RNA-Seq. Nat Methods 5:621–628

Murata J, Roepke J, Gordon H, De Luca V (2008) The leaf epidermome of *Catharanthus roseus* reveals its biochemical specialization. Plant Cell 20:524–542

Nielsen KL, Hogh AL, Emmersen J (2006) DeepSAGE—digital transcriptomics with high sensitivity, simple experimental protocol and multiplexing of samples. Nucleic Acids Res 34:133

Ohlrogge J, Benning C (2000) Unraveling plant metabolism by EST analysis. Curr Opin Plant Biol 3:224–228

Reid JG, Nagaraja AK, Lynn FC, Drabek RB, Muzny DM, Shaw CA, Weiss MK, Naghavi AO, Khan M, Zhu H, Tennakoon J, Gunaratne GH, Corry DB, Miller J, McManus MT, German MS, Gibbs RA, Matzuk MM, Gunaratne PH (2008) Mouse let-7 miRNA populations exhibit RNA editing that is constrained in the 5'-seed/cleavage/anchor regions and stabilize predicted mmu-let-7a: mRNA duplexes. Genome Res 18:1571–1581

Sánchez-León N, Arteaga-Vázquez M, Alvarez-Mejía C, Mendiola-Soto J, Durán-Figueroa N, Rodríguez-Leal D, Rodríguez-Arévalo I, García-Campayo V, García-Aguilar M, Olmedo-Monfil V, Arteaga-Sánchez M, de la Vega OM, Nobuta K, Vemaraju K, Meyers BC, Vielle-Calzada JP (2012) Transcriptional analysis of the *Arabidopsis* ovule by massively parallel signature sequencing. J Exp Bot 63(10):3829–3842

Schadt EE, Turner S, Kasarskis A (2010) A window into third-generation sequencing. Hum Mol Genet 19:227–240

Schena M, Shalon D, Davis RW, Brown PO (1995) Quantitative monitoring of gene expression patterns with a complementary DNA microarray. Science 270:467–470

Sengoelge G, Winnicki W, Kupczok A, von Haeseler A, Schuster M, Pfaller W, Jennings P, Weltermann A, Blake S, Sunder-Plassmann G (2014) A SAGE based approach to human glomerular endothelium: defining the transcriptome, finding a novel molecule and highlighting endothelial diversity. BMC Genom 15:725

Shao Y, Wei J, Wu F, Zhang H, Yang D, Liang Z, Jin W (2016) DsTRD: Danshen transcriptional resource database. PLoS ONE 11(2):e0149747

Shendure J, Porreca GJ, Reppas NB, Lin X, McCutcheon JP, Rosenbaum AM, Wang MD, Zhang K, Mitra RD, Church GM (2005) Accurate multiplex polony sequencing of an evolved bacterial genome. Science 309:1728–1732

Song Z, Guo L, Liu T, Lin C, Wang J, Li X (2017) Comparative RNA-sequence transcriptome analysis of phenolic acid metabolism in *Salvia miltiorrhiza*, a traditional Chinese medicine model plant. Int J Genomics 2017:9364594

Tang JF, Baldwin SJ, Jacobs JME, van der Linden CG, Voorrips RE, Leunissen JA, van Eck H, Vosman B (2008) Large-scale identification of polymorphic microsatellites using an *in silico* approach. BMC Bioinformatics 9:374

Teoh KH, Polichuk DR, Reed DW, Nowak G, Covello PS (2006) *Artemisia annua* L. (Asteraceae) trichome-specific cDNAs reveal CYP71AV1, a cytochrome P450 with a key role in the biosynthesis of the antimalarial sesquiterpene lactone artemisinin. FEBS Lett 580:1411–1416

Tessler LA, Reifenberger JG, Mitra RD (2009) Protein quantification in complex mixtures by solid phase single-molecule counting. Anal Chem 81:7141–7148

Trapnell C, Roberts A, Goff L, Pertea G, Kim D, Kelley DR, Pimentel H, Salzberg SL, Rinn JL, Pachter L (2012) Differential gene and transcript expression analysis of RNA-seq experiments with TopHat and Cufflinks. Nat Protoc 7:562–578

Unamba CI, Nag A, Sharma RK (2015) Next generation sequencing technologies: the doorway to the unexplored genomics of non-model plants. Front Plant Sci 6:1074

Velculescu VE, Zhang L, Vogelstein B, Kinzler KW (1995) Serial analysis of gene expression. Science 270:484–487

Wang BQ (2010) *Salvia miltiorrhiza*: chemical and pharmacological review of a medicinal plant. J Med Plant Res 4(25):2813–2820

Wang Z, Gerstein M, Snyder M (2009) RNA-Seq: a revolutionary tool for transcriptomics. Nat Rev Genet 10:57–63

Wei T, Deng K, Wang H, Zhang L, Wang C, Song W, Zhang Y, Chen C (2018) Comparative transcriptome analyses reveal potential mechanisms of enhanced drought tolerance in transgenic *Salvia miltiorrhiza* plants expressing *AtDREB1A* from *Arabidopsis*. Int J Mol Sci 19:827

Xu X, Jiang Q, Ma X, Ying Q, Shen B, Qian Y, Song H, Wang H (2014) Deep sequencing identifies tissue-specific microRNAs and their target genes involving in the biosynthesis of tanshinones in *Salvia miltiorrhiza*. PLoS ONE 9(11):e111679

Xu Z, Reuben JP, Weirather J, Luo H, Liao B, Zhang X, Zhu Y, Ji A, Zhang B, Hu S, Au KF, Song J, Chen S (2015) Full-length transcriptome sequences and splice variants obtained by a combination of sequencing platforms applied to different root tissues of *Salvia miltiorrhiza* and tanshinone biosynthesis. Plant J 82 (6):951–961

Xu H, Song J, Luo H, Zhang Y, Li Q, Zhu Y, Xu J, Li Y, Song C, Wang B, Sun W, Shen G, Zhang X, Qian J, Ji A, Xu Z, Luo X, He L, Li C, Sun C, Yan H, Cui G, Li X, Li XE, Wei J, Liu J, Wang Y, Hayward A, Nelson D, Ning Z, Peters RJ, Qi X, Chen S (2016)

Analysis of the genome sequence of the medicinal plant *Salvia miltiorrhiza*. Mol Plant 9(6):949–952
Yan Y, Wang Z, Tian W, Dong Z, Spencer DF (2010) Generation and analysis of expressed sequence tags from the medicinal plant *Salvia miltiorrhiza*. Sci China Life Sci 53:273–285
Yang L, Ding G, Lin H, Cheng H, Kong Y, Wei Y, Fang X, Liu R, Wang L, Chen X, Yang C (2013) Transcriptome analysis of medicinal plant *Salvia miltiorrhiza* and identification of genes related to tanshinone biosynthesis. PLoS ONE 8(11):e80464
Zhang X, Luo H, Xu Z, Zhu Y, Ji A, Song J, Chen S (2015a) Genome-wide characterisation and analysis of bHLH transcription factors related to tanshinone biosynthesis in *Salvia miltiorrhiza*. Sci Rep 5:https://doi.org/10.1038/srep11244
Zhang G, Tian Y, Zhang J, Shu L, Yang S, Wang W, Sheng J, Dong Y, Chen W (2015b) Hybrid de novo genome assembly of the Chinese herbal plant danshen (*Salvia miltiorrhiza* Bunge). GigaScience 4:62
Zhang X, Dong J, Liu H, Wang J, Qi Y, Liang Z (2016) Transcriptome sequencing in response to salicylic acid in *Salvia miltiorrhiza*. PLoS ONE 11(1):e0147849
Zhang Y, Xu Z, Ji A, Luo H, Song J (2018) Genomic survey of bZIP transcription factor genes related to tanshinone biosynthesis in Salvia miltiorrhiza. Acta Pharm Sin B 8:295–305
Zhong GX, Li P, Zeng LJ, Guan J, Li DQ, Li SP (2009) Chemical characteristics of *Salvia miltiorrhiza* (Danshen) collected from different locations in China. J Agric Food Chem 57:6879–6887
Zhou L, Zuo Z, Chow MS (2005) Danshen: an overview of its chemistry, pharmacology, pharmacokinetics, and clinical use. J Clin Pharmacol 45:1345–1359
Zhou W, Huang Q, Wu X, Zhou Z, Ding M, Shi M, Huang F, Li S, Wang Y, Kai G (2017) Comprehensive transcriptome profiling of *Salvia miltiorrhiza* for discovery of genes associated with the biosynthesis of tanshinones and phenolic acids. Sci Rep 7:10554

Gene Expression Regulation in *Salvia miltiorrhiza*

8

Caili Li

Abstract
Transcription factors are a group of proteins that modulate many important biological processes by regulating the transcription of downstream target genes. In general, each plant transcription factor contains a DNA-binding domain, an oligomerization domain, a transcription regulation domain and a nuclear localization signal. In the past few years, many progresses have been made in the characterization of *S. miltiorrhiza* transcription factors on a genome-wide scale. It includes cloning of open reading frames (ORFs), sequence analysis, and expression profiling. Many identified transcription factors are members of the MYB, WRKY, bHLH, ZIP and AP2/ERF families. Some of them have been found to play key roles in plant hormone signaling and in the regulation of cell growth and differentiation. Transcription factors have been identified as regulators in the biosynthesis of various secondary metabolites in *S. miltiorrhiza*. In this chapter, I briefly summarized recent research progresses in the regulatory role of transcription factors in *S. miltiorrhiza* growth and differentiation, stresses and metabolic pathways. It will facilitate further characterization of transcription factors in *S. miltiorrhiza*.

C. Li (✉)
Institute of Medicinal Plant Development, Chinese Academy of Medical Sciences & Peking Union Medical College, No. 151 Malianwa North Road, Haidian District, Beijing 100193, China
e-mail: clli@implad.ac.cn

8.1 Introduction

Gene expression regulation at the transcriptional level controls many important biological processes. Transcription factors regulate expression in response to environmental and physiological signals and metabolic pathways. Transcription factors can activate or inhibit the transcription of downstream target genes by directly binding to their promoters in a sequence-specific manner. With the increasing availability of plant genome sequences, it is possible not only to study the function of transcription factors on a genome-wide scale but also to explore the structural and functional similarities and differences among different plant transcription factors. Many progresses have been made in the past few years, including genome-wide cloning and characterization of transcription factor genes, as well as the bioinformatics analysis and dissection of their regulatory mechanisms.

S. miltiorrhiza Bunge, belonging to the *Lamiaceae* family, is a famous and traditional Chinese herbal plant that has been widely used for treating cardiovascular and cerebrovascular

S. Lu (ed.), *The Salvia miltiorrhiza Genome*, Compendium of Plant Genomes,
https://doi.org/10.1007/978-3-030-24716-4_8

diseases (Zhang et al. 2010; Wu and Shi 2008; Zhou et al. 2005). There are mainly two types of bioactive constituents in *S. miltiorrhiza* plants, namely liposoluble tanshinones and water-soluble phenolic acids (Chen et al. 2001; Li 1998). The sequencing of *S. miltiorrhiza* genome has been completed. It was inferred that there were at least 1620 genes encoding transcription factors (Xu et al. 2016). The large number and variation of transcription factors in *S. miltiorrhiza* indicate the complexity and functional significance of transcriptional regulation in *S. miltiorrhiza*. In recent years, it has been found that some transcription factors can regulate the transcription of many genes in metabolic pathways. In this chapter, we briefly summarize recent progresses made in understanding the regulation of metabolic pathways by transcription factors.

8.2 Functional Domains of Transcription Factors

Transcription factors can specifically bind to short DNA sequences located in the promoter of target genes and interact with the transcription pre-initiation complex, thereby enhancing or inhibiting the activity of RNA polymerase II. By doing so, transcription factors modulate the transcription of their target genes. In recent years, a series of transcription factors regulating the response to drought, high salt, low temperature, hormones, pathogen and developmental stimuli have been identified in higher plants (Adam et al. 1994; Adamska et al. 1991; Bansal and Bogorad 1993; Kay 1993; Kozaki et al. 1992; Logemann et al. 1995; Pilgrim et al. 1993; Tonoike et al. 1994).

Typical plant transcription factor consists of a DNA-binding domain, an oligomerization domain, a transcription regulation domain and a nuclear localization signal (NLS). It has been found that some transcription factors lack either the transcription regulation domain or the DNA-binding domain.

8.2.1 DNA-Binding Domain

The DNA-binding domains of plant transcription factors are structurally consistent. They contain amino acid residues that bind to specific DNA sequences in the *cis*-acting elements, thus determining the target specificity. In addition, other residues in this domain may enhance the affinity of transcription factors to DNA by binding to DNA nonspecifically through interacting with the phosphate or deoxyribose moieties (Guilfoyle 1997; Aukerman et al. 1991). Typical DNA-binding domains in plant transcription factors include bZIP domain (Nantel and Quatrano 1996), MYC domain (Abe et al. 1997), MYB domain (Martin 1997), homeodomain (Klinge et al. 1996) and AP2/EREBP domain (Riechmann and Meyerowitz, 1998) (Table 8.1). Moreover, plant transcription factors can be classified according to the conservative regions in the DNA-binding domain and oligomerization domain (Table 8.1). Besides, some of these domains can be divided into subclasses according to the number and location of conserved amino acid residues. For instance, based on the number and arrangement of cysteine (C) and histidine (H) residues, transcription factors containing zinc fingers can be divided into five classes, namely C_2H_2, C_3H and C_2C_2 (GATA finger), C_3HC_4 (RING finger) and C_2HC_5 (LIM finger).

8.2.2 Transcription Regulation Domain

Transcription regulation domain is the key functional domain in transcription factors since it determines the function of transcription factors. Transcription factors in a family generally have distinct actions due to the differences in their transcription regulation domains, which tend to diverge from one to another (Yanagisawa and Sheen 1998). Regulation domains and transcription factors can function as either repressors or activators of the target genes. When several

Table 8.1 Structural feature of conserved domains that are used to classify plant transcription factors

Domain type	Structure	References
Zinc finger	Finger motif(s) each maintained by cysteine and/or histidine residues organized around a zinc ion	Sakamoto et al. (1993)
bZIP	A basic region and a leucine-rich zipper-like motif	Nantel and Quatrano (1996)
MYB-related	A basic region with one to three imperfect repeats each forming a helix-helix-turn-helix	Martin (1997)
WRKY	Approximately 60-amino acids with the highly conserved amino acid sequence WRKYGQK	Xu et al. (2004)
Trihelix	Basic, acidic and proline/glutamine-rich motif which forms a trihelix DNA-binding domain	Kuhn et al. (1993)
Homeodomain	Approximately 60 amino acid residues producing either three or four a-helices and an *N*-terminal arm	Klinge et al. (1996)
MYC b/HLH	A cluster of basic amino acid residues adjacent to a helix-loop-helix motif	Abe et al. (1997)
MADS	Approximately 57 amino acid residues that comprise a long a-helix and two b-strands	Purugganan et al. (1995)
AT-hook motif	A consensus core sequence R(G/P)RGRP with the RGR region contacting the minor groove of A/T-rich DNA	Gupta et al. (1997)
HMG-box	L-shaped domain consisting of three a-helices with an angle of about 808 between the arms	Grasser (1995)
AP2/EREBP	A 68-amino acid region with a conserved domain that constitutes a putative amphiphatic a-helix	Riechmann and Meyerowitz (1998)
B3	A 120 amino acid conserved sequence at the C-termini of VP1 and ABI3	Sakuma et al. (2002)
ARF	A 350 amino acid region similar to B3 in sequence	Ulmasov et al. (1997)

transcription factors target the same *cis*-acting element, repression of the target genes may occur due to the exclusion of the activator from the target promoter. For example, two bZIP proteins from rice, namely osZIP-2a and -2b, could dimerize with wheat bZIP factor mBP1 in vitro to prevent their interaction with the target promoter. While preliminary results showed that these rice bZIP factors can quench other bZIP proteins, a clear conclusion can be drawn only after they are tested more extensively (Nantel and Quatrano 1996). Several observations suggested the existence of repression domains in plant transcription factors, but they remain poorly characterized. PvALF (*Phaseolus vulgaris* ABI-3-like factor) was found to activate the transcription of several selected genes, including the French bean phytohemagglutinin gene *DLEC2*. ROM2 (regulator of maturation-specific protein 2) is a bZIP protein that can bind to the enhancer of *DLEC2* to inhibit PvALF-activated gene transcription (Chern et al. 1996). This indicates that a repression domain in the *N*-terminal half of ROM2 inhibits the PvALF-activated transcription of *DLEC2* (Chern et al. 1996).

Although many studies have shown the existence of transcriptional inhibitory regions in transcription factors, the structural and functional characteristics of transcription factors are not well understood. The mode of action of transcription factor inhibitory regions may be as follows, (1) binding to the functional site of the promoter, thus preventing other transcription factors from binding to the same promoter; (2) inhibiting the transcription of related genes by suppressing other transcription factors; and (3) changing the structure of DNA to disable transcription.

8.2.3 Nuclear Localization Signal (NLS)

The nuclear localization signal region is a region in transcription factors rich in arginine and lysine. Upon synthesis, transcription factors need to be transferred into the nucleus in order to exert their functions, and thus, the function of transcription factors is determined by the nuclear localization signal region. The amino acid sequence, organization and number of nuclear localization signal region vary among different plant species and varieties, and they exhibit irregular distribution in transcription factors. NSL was identified in many transcription factors, such as rice (Dehesh et al. 1995), tomato HSFA1-2 (Lyck et al. 1997), maize O2 (Varagona et al. 1992) and peas PS-IAA4 and PS-IAA6 (Abel et al. 1995).

While in some NLSs the basic residues are closely associated (Lyck et al. 1997), in some other NLSs, the basic residues form two functionally important groups are separated by several non-conserved residues. The structure of the latter group is called the bipartite structure (Lyck et al. 1997). Other transcription factors may contain multiple copies of the NLS that are functionally independent and either clustered or dispersed within the protein (Klinge et al. 1996; Dehesh et al. 1995; Varagona et al. 1992; Boulikas 1994; Dehesh et al. 1995; Boulikas 1994; Meisel and Lam 1996; Klinge et al. 1996).

8.2.4 Oligomerization Domain

The functional domain that enables transcription factors to converge with each other is called the oligomerization domain. Oligomerization domains affect the specificity of DNA binding, the affinity of transcription factors with promoter elements and the nuclear localization of transcription factors (Katagiri et al. 1992; Guiltinan and Miller 1994). These domains have a relatively conserved amino acid sequence, and most of them are connected to the DNA-binding domain to form a specific spatial conformation. The oligomerization domain of the bZIP transcription factors contains several regularly spaced leucine residues and a zipper-like structure (Rook et al. 1998). While the b/HLH transcription factors contain a helix-loop-helix structure (Sainz et al. 1997), the MADS transcription factors have oligomerization domains that form two α-helix and two β-pleated sheets (Purugganan et al. 1995).

There are differences in the length of oligomerization domain among transcription factors in the same family. In most bZIP transcription factors, leucine zipper consists of four to five repeats, but in *Arabidopsis* ATB2, nine repeats were found (Rook et al. 1998). Besides, as observed in MADS transcription factors, the keratin-like domain called K (Purugganan et al. 1995) and an intervening domain designated as I (Mizukami et al. 1996) are necessary for interaction, and the regions outside the oligomerization domain affect the association of different subunits (Mizukami et al. 1996). These variations in the oligomerization domain increase the versatility of the transcription machinery, and they may also affect the regulation of gene transcription in plants.

8.3 Identification of Transcription Factor Families

The *S. miltiorrhiza* genome contained several large transcription factor families, such as MYB, WRKY, bZIP, basic helix-loop-helix (bHLH) and APETALA2 (AP2)/ethylene response element binding protein (EREBP) (Li and Lu 2014; Li et al. 2015; Zhang et al. 2018; Ji et al. 2016; Zhang et al. 2015a). Members of these families have been chosen for expression profiling in recent years.

8.3.1 The MYB Family

The MYB family is the largest transcription factor family in plants. It is divided into three subfamilies according to the number and position of the MYB repeats, namely R2R3, R1R2R3 and MYB-related. Recently, 111 *S. miltiorrhiza R2R3-MYB* genes were cloned and characterized systematically (Li and Lu 2014; Li et al. 2018). The *MYB* genes in *S. miltiorrhiza* were more conserved and diversified than those in *Arabidopsis* (Li and Lu 2014). *SmMYBs* were usually involved in a variety of physiological processes, and their expression was induced by developmental or environmental stimuli (Li and Lu 2014; Li et al. 2018).

8.3.2 The WRKY Family

WRKY is a large transcription factor family specific to the green lineage, including green algae and land plants. The first *WRKY* gene *SPF1* was cloned from *Ipomoea batatas* about twenty-five years ago (Ishiguro and Nakamura 1994). Since then, extensive progresses have been made in the identification and functional characterization of the *WRKY* genes. The defining feature for WRKY transcription factors is their DNA-binding domain, known as the WRKY domain. A total of 69 *SmWRKYs* were identified from the transcriptome of *S. miltiorrhiza*. Phylogenetic analysis indicated that some *SmWRKYs* had close phylogenetic relationships with *WRKYs* from other plants known to be regulators of terpene metabolism (Li et al. 2015; Yu et al. 2018).

8.3.3 The AP2/ERF Family

The AP2/ERF family, one of the largest transcription factor families in the plant kingdom, consists of five subfamilies, namely AP2 (APETALA2), ERF (ethylene-responsive factor), DREB (dehydration-responsive element binding proteins), RAV (related to ABI3/VP1) and Soloist (Sakuma et al. 2002; Yamasaki et al. 2013; Licausi et al. 2013). They are mainly involved in the regulation of metabolism and developmental processes. Members of the APETALA2/ethylene response factors (AP2/ERF) subfamilies are characterized by the presence of the AP2 DNA-binding domain. In total, 170 *AP2/ERF* genes were identified from the *S. miltiorrhiza* genome and divided into five relatively conserved subfamilies. The phylogeny, gene structure, conserved structure and expression profile of AP2/ERF family members in *S. miltiorrhiza* have been comprehensively surveyed (Ji et al. 2016).

8.3.4 The bHLH Family

The basic helix-loop-helix (bHLH) transcription factor family is also one of the largest transcription factor families in plants. According to genomic analyses, more than 630 bHLH transcription factors have been found in several important food crops. In total, 167, 177, 190 and at least 191 *bHLH* transcription factor genes have been predicted in *A. thaliana* (Carretero-Paulet et al. 2010), *Oryza sativa* (Carretero-Paulet et al. 2010), *Nicotiana tabacum* (Rushton et al. 2008) and *Vitis vinifera* (Jaillon et al. 2007), respectively. The availability of the *S. miltiorrhiza* genome has facilitated the studies of bHLH transcription factors. In the genome of *S. miltiorrhiza*, 127 *bHLH* transcription factor genes were identified. Phylogenetic analysis indicated that these *SmbHLHs* could be classified into 25 subfamilies (Zhang et al. 2015a).

8.3.5 The bZIP Family

The bZIP family is one of the largest and most conserved transcription factor families in eukaryotic species. The bZIP transcription factors have a highly conserved bZIP domain consisting of a basic DNA-binding region and a leucine zipper region that confers dimerization specificity (Hurst 1995). The basic region preferentially binds to the DNA sequences with an ACGT core, particularly G-box (CACGTG),

C-box (GACGTC) and A-box sequences (TACGTA) (Izawa et al. 1993; Foster et al. 1994; Sibéril et al. 2001). bZIP transcription factors have been extensively studied in plants. A total of 70 *SmbZIP* transcription factor genes were identified in *S. miltiorrhiza*. The expression profiles of these *SmbZIP* genes identified from the transcriptome showed that seven *SmbZIP* genes were highly expressed in the root and periderm, where tanshinone biosynthesis and accumulation take place (Zhang et al. 2018).

8.4 Regulation of Plant Processes by Transcription Factors

8.4.1 Transcription Factors in Cell Growth and Differentiation Regulation

Many transcription factors are known to play key roles in various regulatory networks that govern the growth and differentiation of plant cells. In *Arabidopsis*, *AtMYB23* is functionally equivalent to *GL1* during trichome initiation but plays a different role during trichome branching (Tominaga-Wada et al. 2012). It was found that a protein complex consisted of bHLH GLABRA3 (GL3), ENHANCER OF GLABRA3 (EGL3) and several MYB members associated with a WD40-repeat protein (TRANSPARENT TESTA GLABRA1 [TTG1]) controlled the identity of epidermal cells (Bernhardt et al. 2003). This complex was also found to be involved in the differentiation of root hairs, stomata and trichomes by controlling the transcription of a downstream regulator (Bernhardt et al. 2003).

In *S. miltiorrhiza*, the ternary transcription complex SmTTG1-SmMYB111-SmbHLH51 was found to be involved in the regulation of root hair and trichome formation (Li et al. 2018). Through functional characterization, it was found that SmMYB62, SmMYB78, SmMYB80 and SmMYB99 may be involved in the regulation of flower growth and development in *S. miltiorrhiza* (Li and Lu 2014). Besides, the SPL transcription factors in *S. miltiorrhiza* affect the transition from vegetative growth to reproductive growth in *S. miltiorrhiza* (Zhang et al. 2014). The expression of SPLs increased gradually with the maturation of *S. miltiorrhiza* plants (Zhang et al. 2014). The SPL transcription factors can further activate a series of downstream transcription factors related to flowering, thereby promoting flower development (Zhang et al. 2014).

8.4.2 Transcription Factors in Response to Biotic and Abiotic Stresses

The stresses that plant faced during growth, including biotic and abiotic stresses, often seriously affect growth and development. Transcription factors play a key role in plant stress signal transduction. In the presence of biotic and abiotic stresses, transcription factors are constantly synthesized to transmit and amplify the signal to regulate the transcription of downstream genes to improve stress resistance.

In *A. thaliana*, the expression of *AtMYB30* was found to be closely associated with cell death in the allergic reaction induced by pathogen attack or elicitor treatment (Froidure et al. 2010). Overexpression of *GhWRKY27a* reduces tolerance to drought and resistance to *Rhizoctonia solani* infection in transgenic tobacco (Yan et al. 2015). Overexpression of *OsWRKY11* under the control of *HSP101* promoter led to enhanced drought tolerance, as evidenced by the slower leaf-wilting and increased survival rate of the green parts (Lee et al. 2018).

Studies have shown that many transcription factors are involved in stress response of *S. miltiorrhiza*. Upon treatment with yeast extract and Ag^+, the transcription of 412 transcription factor genes was found to be modulated in the hairy roots of *S. miltiorrhiza* (Gao et al. 2014). The majority of *SmWRKY* genes were MeJA-, yeast extract- and Ag^+-responsive (Li et al. 2015). *SmbZIP7* and *SmbZIP20* were responsive to Ag^+ treatment (Zhang et al. 2018). Overexpression of *SmNAC1* enhanced the tolerance to high zinc in plants, and zinc was enriched in the shoot tissues (Zhu et al. 2019). *SmERF1L1* is responsive to methyl jasmonate (MeJA), yeast extraction (YE), salicylic acid (SA) and ethylene treatments (Huang et al. 2019). These results demonstrate that transcription factors are involved in the stress response in *S. miltiorrhiza*. Since many medicinal active components are specialized metabolites that play an important role in plant defense responses, one possible mechanism underlying the involvement of transcription factors in the stress response of *S. miltiorrhiza* may be that transcription factors regulate the biosynthesis of these components.

8.5 Transcription Factors in Regulation of Secondary Metabolism

S. miltiorrhiza Bunge is a widely used Chinese medicinal plant. The main bioactive compounds in *S. miltiorrhiza* are phenolic acids (Ma et al. 2013) and terpenoid tanshinones (Ma et al. 2015). Phenolic acids extracted from *S. miltiorrhiza* have drawn considerable attention in recent years due to their remarkable pharmacological activities. Phenolic acids derived from the rosmarinic acid pathway, such as rosmarinic acid and salvianolic acid B, are important bioactive components in *S. miltiorrhiza* (Zhao et al. 2008). The biosynthetic pathway of phenolic acids is believed to include both phenylpropanoid and tyrosine-derived pathways (Di et al. 2013; Ma et al. 2013; Wang et al. 2015). The former is a general pathway for the synthesis of phenolic acids, flavonoids (e.g., anthocyanins and flavones) and lignins.

Terpenoid tanshinones contain mainly cryptotanshinone (CT), tanshinone I (T1), tanshinone IIA (TIIA) and dihydrotanshinone (HT) (Zhou et al. 2017). As a kind of diterpenes, tanshinone is derived from isopentenyl diphosphate (IPP), and its isomer dimethylallyl diphosphate (DMAPP) is synthesized through two different pathways in separate compartments, namely the MEP pathway (the methylerythritol phosphate pathway in the plastids) and the MVA pathway (the mevalonate pathway in the cytosol) (Xu et al. 2010; Kai et al. 2011). In recent years, many genes encoding key enzymes in tanshinone biosynthesis, such as *SmGGPPS* (*geranylgeranyl diphosphate synthase*), *SmDXR* (*1-deoxy-D-xylulose 5-phosphate eductoisomerase*), *SmCPS1* (*copalyl diphosphate synthase*), *SmKSL1* (*kaurene synthase-like*) and *SmCYP76AH1* have been cloned from *S. miltiorrhiza* and characterized (Kai et al. 2011; Wu et al. 2009; Gao et al. 2009; Guo et al. 2013; 2016). Overexpression or RNA interference of one or two of these genes could significantly modulate the production of metabolites.

Transcription factors are sequence-specific DNA-binding proteins that interact with the promoter regions of target genes and regulate the transcription of genetic information from DNA to mRNA. A single transcription factor could regulate the transcription of several genes at the same time. Therefore, it is an effective strategy to use transcription factors to improve the production of secondary metabolites as transcription factors are often capable of coordinately regulating multiple biosynthetic pathways (Zhang et al. 2015b). Generally, overexpression of transcription factors is used for engineering to increase the level of metabolites. It has been reported that members of the MYBs, AP2/ERF, bHLH, WRKY and Basic Leucine Zipper (bZIP) families have global regulatory functions in specialized metabolism in *S. miltiorrhiza* (Li and Lu 2014; Li et al. 2015; Zhang et al. 2018; Ji et al. 2016; Zhang et al. 2015a).

8.5.1 Regulation of MYB in Secondary Metabolism

As members of one of the largest transcription factor families, the R2R3-MYB family, which is characterized by the presence of a highly conserved MYB domain at N-terminus and an activation or repression domain at the C-terminus, plays important roles in various physiological and biochemical processes in plants (Dubos et al. 2010). Various R2R3-MYBs have been found to be involved in the regulation of terpenoid and phenylpropanoid metabolism. For instance, strawberry FaMYB2 has been shown to suppress anthocyanin and flavonol accumulation in transgenic tobacco plants (Aharoni et al. 2001). AtMYB62 included in S20 was involved in gibberellin metabolism and signaling and affected root architecture, Pi uptake and acid phosphatase activity (Devaiah et al. 2009). Grapevine VvMYB5b belonging to S5 was involved in flavonoid and terpenoid metabolism. Overexpression of *VvMYB5b* resulted in the downregulation of phenylpropanoid metabolism and upregulation of carotenoid metabolism (Mahjoub et al. 2009). Loblolly pine PtMYB14 belonging to S4 was a putative regulator of the isoprenoid-oriented response, which leads to the accumulation of sesquiterpene in conifers (Bedon et al. 2010).

In *S. miltiorrhiza*, 111 R2R3-MYBs were identified. While some R2R3-MYBs belonging to subgroups S3, S4, S5, S6, S7, S13 and S21 were predicted to be involved in the biosynthesis of salvianolic acids (SAs), some in subgroups S4, S5 and S20 were assumed to be potential regulators of terpenoid biosynthesis (Li and Lu 2014). In *S. miltiorrhiza*, only a few endogenous MYBs have been functionally verified to be involved in the regulation of phenolic acids biosynthesis. Previous research reported that *Arabidopsis* transcription factor production of anthocyanin pigment 1 (AtPAP1) exhibited strong capability to promote the production of phenolic acids in *S. miltiorrhiza* (Zhang et al. 2010). The repressor role of *SmMYB39* in the rosmarinic acid pathway has been characterized through overexpression and RNAi-mediated silencing. Overexpression/RNAi-mediated silencing of *SmMYB39* in *S. miltiorrhiza* plantlets dramatically affected the accumulation of 4-coumaric acid, rosmarinic acid, salvianolic acid B and salvianolic acid A by regulating the transcription of genes encoding 4-hydroxylase (C4H) and tyrosine aminotransferase (TAT) and the enzyme activities of C4H and TAT (Zhang et al. 2013).

Overexpression of *ZmC1* increased the yield of total tanshinones by 2.4-fold through upregulating several genes, such as *SmMDC* and *SmPMK* in *S. miltiorrhiza* hairy roots (Zhao et al. 2015). Overexpression of *SmMYB36* promoted tanshinone accumulation by increasing the activity of MYB binding elements in *DXR*, *MCT* and *GGPPS1* promoters (Ding et al. 2017). Overexpression of another MYB transcription factor gene, *SmMYB9b*, in *S. miltiorrhiza* hairy roots enhanced tanshinone concentration by 1.2-fold through stimulating the transcription of terpenoid biosynthetic genes *SmDXS2*, *SmDXR*, *SmGGPPS* and *SmKSL1* (Zhang et al. 2017).

8.5.2 Regulatory Function of bHLH in Phenolic Acid and Tanshinone Biosynthesis

The bHLH transcription factors also play an important role in the regulation of metabolism. An *A. thaliana* bHLH, AtMYC2, could directly bind to the promoters of sesquiterpene synthase genes, namely *TPS21* and *TPS11*, to activate their transcription, thus increasing the production of sesquiterpene (Hong et al. 2012). Another bHLH transcription factor in *A. thaliana*, phytochrome-interacting factor 5 (PIF5), functioned as a positive regulator of the MEP pathway and enhanced isopentyl diphosphate (IPP) metabolism (Mannen et al. 2014). In *Artemisia annua*, overexpression of *AaMYC2* significantly increased the transcription of *CYP71AV1* and *DBR2*, resulting in the increase of artemisinin content (Shen et al. 2016).

The bHLH transcription factors play an important role in the regulation of phenolic acid biosynthesis. Overexpression of *SmbHLH148* activated the expression of genes involved in the biosynthesis of phenolic acids and thus significantly increased the accumulation of caffeic acid, rosmarinic acid and salvianolic acid B (Xing et al. 2018a).

SmbHLH10 could positively regulate tanshinone biosynthesis in *S. miltiorrhiza* hairy roots. In *SmbHLH10* overexpression roots, the contents of four major tanshinones were increased (Xing et al. 2018b). SmbHLH10 could directly bind to the G-box in the promoter of several genes, thus activating their transcription and upregulating tanshinone biosynthesis (Xing et al. 2018b). In addition, a total of seven *bHLH* genes were found to be potentially involved in the regulation of tanshinone biosynthesis (Zhang et al. 2015a, b).

MYC is a subgroup of the bHLH family that has been reported to positively regulate the yield of tanshinones in *S. miltiorrhiza*. *SmMYC2a* and *SmMYC2b*, two orthologs of *MYC2*, were found to interact with *SmJAZs* and positively regulate the biosynthesis of tanshinones and salvianolic acid B in *S. miltiorrhiza* hairy roots through modulating the transcription of multiple genes in the tanshinone biosynthetic pathway (Zhou et al. 2016). Overexpression of *SmMYC2* increases the production of phenolic acids in *S. miltiorrhiza* (Yang et al. 2017). *SmbHLH37* negatively regulates JA signaling and functions antagonistically with *SmMYC2* in regulating salvianolic acid B biosynthesis in *S. miltiorrhiza* (Du et al. 2018).

8.5.3 Regulation of AP2/ERF in Bioactive Component Biosynthesis

The AP2/ERF family is one of the most important transcription factor families involved in the regulation of metabolism (Licausi et al. 2013). In *Catharanthus roseus*, octadecanoid-responsive *Catharanthus* AP2/ERF-domain 3 (ORCA3) and ORCA4 played key roles in TIA biosynthesis (Van der Fits and Memelink 2000, 2001; Paul et al. 2017). In tobacco, ERF189, ERF221, ORC1 and NtERF32 are related to nicotine biosynthesis (Shoji et al. 2010; De Boer et al. 2011; Sears et al. 2014). The introduction of *AaERF1*, *AaERF2* and *AaORA* has been shown to promote the transcription of *AaADS* and *AaCYP71AV1*, respectively, resulting in enhanced accumulation of artemisinin and artemisinic acid in *Artemisia annua* (Yu et al. 2012; Lu et al. 2013; Tan et al. 2015).

SmERF115 is a positive regulator of phenolic acid biosynthesis. Transcription of the key biosynthetic gene *SmRAS1* was upregulated in *SmERF115* overexpression lines and downregulated in *SmERF115*-RNAi lines (Sun et al. 2019). SmERF115 could directly bind to the promoter of *SmRAS1* and activate the transcription of *SmRAS1* in vivo. Accordingly, the production of phenolic acid was increased and reduced in hairy roots with *SmERF115* overexpressed and silencing, respectively (Sun et al. 2019).

SmERF128 activated the transcription of *SmCPS1*, *SmKSL1* and *SmCYP76AH1* by binding to the GCC box and to the CRTDREHVCBF2 (CBF2) and RAV1AAT (RAA) motifs within their promoters to regulate tanshinone biosynthesis. Overexpression of *SmERF128* increased the transcription levels of *SmCPS1*, *SmKSL1* and *SmCYP76AH1*, and their transcription was decreased when *SmERF128* was silenced. Accordingly, the content of tanshinone was reduced in *SmERF128* RNA interference (RNAi) hairy roots and dramatically increased in *SmERF128* overexpression hairy roots. *SmERF1L1*, encoding AP2/ERF transcription factor in *S. miltiorrhiza*, is a jasmonic acid (JA)-responsive gene. Overexpression of *SmERF1L1* significantly increased the content of tanshinones by upregulating the transcription of genes involved in tanshinone biosynthesis, especially *SmDXR*. Meanwhile, the level of phenolic acids was decreased in *SmERF1L1* overexpressed *S. miltiorrhiza* (Huang et al. 2019).

8.5.4 Regulatory Function of WRKY in Tanshinones Biosynthesis

Increasing evidence has demonstrated the importance of *WRKY* genes in regulating the biosynthesis of specialized metabolites (Suttipanta et al. 2011). For example, in *Gossypium arboreum*, *GaWRKY1* participates in sesquiterpene biosynthesis by modulating (+)-δ-cadinene synthase (CAD1) activity (Xu et al. 2004). Recently, *WRKYs* have been isolated from some medicinal plants, including *C. roseus* (Suttipanta et al. 2011), *Artemisia annua* (Chen et al. 2017) and *Withania somnifera* (Singh et al. 2017). Overexpression of *CrWRKY1* increased the content of serpentine in hairy roots through binding to the W-box elements in the promoters of *tryptophan decarboxylase* (*TDC*) and *TDC* genes, which were involved in the indole alkaloid (TIA) biosynthetic pathway (Suttipanta et al. 2011). The WRKY transcription factor GLANDULAR TRICHOME-SPECIFIC WRKY 1 (AaGSW1) positively regulates the transcription of *AaCYP71AV1* and *AaORA* by directly binding to the W-box motifs in their promoters (Chen et al. 2017). Overexpression of *AaGSW1* in *A. annua* significantly improved artemisinin and dihydroartemisinic acid contents (Chen et al. 2017). A WRKY transcription factor from *W. somnifera* binds to the W-box motif in the promoters of *squalene synthase* and *squalene epoxidase*, regulating the accumulation of triterpenoids, including phytosterols and withanolides, in *W. somnifera* (Singh et al. 2017).

SmWRKY1 participates in the regulation of tanshinone biosynthesis. Overexpression of *SmWRKY1* significantly elevated the transcription of genes encoding enzymes in the MEP pathway, especially *SmDXR*, resulting in increased tanshinone production (Cao et al. 2018). SmWRKY1 activated the transcription of *SmDXR* by directly binding to its promoter (Cao et al. 2018). SmWRKY2 was a positive regulator of tanshinone biosynthesis by binding to the W-box in *SmCPS* promoter. Overexpression of *SmWRKY2* in *S. miltiorrhiza* hairy roots significantly increased the transcription of *SmCPS*, resulting in increased accumulation of tanshinones (Deng et al. 2019). Besides, tanshinone production was slightly reduced in the *antisense-SmWRKY2* line (Deng et al. 2019). Our group identified 61 *SmWRKY*s from *S. miltiorrhiza* genome and characterized them in detail. We found that *SmWRKY3* and *SmWRKY9* were involved in the regulation of tanshinone biosynthesis (Li et al. 2015).

8.5.5 Complexes of Transcription Factors

The phenylpropanoid pathway in higher plants is under the control of specific R2R3-MYB and bHLH transcription factors (Dubos et al. 2010). PtrMYB57-bHLH131-PtrTTG1 negatively regulates the biosynthesis of anthocyanins and proanthocyanidins in *Populus trichocarpa* (Wan et al. 2017). SmTTG1 was found to be necessary for salvianolic acid B biosynthesis (Li et al. 2018). Using SmTTG1 as bait, SmMYB111 was identified from the *S. miltiorrhiza* cDNA library by yeast two-hybrid (Y2H) assays. Transgenic plants overexpressing and silencing *SmMYB111* significantly induced and reduced the accumulation of salvianolic acid B, respectively (Li et al. 2018). In addition, SmMYB111 interacts with SmTTG1 and SmbHLH51. SmMYB111 was found to positively regulate the synthesis of phenolic acids, salvianolic acid B and rosmarinic acid (RA) in *S. miltiorrhiza* by forming the SmTTG1-SmMYB111-SmbHLH51 ternary transcription complex (Li et al. 2018).

Recently, transcriptome analysis has been used to identify candidate genes in the specialized metabolic pathways and explore the potential factors associated with important traits in *S. miltiorrhiza* (Yang et al. 2013; Song et al. 2017; Wei et al. 2017; Pei et al. 2018). Manipulating a single transcription factor can reprogram the transcriptome profile and is an effective way to modulate the production of some specialized metabolites. Since ectopic expression of *AtPAP1* in *S. miltiorrhiza* greatly improved the level of salvianolic acid B, the expression level of many genes associated with the salvianolic acid B biosynthetic pathway must be changed (Zhang et al. 2010). By comparing the transcriptome of transgenic *S. miltiorrhiza* plants overexpressing *AtPAP1* and that of the wild-type (WT) plants, some novel genes associated with salvianolic acid B biosynthesis and specific target genes of *AtPAP1* were identified (Wu et al. 2018). SmbHLH51, encoded by a novel *bHLH* gene, was identified as a positive transcriptional regulator of the phenolic pathway by overexpressing and silencing *SmbHLH51* in *S. miltiorrhiza* (Wu et al. 2018). AtPAP1 stimulates phenolic acid biosynthesis in *S. miltiorrhiza* by activating the transcription of *SmbHLH51* and probably forming a transcription complex with SmbHLH51 (Wu et al. 2018).

8.6 Conclusion

Transcription factors play multifaceted and pivotal roles in the regulation of biological processes in all living organisms. With the decoding of *S. miltiorrhiza* genome and the increasing availability of transcriptome data, progress in the research of transcription factors of *S. miltiorrhiza* is being made rapidly. To date, several MYB, WRKY, AP2/ERF, bZIP and bHLH transcription factors have been identified as regulators in the biosynthesis of different groups of secondary metabolites in *S. miltiorrhiza*. These transcription factors positively or negatively regulate the expression of multiple genes encoding key enzymes and thereby control biosynthetic pathways.

Transcription factors are tools that can be used to improve the yield and diversity of metabolites. As the metabolic flow is regulated at multiple steps and by multiple transcription factors, overexpression of a single transcription factor is often insufficient. In order to understand the regulation of metabolic network, it is still important to find more transcription factors, determine their functionality and their interaction with other transcription factors.

Regulation of transcription factors at the protein level is the critical key to effectively and accurately control of their activity. Protein–protein interaction has resulted in the identification of different classes of transcription factor interacting proteins. Posttranslational modifications of transcription factors can affect their transcriptional activity. Despite increasing knowledge of protein–protein interaction and posttranslational modifications of proteins, we still know very little about them for transcription factors. More knowledge about protein–protein interactions and posttranslational modifications will be crucial to understand how transcription factors control the expression of their target genes.

Acknowledgements This work was supported by the CAMS Innovation Fund for Medical Sciences (CIFMS) (2016-I2 M-3-016) and the National Natural Science Foundation of China (81603225).

References

Abe H, Yamaguchi-Shinozaki K, Urao T, Iwasaki T, Hosokawa D, Shinozaki K (1997) Role of *arabidopsis* MYC and MYB homologs in drought- and abscisic acid-regulated gene expression. Plant Cell 9:1859–1868

Abel S, Nguyen MD, Theologis A (1995) The PS-IAA4/5-like family of early auxin-inducible mRNAs in *Arabidopsis thaliana*. J Mol Biol 251:533–549

Adam E, Szell M, Szekeres M, Schaefer E, Nagy F (1994) The developmental and tissue-specific expression of tobacco phytochrome-A genes. Plant J 6:283–293

Adamska I, Scheel B, Kloppstech K (1991) Circadian oscillations of nuclear-encoded chloroplast proteins in pea (*Pisum sativum*). Plant Mol Biol 17:1055–1065

Aukerman MJ, Schmidt RJ, Burr B, Burr FA (1991) An arginine to lysine substitution in the bZIP domain of an *opaque-2* mutant in maize abolishes specific DNA binding. Genes Dev 5:310–320

Aharoni A, De Vos CH, Wein M, Sun Z, Greco R, Kroon A, Mol JN, O'Connell AP (2001) The strawberry FaMYB1 transcription factor suppresses anthocyanin and flavonol accumulation in transgenic tobacco. Plant J 28:319–332

Bansal KC, Bogorad L (1993) Cell type-preferred expression of maize cab-m1: repression in bundle sheath cells and enhancement in mesophyll cells. Proc Natl Acad Sci USA 90:4057–4061

Bedon F, Bomal C, Caron S, Levasseur C, Boyle B, Mansfield SD, Schmidt A, Gershenzon J, Grima-Pettenati J, Séguin A, MacKay J (2010) Subgroup 4 R2R3-MYBs in conifer trees: gene family expansion and contribution to the isoprenoidand flavonoid-oriented responses. J Exp Bot 61:3847–3864

Bernhardt C, Lee MM, Gonzalez A, Zhang F, Lloyd A, Schiefelbein J (2003) The bHLH genes *GLABRA3 (GL3)* and *ENHANCER OF GLABRA3 (EGL3)* specify epidermal cell fate in the *Arabidopsis* root. Development 130:6431–6439

Boulikas T (1994) Putative nuclear localization signals (NLS) in protein transcription factors. J Cell Biochem 55:32–58

Cao W, Wang Y, Shi M, Hao X, Zhao W, Wang Y, Ren J, Kai G (2018) Transcription Factor SmWRKY1 Positively Promotes the Biosynthesis of Tanshinones in *Salvia miltiorrhiza*. Front Plant Sci 9:554

Carretero-Paulet L, Galstyan A, Roig-Villanova I, Martínez-García JF, Bilbao-Castro JR, Robertson DL (2010) Genome-wide classification and evolutionary analysis of the bHLH family of transcription factors in *Arabidopsis*, poplar, rice, moss, and algae. Plant Physiol 153:1398–1412

Chen H, Chena F, Chiu FCK, Lo CMY (2001) The effect of yeast elicitor on the growth and secondary metabolism of hairy root cultures of *Salvia miltiorrhiza*. Enzyme Microb Tech 28:100–105

Chen M, Yan T, Shen Q, Lu X, Pan Q, Huang Y, Tang Y, Fu X, Liu M, Jiang W, Lv Z, Shi P, Ma Y, Hao X, Zhang L, Li L, Tang K (2017) GLANDULAR TRICHOME-SPECIFIC WRKY 1 promotes artemisinin biosynthesis in *Artemisia annua*. New Phytol 214:304–316

Chern MS, Bobb AJ, Bustos MM (1996) The regulator of *MAT2* (ROM2) protein binds to early maturation promoters and represses PvALF-activated transcription. Plant Cell 8:305–321

De Boer K, Tilleman S, Pauwels L, Vanden Bossche R, De Sutter V, Vanderhaeghen R, Hilson P, Hamill JD, Goossens A (2011) APETALA2/ETHYLENE RESPONSE FACTOR and basic helix-loop-helix tobacco transcription factors cooperatively mediate jasmonate-elicited nicotine biosynthesis. Plant J 66:1053–1065

Dehesh K, Smith LG, Tepperman JM, Quail PH (1995) Twin autonomous bipartite nuclear localization signals direct nuclear import of GT-2. Plant Journal 8:25–36

Deng C, Hao X, Shi M, Fu R, Wang Y, Zhang Y, Zhou W, Feng Y, Makunga NP, Kai G (2019) Tanshinone production could be increased by the expression of SmWRKY2 in *Salvia miltiorrhiza* hairy roots. Plant Sci 284:1–8

Devaiah BN, Madhuvanthi R, Karthikeyan AS, Raghothama KG (2009) Phosphate starvation responses and gibberellic acid biosynthesis are regulated by the MYB62 transcription factor in *Arabidopsis*. Mol Plant 2:43–58

Di P, Zhang L, Chen J, Tan H, Xiao Y, Dong X, Zhou X, Chen W (2013) 13C Tracer reveals phenolic acids biosynthesis in hairy root cultures of *Salvia miltiorrhiza*. ACS Chem Biol 8:1537–1548

Ding K, Pei T, Bai Z, Jia Y, Ma P, Liang Z (2017) SmMYB36, a novel R2R3-MYB transcription factor, enhances tanshinone accumulation and decreases phenolic acid content in *Salvia miltiorrhiza* hairy roots. Sci Rep 7:5104

Du T, Niu J, Su J, Li S, Guo X, Li L, Cao X, Kang J (2018) SmbHLH37 functions antagonistically with SmMYC2 in regulating Jasmonate-Mediated biosynthesis of phenolic acids in *Salvia miltiorrhiza*. Front Plant Sci 9:1720

Dubos C, Stracke R, Grotewold E, Weisshaar B, Martin C, Lepiniec L (2010) MYB transcription factors in *Arabidopsis*. Trends Plant Sci 15:573–581

Foster R, Izawa T, Chua NH (1994) Plant bZIP proteins gather at ACGT elements. FASEB J 8:192–200

Froidure S, Roby D, Rivas S (2010) Expression of the Arabidopsis transcription factor AtMYB30 is post-transcriptionally regulated. Plant Physiol Biochem 48:735–739

Gao W, Hillwig ML, Huang L, Cui G, Wang X, Kong J, Yang B, Peters RJ (2009) A functional genomics approach to tanshinone biosynthesis provides stereochemical insights. Org Letters 11:5170–5173

Gao W, Sun HX, Xiao H, Cui G, Hillwig ML, Jackson A, Wang X, Shen Y, Zhao N, Zhang L, Wang XJ, Peters RJ, Huang L (2014) Combining metabolomics and transcriptomics to characterize tanshinone biosynthesis in *Salvia miltiorrhiza*. BMC Genom 15:73

Grasser KD (1995) Plant chromosomal high mobility group (HMG) proteins. Plant J 7:185–192

Guilfoyle TJ (1997) The structure of plant gene promoters. Genet Engin 19:15–47

Guiltinan MJ, Miller L (1994) Molecular characterization of the DNA-binding and dimerization domains of the bZIP transcription factor, EmBP-1. Plant Mol Biol 26:1041–1053

Guo J, Ma X, Cai Y, Ma Y, Zhan Z, Zhou YJ, Liu W, Guan M, Yang J, Cui G, Kang L, Yang L, Shen Y, Tang J, Lin H, Ma X, Jin B, Liu Z, Peters RJ, Zhao ZK, Huang L (2016) Cytochrome P450 promiscuity leads to a bifurcating biosynthetic pathway for tanshinones. New Phytol 210:525–534

Guo J, Zhou YJ, Hillwig ML, Shen Y, Yang L, Wang Y, Zhang X, Liu W, Peters RJ, Chen X, Zhao ZK, Huang L (2013) CYP76AH1 catalyzes turnover of miltiradiene in tanshinones biosynthesis and enables heterologous production of ferruginol in yeasts. P Natl A Sci 110:12108–12113

Gupta R, Webster CI, Walker AR, Gray JC (1997) Chromosomal location and expression of the single-copy gene encoding high-mobility-group protein HMG-I/Y in Arabidopsis thaliana. Plant Mol Biol 34:529–536

Hong GJ, Xue XY, Mao YB, Wang LJ, Chen XY (2012) *Arabidopsis* MYC2 interacts with DELLA proteins in regulating sesquiterpene synthase gene expression. Plant Cell 24:2635–2648

Huang Q, Sun M, Yuan T, Wang Y, Shi M, Lu S, Tang B, Pan J, Wang Y, Kai G (2019) The AP2/ERF transcription factor SmERF1L1 regulates the biosynthesis of tanshinones and phenolic acids in *Salvia miltiorrhiza*. Food Chem 274:368–375

Hurst HC (1995) Transcription factors 1: bZIP proteins. Protein Profile 2:105–168

Ishiguro S, Nakamura K (1994) Characterization of a cDNA encoding a novel DNA-binding protein, SPF1, that recognizes SP8 sequences in the 5' upstream regions of genes coding for sporamin and beta-amylase from sweet potato. Mol Gen Genet 244:563–571

Izawa T, Foster R, Chua NH (1993) Plant bZIP protein DNA binding specificity. J Mol Biol 230:1131–1144
Jaillon O, Aury JM, Noel B, Policriti A, Clepet C, Casagrande A, Choisne N, Aubourg S, Vitulo N, Jubin C, Vezzi A, Legeai F, Hugueney P, Dasilva C, Horner D, Mica E, Jublot D, Poulain J, Bruyère C, Billault A, Segurens B, Gouyvenoux M, Ugarte E, Cattonaro F, Anthouard V, Vico V, Del Fabbro C, Alaux M, Di Gaspero G, Dumas V, Felice N, Paillard S, Juman I, Moroldo M, Scalabrin S, Canaguier A, Le Clainche I, Malacrida G, Durand E, Pesole G, Laucou V, Chatelet P, Merdinoglu D, Delledonne M, Pezzotti M, Lecharny A, Scarpelli C, Artiguenave F, Pè ME, Valle G, Morgante M, Caboche M, Adam-Blondon AF, Weissenbach J, Quétier F, Wincker P (2007) The grapevine genome sequence suggests ancestral hexaploidization in major angiosperm phyla. Nature 449:463–467
Ji AJ, Luo HM, Xu ZC, Zhang X, Zhu YJ, Liao BS, Yao H, Song JY, Chen SL (2016) Genome-wide identification of the *AP2/ERF* gene family involved in active constituent biosynthesis in *Salvia miltiorrhiza*. Plant Genome 9
Kai GY, Xu H, Zhou C, Liao P, Xiao J, Luo X, You L, Zhang L (2011) Metabolic engineering tanshinone biosynthetic pathway in *Salvia miltiorrhiza* hairy root cultures. Metab Eng 13:319–327
Katagiri F, Seipel K, Chua NH (1992) Identification of a novel dimer stabilization region in a plant bZIP transcription activator. Mol Cell Biol 12:4809–4816
Kay SA (1993) Shedding light on clock controlled cab gene transcription in higher plants. Semin Cell Biol 4:81–86
Klinge B, Uberlacker B, Korfhage C, Werr W (1996) *ZmHox*: a noval class of maize homeobox genes. Plant Mol Biol 30:439–453
Kozaki A, Sakamoto A, Takeba G (1992) The promoter of the gene for plastidic glutamine synthetase (GS2) from rice is developmentally regulated and exhibits substrate-induced expression in transgenic tobacco plants. Plant Cell Physiol 33:233–238
Kuhn RM, Caspar T, Dehesh K, Quail PH (1993) DNA binding factor GT-2 from Arabidopsis. Plant Mol Biol 23:337–348
Lee H, Cha J, Choi C, Choi N, Ji HS, Park SR, Lee S, Hwang DJ (2018) Rice *WRKY11* plays a role in pathogen defense and drought tolerance. Rice (N Y) 11:5
Li C, Lu S (2014) Genome-wide characterization and comparative analysis of R2R3-MYB transcription factors shows the complexity of MYB-associated regulatory networks in *Salvia miltiorrhiza*. BMC Genom 15:277
Li C, Li D, Shao F, Lu S (2015) Molecular cloning and expression analysis of WRKY transcription factor genes in *Salvia miltiorrhiza*. BMC Genom 16:200
Li LN (1998) Biologically active components from traditional Chinese medicines. Pure Appl Chem 70:547–554
Li S, Wu Y, Kuang J, Wang H, Du T, Huang Y, Zhang Y, Cao X, Wang Z (2018) SmMYB111 is a key factor to phenolic acid biosynthesis and interacts with both SmTTG1 and SmbHLH51 in *Salvia miltiorrhiza*. J Agric Food Chem 66:8069–8078
Licausi F, Ohme-Takagi M, Perata P (2013) APETALA2/Ethylene Responsive Factor (AP2/ERF) transcription factors: mediators of stress responses and developmental programs. New Phytol 199:639–649
Logemann E, Parniske M, Hahlbrock K (1995) Modes of expression and common structural features of the complete phenylalanine ammonia-lyase gene family in parsley. Proc Natl Acad Sci USA 92:5905–5909
Lu X, Zhang L, Zhang F et al (2013) AaORA, a trichome-specific AP2/ERF transcription factor of *Artemisia annua*, is a positive regulator in the artemisinin biosynthetic pathway and in disease resistance to *Botrytis cinerea*. New Phytol 198:1191–1202
Lyck R, Harmening U, Höhfeld I, Treuter E, Scharf KD, Nover L (1997) Intracellular distribution and identification of the nuclear localization signals of two plant heat-stress transcription factors. Planta 202:117–125
Ma P, Liu J, Osbourn A, Dong J, Liang Z (2015) Regulation and metabolic engineering of tanshinone biosynthesis. RSC Adv 5:18137–18144
Ma P, Liu J, Zhang C, Liang Z (2013) Regulation of water-soluble phenolic acid biosynthesis in *Salvia miltiorrhiza* Bunge. Appl Biochem Biotech 170:1253–1262
Mahjoub A, Hernould M, Joubès J, Decendit A, Mars M, Barrieu F, Hamdi S, Delrot S (2009) Overexpression of a grapevine R2R3-MYB factor in tomato affects vegetative development flower morphology and flavonoid and terpenoid metabolism. Plant Physiol Biochem 47:551–561
Mannen K, Matsumoto T, Takahashi S, Yamaguchi Y, Tsukagoshi M, Sano R, Suzuki H, Sakurai N, Shibata D, Koyama T, Nakayama T (2014) Coordinated transcriptional regulation of isopentenyl diphosphate biosynthetic pathway enzymes in plastids by phytochrome-interacting factor 5. Biochem Biophys Res Commun 443:768–774
Martin C (1997) MYB transcription factors in plants. Trends Gene 13:67–73
Meisel L, Lam E (1996) The conserved ELK-homeodomain of KNOTTED-1 contains two regions that signal nuclear localization. Plant Mol Biol 30:1–14
Mizukami Y, Huang H, Tudor M, Hu Y, Ma H (1996) Functional domains of the floral regulator AGAMOUS: characterization of the DNA binding domain and analysis of dominant negative mutations. Plant Cell 8:831–845
Nantel A, Quatrano RS (1996) Characterization of three rice basic/leucine zipper factors, including two inhibitors of EmBP-1 DNA binding activity. J Biol Chem 271:31296–31305
Paul P, Singh SK, Patra B, Sui X, Pattanaik S, Yuan L (2017) A differentially regulated AP2/ERF

transcription factor gene cluster acts downstream of a MAP kinase cascade to modulate terpenoid indole alkaloid biosynthesis in *Catharanthus roseus*. New Phytol 213:1107–1123
Pei T, Ma P, Ding K, Liu S, Jia Y, Ru M, Dong J, Liang Z (2018) SmJAZ8 acts as a core repressor regulating JA-induced biosynthesis of salvianolic acids and tanshinones in *Salvia miltiorrhiza* hairy roots. J Exp Bot 69:1663–1678
Pilgrim ML, Caspar T, Quail PH, McClung CR (1993) Circadian and light-regulated expression of nitrate reductase in *Arabidopsis*. Plant Mol Biol 23:349–364
Purugganan MD, Rounsley SD, Schmidt RJ, Yanofsky MF (1995) Molecular evolution of flower development: diversification of the plant MADS-box regulatory gene family. Genetics 140:345–356
Riechmann JL, Meyerowitz EM (1998) The AP2/EREBP family of plant transcription factors. Biol Chem 379:633–646
Rook F, Weisbeek P, Smeekens S (1998) The light-regulated *Arabidopsis* bZIP transcription factor gene *ATB2* encodes a protein with an unusually long leucine zipper domain. Plant Mol Biol 37:171–178
Rushton PJ, Bokowiec MT, Han S, Zhang H, Brannock JF, Chen X, Laudeman TW, Timko MP (2008) Tobacco transcription factors: novel insights into transcriptional regulation in the *Solanaceae*. Plant Physiol 147:280–295
Sainz MB, Grotewold E, Chandler VL (1997) Evidence for direct activation of an anthocyanin promoter by the maize C1 protein and comparison of DNA binding by related Myb domain proteins. Plant Cell 9:611–625
Sakamoto A, Minami M, Huh GH, Iwabuchi M (1993) The putative zinc-finger protein WZF1 interacts with a cis-acting element of wheat histone genes. Eur J Biochem 217:1049–1056
Sakuma Y, Liu Q, Dubouzet JG, Abe H, Shinozaki K, Yamaguchi-Shinozaki K (2002) DNA-binding specificity of the ERF/AP2 domain of *Arabidopsis* DREBs, transcription factors involved in dehydration and cold-inducible gene expression. Biochem Biophys Res Commun 290:998–1009
Sears MT, Zhang H, Rushton PJ, Wu M, Han S, Spano AJ, Timko MP (2014) NtERF32: a non-NIC2 locus AP2/ERF transcription factor required in jasmonate-inducible nicotine biosynthesis in tobacco. Plant Mol Biol 84:49–66
Shen Q, Lu X, Yan T, Fu X, Lv Z, Zhang F, Pan Q, Wang G, Sun X, Tang K (2016) The jasmonate-responsive AaMYC2 transcription factor positively regulates artemisinin biosynthesis in *Artemisia annua*. New Phytol 210:1269–1281
Shoji T, Kajikawa M, Hashimoto T (2010) Clustered transcription factor genes regulate nicotine biosynthesis in tobacco. Plant Cell 22:3390–3409
Sibéril Y, Benhamron S, Memelink J, Giglioli-Guivarc'h N, Thiersault M, Boisson B, Doireau P, Gantet P (2001) *Catharanthus roseus* G-box binding factors 1 and 2 act as repressors of strictosidine synthase gene expression in cell cultures. Plant Mol Biol 45:477–488
Singh AK, Kumar SR, Dwivedi V, Rai A, Pal S, Shasany AK, Nagegowda DA (2017) A WRKY transcription factor from Withania somnifera regulates triterpenoid withanolide accumulation and biotic stress tolerance through modulation of phytosterol and defense pathways. New Phytol 215:1115–1131
Song Z, Guo L, Liu T, Lin C, Wang J, Li X (2017) Comparative RNA-sequence transcriptome analysis of phenolic acid metabolism in *Salvia miltiorrhiza*, a traditional Chinese medicine model plant. Int J Genomics 2017:9364594
Sun M, Shi M, Wang Y, Huang Q, Yuan T, Wang Q, Wang C, Zhou W, Kai G (2019) The biosynthesis of phenolic acids is positively regulated by the JA-responsive transcription factor ERF115 in *Salvia miltiorrhiza*. J Exp Bot 70:243–254
Suttipanta N, Pattanaik S, Kulshrestha M, Patra B, Singh SK, Yuan L (2011) The transcription factor CrWRKY1 positively regulates the terpenoid indole alkaloid biosynthesis in *Catharanthus roseus*. Plant Physiol 157:2081–2093
Tan H, Xiao L, Gao S, Li Q, Chen J, Xiao Y, Ji Q, Chen R, Chen W, Zhang L (2015) *TRICHOME AND ARTEMISININ REGULATOR 1* is required for trichome development and artemisinin biosynthesis in *Artemisia annua*. Mol Plant 8:1396–1411
Tominaga-Wada R, Nukumizu Y, Sato S, Kato T, Tabata S, Wada T (2012) Functional divergence of MYB-related genes *WEREWOLF* and *AtMYB23* in *Arabidopsis*. Biosci Biotechnol Biochem 76:883–887
Tonoike H, Han IS, Jongewaard I, Doyle M, Guiltinan M, Fosket DE (1994) Hypocotyl expression and light down regulation of the soybean *tubulin* gene, *tubB1*. Plant J 5:343–351
Ulmasov T, Hagen G, Guilfoyle TJ (1997) ARF1, a transcription factor that binds to auxin response elements. Science 276:1865–1868
Van der Fits L, Memelink J (2000) ORCA3, a jasmonate-responsive transcriptional regulator of plant primary and secondary metabolism. Science 289:295–297
Van der Fits L, Memelink J (2001) The jasmonate-inducible AP2/ERF domain transcription factor ORCA3 activates gene expression via interaction with a jasmonate-responsive promoter element. Plant J 25:43–53
Varagona MJ, Schmidt RJ, Raikhel NV (1992) Nuclear localization signal(s) required for nuclear targeting of the maize regulatory protein Opaque-2. Plant Cell 4:1213–1227
Wan S, Li C, Ma X, Luo K (2017) *PtrMYB57* contributes to the negative regulation of anthocyanin and proanthocyanidin biosynthesis in poplar. Plant Cell Rep 36:1263–1276
Wang B, Sun W, Li Q, Li Y, Luo H, Song J, Sun C, Qian J, Zhu Y, Hayward A, Xu H, Chen S (2015) Genome-wide identification of phenolic acid biosynthetic genes in *Salvia miltiorrhiza*. Planta 241:711–725
Wei T, Deng K, Zhang Q, Gao Y, Liu Y, Yang M, Zhang L, Zheng X, Wang C, Liu Z, Chen C, Zhang Y

(2017) Modulating *AtDREB1C* expression improves drought tolerance in *Salvia miltiorrhiza*. Front Plant Sci 8:52

Wu JY, Shi M (2008) Ultrahigh diterpenoid tanshinone production through repeated osmotic stress and elicitor stimulation in fed-batch culture of *Salvia miltiorrhiza* hairy roots. Appl Microbiol Biot 78:441–448

Wu SJ, Shi M, Wu JY (2009) Cloning and characterization of the 1-deoxy-D-xylulose 5-phosphate reductoisomerase gene for diterpenoid tanshinone biosynthesis in *Salvia miltiorrhiza* (Chinese sage) hairy roots. Biotechnol Appl Bioc 52:89

Wu Y, Zhang Y, Li L, Guo X, Wang B, Cao X, Wang Z (2018) AtPAP1 Interacts with and activates SmbHLH51, a positive regulator to phenolic acids biosynthesis in *Salvia miltiorrhiza*. Front Plant Sci 9

Xing B, Liang L, Liu L, Hou Z, Yang D, Yan K, Zhang X, Liang Z (2018a) Overexpression of SmbHLH148 induced biosynthesis of tanshinones as well as phenolic acids in *Salvia miltiorrhiza* hairy roots. Plant Cell Rep 37:1681–1692

Xing B, Yang D, Yu H, Zhang B, Yan K, Zhang X, Han R, Liang Z (2018b) Overexpression of SmbHLH10 enhances tanshinones biosynthesis in *Salvia miltiorrhiza* hairy roots. Plant Sci 276:229–238

Xu H, Song J, Luo H, Zhang Y, Li Q, Zhu Y, Xu J, Li Y, Song C, Wang B, Sun W, Shen G, Zhang X, Qian J, Ji A, Xu Z, Luo X, He L, Li C, Sun C, Yan H, Cui G, Li X, Li X, Wei J, Liu J, Wang Y, Hayward A, Nelson D, Ning Z, Peters RJ, Qi X, Chen S (2016) Analysis of the genome sequence of the medicinal plant *Salvia miltiorrhiza*. Mol Plant 6:949–952

Xu H, Zhang L, Zhou CC, Xiao JB, Liao P, Kai GY (2010) Metabolic regulation and genetic engineering of pharmaceutical component tanshinone biosynthesis in *Salvia miltiorrhiza*. J Med Plant Res 4:2591–2597

Xu YH, Wang JW, Wang S, Wang JY, Chen XY (2004) Characterization of GaWRKY1, a cotton transcription factor that regulates the sesquiterpene synthase gene (C)-d-cadinene synthase-A. Plant Physiol 135:507–515

Yamasaki K, Kigawa T, Seki M, Shinozaki K, Yokoyama S (2013) DNA-binding domains of plant-specific transcription factors: structure, function, and evolution. Trends Plant Sci 18:267–276

Yan Y, Jia H, Wang F, Wang C, Liu S, Guo X (2015) Overexpression of *GhWRKY27a* reduces tolerance to drought stress and resistance to *Rhizoctonia solani* infection in transgenic *Nicotiana benthamiana*. Front Physiol 6:265

Yanagisawa S, Sheen J (1998) Involvement of maize *Dof* zinc finger proteins in tissue-specific and light-regulated gene expression. Plant Cell 10:75–89

Yang L, Ding G, Lin H, Cheng H, Kong Y, Wei Y, Fang X, Liu R, Wang L, Chen X, Yang C (2013) Transcriptome analysis of medicinal plant *Salvia miltiorrhiza* and identification of genes related to tanshinone biosynthesis. PLoS ONE 8:e80464

Yang N, Zhou W, Su J, Wang X, Li L, Wang L, Cao X, Wang Z (2017) Overexpression of SmMYC2 increases the production of phenolic acids in *Salvia miltiorrhiza*. Front Plant Sci 8:1804

Yu H, Guo W, Yang D, Hou Z, Liang Z (2018) Transcriptional profiles of *SmWRKY* family genes and their putative roles in the biosynthesis of tanshinone and phenolic acids in *Salvia miltiorrhiza*. Int J Mol Sci 19:E1593

Yu ZX, Li JX, Yang CQ, Hu WL, Wang LJ, Chen XY (2012) The jasmonate-responsive AP2/ERF transcription factors AaERF1 and AaERF2 positively regulate artemisinin biosynthesis in *Artemisia annua* L. Mol Plant 5:353–365

Zhang J, Zhou L, Zheng X, Zhang J, Yang L, Tan R, Zhao S (2017) Overexpression of SmMYB9b enhances tanshinone concentration in *Salvia miltiorrhiza* hairy roots. Plant Cell Rep 36:1297–1309

Zhang L, Wu B, Zhao D, Li C, Shao F, Lu S (2014) Genome-wide analysis and molecular dissection of the SPL gene family in *Salvia miltiorrhiza*. J Integr Plant Biol 56:38–50

Zhang S, Ma P, Yang D, Li W, Liang Z, Liu Y, Liu F (2013) Cloning and characterization of a putative R2R3 MYB transcriptional repressor of the rosmarinic acid biosynthetic pathway from *Salvia miltiorrhiza*. PLoS ONE 8:e73259

Zhang X, Luo H, Xu Z, Zhu Y, Ji A, Song J, Chen S (2015a) Genome-wide characterisation and analysis of bHLH transcription factors related to tanshinone biosynthesis in *Salvia miltiorrhiza*. Sci Rep 5:11244

Zhang Y, Butelli E, Alseekh S, Tohge T, Rallapalli G, Luo J, Kawar PG, Hill L, Santino A, Fernie AR, Martin C (2015b) Multi-level engineering facilitates the production of phenylpropanoid compounds in tomato. Nat Commun 6:8635

Zhang Y, Xu Z, Ji A, Luo H, Song J (2018) Genomic survey of bZIP transcription factor genes related to tanshinone biosynthesis in *Salvia miltiorrhiza*. Acta Pharm Sin B 8:295–305

Zhang Y, Yan Y, Wang Z (2010) The *Arabidopsis* PAP1 transcription factor plays an important role in the enrichment of phenolic acids in *Salvia miltiorrhiza*. J Agric Food Chem 58:12168–12175

Zhao GR, Zhang HM, Ye TX, Xiang ZJ, Yuan YJ, Guo ZX, Zhao LB (2008) Characterization of the radical scavenging and antioxidant activities of danshensu and salvianolic acid B. Food Chem Toxico l 46:73–81

Zhao SJ, Zhang JJ, Tan RH, Yang L, Zheng XY (2015) Enhancing diterpenoid concentration in *Salvia miltiorrhiza* hairy roots through pathway engineering with maize C1 transcription factor. J Exp Bot 66:7211–7226

Zhou L, Zuo Z, Chow MSS (2005) Danshen: an overview of its chemistry, pharmacology, pharmacokinetics, and clinical use. J Clin Pharmacol 45:1345–1359

Zhou W, Huang Q, Wu X, Zhou Z, Ding M, Shi M, Huang F, Li S, Wang Y, Kai G (2017) Comprehensive transcriptome profiling of *Salvia miltiorrhiza* for discovery of genes associated with the biosynthesis of tanshinones and phenolic acids. Sci Rep 7:10554

Zhou Y, Sun W, Chen J, Tan H, Xiao Y, Li Q, Ji Q, Gao S, Chen L, Chen S, Zhang L, Chen W (2016) SmMYC2a and SmMYC2b played similar but irreplaceable roles in regulating the biosynthesis of tanshinones and phenolic acids in *Salvia miltiorrhiza*. Sci Rep 6:22852

Zhu B, Huo DA, Hong XX, Guo J, Peng T, Liu J, Huang XL, Yan HQ, Weng QB, Zhang XC, Du XY (2019) The *Salvia miltiorrhiza* NAC transcription factor SmNAC1 enhances zinc content in transgenic *Arabidopsis*. Gene 688:54–61

Bioinformatic Tools for *Salvia miltiorrhiza* Functional Genomics

9

Liqiang Wang and Chang Liu

Abstract
For *Salvia miltiorrhiza*, bioinformatic tools have been used to analyze its DNA sequences, protein sequences as well as next-generation DNA sequencing (NGS) data studying the transcriptome, circular RNAs, etc. Here, we first described the basic file formats used in studying *S. miltiorrhiza*. FASTA is the standard format describing nucleotide and amino acid sequences. FASTQ format adds the quality scores to the FASTA file. GFF/GTF is used to describe the genome and gene structures, such as genes, CDS, proteins, exons, and introns. SAM format is widely used to describe the mapping of NGS reads to the reference sequences. For various analysis tasks, many commercial software tools have been developed. However, for users who prefer to use free software tools or develop computational pipelines, the basic function of EMBOSS software is introduced. Mapping of reads to reference sequences is the most widely used method for analyzing NGS data, and we introduced several frequently used tools. Bowtie2 is one of the most widely used tools for read mapping. SAMtools can be used to analyze the mapping data. As an example, several software packages identifying circRNAs are described. This chapter thus provides an introduction to the most fundamental bioinformatic tools.

L. Wang · C. Liu (✉)
Key Laboratory of Bioactive Substances and Resources Utilization of Chinese Herbal Medicine, Ministry of Education, Institute of Medicinal Plant Development, Chinese Academy of Medical Sciences and Peking Union Medical College, No. 151 Malianwa North Road, Haidian District, 100193 Beijing, People's Republic of China
e-mail: cliu@implad.ac.cn

L. Wang
e-mail: lqwang@implad.ac.cn

9.1 Introduction

Functional genomics is an area of molecular biology that scientists use to understand the relationship between an organism's genome (DNA sequence) and its phenotype. It captures information on changes in the way that genes are expressed at the level of RNA and protein, and how that might reflect functions of genes or differences in how an organism appears, functions, and adapts to its environment. It is used to understand how small variations in DNA sequence might result in a change in phenotype that affects survival or fitness. In this chapter, we focus on the bioinformatic tools for *S. miltiorrhiza* functional genomics research. Below, we will discuss the various file formats as well as some software tools associated with the studies on functional genomics. Functional genomics is a large research area, and various bioinformatic tools have been developed to face novel

S. Lu (ed.), *The Salvia miltiorrhiza Genome*, Compendium of Plant Genomes,
https://doi.org/10.1007/978-3-030-24716-4_9

experimental results. Due to the space limitation, it is impossible to describe all bioinformatic tools used for the functional genomic studies of *Salvia miltiorrhiza*. The tools described here serve as good examples. Users of other tools are advised to refer to recent review articles for particular types of analyses.

9.2 DNA Sequence Format

9.2.1 IUPAC

The International Union of Pure and Applied Chemistry (IUPAC) is an international federation of National Adhering Organizations that represents chemists in individual countries. IUPAC's Inter-divisional Committee on Nomenclature and Symbols (IUPAC nomenclature) is the recognized world authority in developing standards for the naming of the chemical elements and compounds. IUPAC is best known for its works standardizing nomenclature in chemistry and other fields of science, but IUPAC has publications in many fields including chemistry, biology, and physics. IUPAC also has a system for giving single-letter codes to identify amino acids and nucleotide bases. "A", "G", "C", and "T" have been used to describe the four bases representing purines (adenine and guanine) and pyrimidines (cytosine and thymine or uracil). Some degenerate codes have been used to describe various combinations of purines or pyrimidines (Table 9.1). Similarly, degenerate codes have been used to describe various combinations of amino acids (Table 9.2). Using these

Table 9.1 Degenerate codes for nucleotides

Nucleic acid code	Meaning	Mnemonic
R	A or G	Purine
Y	C, T or U	Pyrimidines
K	G, T or U	Bases which are ketones
M	A or C	Bases with amino groups
S	C or G	Strong interaction
W	A, T or U	Weak interaction
B	Not A (i.e. C, G, T or U)	B comes after A
D	Not C (i.e. A, G, T or U)	D comes after C
H	Not G (i.e. A, C, T or U)	H comes after G
V	Neither T nor U (i.e. A, C or G)	V comes after U
N	A, C, G, T or U	Nucleic acid
X	Masked	
–	Gap of indeterminate length	

Table 9.2 Degenerate codes for amino acids

Amino acid code	Meaning
B	Aspartic acid or asparagine
J	Leucine or isoleucine
Z	Glutamic acid or glutamine
X	Any
*	Translation stop
–	Gap of indeterminate length

degenerate codes allows the representation of ambiguous bases or amino acids (https://en.wikipedia.org/wiki/International_Union_of_Pure_and_Applied_Chemistry).

9.2.2 FASTA Format

The FASTA format (https://www.ebi.ac.uk/Tools/sss/fasta) is a text-based format for representing either nucleotide sequences or amino acid (protein) sequences, in which nucleotides or amino acids are represented using single-letter IUPAC codes. The format also allows for sequence names and comments to precede the sequences.

The first line in a FASTA file started either with a ">" (greater-than) symbol or, less frequently, a ";" (semicolon) was taken as a comment. Subsequent lines starting with a semicolon would be ignored by software. For these days, the first line often starts with a unique library accession number and followed with a summary description. Following the initial line (used for a unique description of the sequence) is the actual sequence itself in standard one-letter character string. Anything other than a valid character would be ignored (including spaces, tabulators, asterisks). Originally, it was also common to end the sequence with an "*" (asterisk) character (in analogy with use in PIR formatted sequences) and, for the same reason, to leave a blank line between the description and the sequence. Two sequences are provided below as examples for FASTA sequences. The first one is for a DNA sequence, and the second one is for a protein sequence:

```
>GU180369.1_Alpinia_oxyphylla_voucher_PS0533MT02_ITS2
GCATCGTCGCCTTTGCTCCTTGCTCTGTT
GGTGCCAAGCGCGGAAATTGGCCCCGTGT
GCCCTCGGGCACAGTCGGCTGAAGAGTG
GGTAATCTCGGCAGTCGTCGGGCGCGAT
GGGTGTTGGTCGCCCTGTGCGTGAACTG
AGCGTCGTCCCCGTCGTGCTGGGATGAG
TCCCAAAGAGACCCTGTGTGATTGCGGC
GTCGCATGAAAGTGCCGTGTCCATCAGA
TTGTGGCCCCAAG*
>matK_[1769:3316](-)
MEELQGYLEEYRSRQQQFLYPLLFQEYIYV
FAYDHGLNSSIFYEPQNSLGYDNKFSSVLV
KRLIIRMYQKNYWIYSVNDIYQNIFVGHNN
YFYFHFFSQIISEGFAVIVEIPFSLQLISSLEE
KEIPKSHNLQSSHSIFPFLEDKLLHLNYLSD
ILIPYPAHMEILVQMLQSWIQDVLSLHLLQ
FLLHEYYNWNSLIIPNKSIYVFSKDNKRLFC
FLYNLYIYEYEFLLVFPCKQSSFLRLISSGVL
LERIHFYVKIERLGVCRIFCQKTLWIFKDPF
IHYIRYQGKSILGSRGTHFLMKKWKYHLV
HFWQYHFHFWSQPYRIDIKKLSNYSFYFLG
YFSSVQMNSSIVRNQMLENSFLVDTLTKKF
DTRIPIIPLIRSLSKAQFCTVYGYPISKPILTD
LADCDIINRFGRICRKLSHYHSGSSKKQSLY
RMKYILRLSCARTLARKHKSSARSFLQRLS
SGLLEEFFTEEEQVISLIFPKITSFYLYGSYR
ERIWYFDIIRINDLVNSLLVTT*
```

9.2.3 FASTQ Format

FASTQ format (https://en.wikipedia.org/wiki/FASTQ_format) is a text-based format for storing both a biological sequence (usually nucleotide sequence) and its corresponding quality scores. Both the sequence letter and the quality score are each encoded with a single ASCII character for brevity. It was originally developed at the Wellcome Trust Sanger Institute to bundle a FASTA formatted sequence and its quality data, but has recently become the de facto standard for storing the output of high-throughput sequencing instruments such as the various models of next generation and third-generation DNA sequencers.

A FASTQ file normally uses four lines per sequence. Line 1 begins with a "@" character and is followed by a sequence identifier and an optional description (like a FASTA title line). Line 2 is the raw sequence letters. Line 3 begins with a "+" character and is optionally followed by the same sequence identifier (and any description) again. Line 4 encodes the quality values for the sequence in Line 2 and must contain the same number of symbols as letters in the sequence. A FASTQ file containing a single sequence might look like this:

```
@E00514:44:HC2LJALXX:6:1101:5172:2329
1:N:0:GGCTACAT
CGCTCTACGAGAACGACGTCGACGAGAA
GGTGCGGCTCGCCGGCCCG
+
<<F-AJJJFAF7FJJJ<JJJ–7AJJ<<-F7AJF–A-AF<
AJF–A
```

The byte representing quality runs from 0 × 21 (lowest quality; "!" in ASCII) to 0 × 7e (highest quality; "~" in ASCII). Here are the quality value characters in left-to-right increasing order of quality (ASCII): !"#$%&'()* + ,-./ 0123456789:; <=>?@ ABCDEFGHIJKLMNOP QRSTUVWXYZ[\]^_`abcdefghijklmnopqrstuvw xyz{|}~. Using these codes rather than the actual quality score allows the alignment of the nucleotides and the quality scores.

9.2.4 SAM Format

Sequence Alignment Map (SAM) is a text-based format originally for storing biological sequences aligned to a reference sequence developed by Li et al. (2009). It is a tab-delimited text format consisting of a header section, which is optional, and an alignment section. If present, the header must be prior to the alignments. In the SAM format, each alignment line typically represents the linear alignment of a segment. Each line consists of 11 or more tab-separated fields. Header lines start with "@", while alignment lines do not. The first eleven fields are always present and in the order shown below; if the information represented by any of these fields is unavailable, that field's value will be a placeholder, either "0" or "*" as determined by the field's type. A standard SAM files contain data arranged in eleven different fields (Table 9.3). The names and their meaning for each field are provided. It should be pointed out that the Binary Alignment Map (BAM) format is the same as the SAM format in contents, but otherwise the files are smaller than that of the SAM format.

All mapped segments in alignment lines are represented on the forward genomic strand. For

Table 9.3 Mandatory fields in the SAM format

Col	Field	Type	Regexp/range	Brief description
1	QNAME	String	[!-?A-~]{1,254}	QNAME: Query template NAME. Reads/segments having identical QNAME are regarded to come from the same template. A QNAME "*" indicates the information is unavailable. In a SAM file, a read may occupy multiple alignment lines, when its alignment is chimeric or when multiple mappings are given
2	FLAG	Int	$[0, 2^{16} - 1]$	Combination of bitwise FLAGs
3	RNAME	String	*|[:rname:^*=][:rname:]*	Reference sequence NAME of the alignment. If @SQ header lines are present, RNAME (if not "*") must be present in one of the SQ-SN tags. An unmapped segment without coordinate has a "*" at this field. However, an unmapped segment may also have an ordinary coordinate such that it can be placed at a desired position after sorting. If RNAME is "*", no assumptions can be made about POS and CIGAR.
4	POS	Int	$[0, 2^{31-1}]$	1-based leftmost mapping POSition of the first matching base. The first base in a reference sequence has coordinate 1. POS is set as 0 for an unmapped read without coordinate. If POS is 0, no assumptions can be made about RNAME and CIGAR
5	MAPQ	Int	$[0, 2^8 - 1]$	MAPping Quality. It equals −10 log10 Pr{mapping position is wrong}, rounded to the nearest integer. A value 255 indicates that the mapping quality is not available

(continued)

Table 9.3 (continued)

Col	Field	Type	Regexp/range	Brief description
6	CIGAR	String	*|([0-9]+[MIDNSHPX=])+	Concise Idiosyncratic Gapped Alignment Report (CIGAR) string
7	RNEXT	String	*|=|[:rname:^*=][:rname:]*	Reference sequence name of the primary alignment of the NEXT read in the template. For the last read, the next read is the first read in the template. If @SQ header lines are present, RNEXT (if not "*" or "=") must be present in one of the SQ-SN tags. This field is set as "*" when the information is unavailable, and set as "=" if RNEXT is identical to RNAME. If not "=" and the next read in the template has one primary mapping (see also bit 0 × 100 in FLAG), this field is identical to RNAME at the primary line of the next read. If RNEXT is "*", no assumptions can be made on PNEXT and bit 0x20
8	PNEXT	Int	$[0, 2^{31-1}]$	Position of the primary alignment of the NEXT read in the template. Set as 0 when the information is unavailable. This field equals POS at the primary line of the next read. If PNEXT is 0, no assumptions can be made on RNEXT and bit 0x20.
9	TLEN	Int	$[-2^{31}+1, 2^{31-1}]$	Signed observed template LENgth. If all segments are mapped to the same reference, the unsigned observed template length equals the number of bases from the leftmost mapped base to the rightmost mapped base. The leftmost segment has a plus sign, and the rightmost has a minus sign. The sign of segments in the middle is undefined. It is set as 0 for single-segment template or when the information is unavailable
10	SEQ	String	*|[A-Za-z=.]+	Segment SEQuence. This field can be a "*" when the sequence is not stored. If not a "*", the length of the sequence must equal the sum of lengths of M/I/S/=/X operations in CIGAR. An "=" denotes the base is identical to the reference base. No assumptions can be made on the letter cases
11	QUAL	String	[!-~]+	ASCII of base QUALity plus 33 (same as the quality string in the Sanger FASTQ format). A base quality is the phred-scaled base error probability which equals −10 log10 Pr{base is wrong}. This field can be a "*" when quality is not stored. If not a "*", SEQ must not be a "*" and the length of the quality string ought to equal the length of SEQ

segments that have been mapped to the reverse strand, the recorded SEQ is reverse complemented from the original unmapped sequence and CIGAR, QUAL, and strand-sensitive optional fields are reversed and thus recorded consistently with the sequence bases as represented. The FASTQ and SAM files are the main data format describing the NGS data. Tile contents are usually handled by various software tools, and human readability is less of a concern. However, an in-depth understanding of these formats allows the identification of novel molecules such as circular RNAs. Below is an exemplar text in a SAM file:

```
@HD	VN:1.0	SO:coordinate
@SQSN:GL30704-R1	LN:4936
@PGID:bowtie2	PN:bowtie2	VN:2.3.4	CL:"./1.fasta.DB -S 1.fasta.sam -U sjjMD1.fq"
E00514:44:HC2LJALXX:6:1108:18456:54770	0	GL30704-R1	131	42	47M	*	0	0
	TCGTCTAATTGGCTCTTTGTAGTCTTGCTCTTGACCTTTCGTGACTA
	FFJFFFJJJJJJFJJJJJJJJJJJJJJFJJJJJJJJJJJJJJJ	AS:i:0	XN:i:0	XM:i:0	XO:i:0	XG:i:0
	NM:i:0	MD:Z:47	YT:Z:UU
E00514:44:HC2LJALXX:6:1113:12327:72526	16	GL30704-R1	132	42	47M	*	0	0
	CGTCTAATTGGCTCTTTGTAGTCTTGCTCTTGACCTTTCGTGACTAA
	JJJJJJJJJJJJJJJJJJJJJJJJJJJJJJJJJJJJJJJJJJJJJFF	AS:i:0	XN:i:0	XM:i:0	XO:i:0	XG:i:0
	NM:i:0	MD:Z:47	YT:Z:UU
E00514:44:HC2LJALXX:6:2212:17168:20084	16	GL30704-R1	132	42	47M	*	0	0
	CGTCTAATTGGCTCTTTGTAGTCTTGCTCTTGACCTTTCGTGACTAA
	JJJJJJJJJJJJJJJJJJJJJJJJJJJJJJJJJJJJJJJJJJJJJFF	AS:i:0	XN:i:0	XM:i:0	XO:i:0	XG:i:0
	NM:i:0	MD:Z:47	YT:Z:UU
```

9.2.5 GFF/GTF

The General Feature Format (GFF) (https://github.com/The-Sequence-Ontology/Specifications/blob/master/gff3.md) is used to describe the genome or gene structures, for example, the start and end position of gene, and the boundary of exons and introns. A GFF file consists of one line per feature, each containing nine columns of data (Table 9.4), plus optional track definition lines. The General Transfer Format (GTF) is identical to GFF version 2 (http://mblab.wustl.edu/GTF22.html). Fields must be tab-separated. Also, all but the final field in each feature line must contain a value; "empty" columns should be denoted with a ".".

An exemplar GFF file might look like below:

Table 9.4 Mandatory fields in the GFF/GTF

Col	Field	Brief description
1	Sequence	The name of the sequence where the feature is located
2	Source	Name of the program that generated this feature or the data source (database or project name)
3	Feature	Feature type name, e.g., gene, variation, similarity
4	Start	Start position of the feature, with sequence numbering starting at 1
5	End	End position of the feature, with sequence numbering starting at 1
6	Score	A floating-point value
7	Strand	Defined as + (forward) or - (reverse)
8	Frame	One of "0", "1", or "2". "0" indicates that the first base of the feature is the first base of a codon, "1" that the second base is the first base of a codon, and so on
9	Attribute	A semicolon-separated list of tag-value pairs, providing additional information about each feature

Table 9.6 Most important parameters used in SAMtools commands

Parameters	Value	Function
–Indexing	dict	Create a sequence dictionary file
	faidx	Index/extract FASTA
	index	Index alignment
–Editing	calmd	Recalculate MD/NM tags and "=" bases
	fixmate	Fix mate information
	reheader	Replace BAM header
	rmdup	Remove PCR duplicates
	targetcut	Cut fosmid regions (for fosmid pool only)
	addreplacerg	Add or replace RG tags
–File operations	collate	Shuffle and group alignments by name
	cat	Concatenate BAMs
	merge	Merge sorted alignments
	mpileup	Multi-way pileup
	sort	Sort alignment file
	split	Split a file by read group
	quickcheck	Quickly check if SAM/BAM/CRAM file appears intact
	FASTQ	Convert a BAM to a FASTQ
	fasta	Convert a BAM to a FASTA
–Statistics	bedcov	Read depth per BED region
	depth	Compute the depth
	flagstat	Simple stats
	idxstats	BAM index stats
	phase	Phase heterozygotes
	stats	Generate stats (former bamcheck)
–Viewing	flags	Explain BAM flags
	tview	Text alignment viewer
	view	SAM <->BAM <->CRAM conversion
	depad	Convert padded BAM to unpadded BAM

Below please find several examples of SAMtools commands. Explanation of these commands is provided afterward in some cases.

(1) Using "view" to convert a file in BAM format to SAM format
>samtools view sample.bam > sample.sam
(2) Using "view" to convert a file in SAM format to BAM format
>samtools view -bS sample.sam > sample.bam
(3) Using "view" to extract all reads mapped to a particular region
>samtools view -h -b sample_sorted.bam "chr1:10-13" > tiny_sorted.bam

The above command will extract the same reads as above, but instead of displaying them, write them to a new BAM file, tiny_sorted.bam. The -b option makes the output compressed, and the -h option causes the SAM headers to be outputed also. These headers include a description of the reference that the reads in sample_sorted.bam were aligned to and will be needed if the tiny_-sorted.bam file is to be used with some of the more advanced SAMtools commands. The order of extracted reads is preserved.

(4) Using "tview" to start an interactive viewer
>samtools tview sample_sorted.bam

This command will start an interactive viewer to visualize a small region of the reference, the reads aligned, and mismatches. Within the view, can jump to a new location by typing g: and a location, like g:chr1:10,000,000. If the reference element name and following colon are replaced with {{{1}}}, the current reference element is used; i.e., if {{{1}}} is typed after the previous "goto" command, the viewer jumps to the region 200 base pairs down on chr1. Typing "?" brings up help information.

(5) Using "sort" to sort the alignment in BAM format
>samtools sort unsorted_in.bam sorted_out

This command will read the specified unsorte-d_in.bam as input, sort it by aligned read position, and write it out to sorted_out.bam, the bam file whose name (without extension) was specified.

(6) Using "index" to build an index for an alignment
>samtools index sorted.bam

This command will create an index file, sor-ted.bam.bai for the sorted.bam file.

9.3.2 EMBOSS

EMBOSS is an acronym for European Molecular Biology Open Software Suite. EMBOSS is a free open-source software analysis package developed for the needs of the molecular biology and bioinformatic user community (Rice et al. 2000). The EMBOSS package contains a variety of applications for sequence alignment, rapid database searching with sequence patterns, protein motif identification (including domain analysis), and much more. Herein, several examples are provided to give readers a general sense how these programs can be used in analyzing the genome of *S. miltiorrhiza*. Fortunately, all commands in the EMBOSS were designed to use consistent types of parameter and names. All other programs work in pretty much the same matter.

(1) extractseq. This command is under the category "SEQUENCE MANIPULATION" and is used to extract regions from a sequence. The parameters for "extractseq" are described in Table 9.7.
(2) seqret. This command is under the category "SEQUENCE RETRIEVAL" and is a frequently used tool to read and write (return) sequences. The standard (mandatory) qualifiers of seqret from EMBOSS are shown in Table 9.8.
(3) einverted. This command is under the category "REPEAT" and is used to find inverted repeats in nucleotide sequences. The commonly used parameters are described in Table 9.9.

Table 9.7 Standard (mandatory) qualifiers of extractseq from EMBOSS

Parameters	Value	Description
-sequence	sequence	Sequence filename and optional format, or reference (input USA)
-sequence	sequence	Sequence filename and optional format, or reference (input USA)
-regions	range	[Whole sequence] regions to extract. A set of regions is specified by a set of pairs of positions. The positions are integers. They are separated by any non-digit, non-alpha character. Examples of region specifications are: 24-45, 56-78; 1:45, 67 = 99;765..888; 1,5,8,10,23,45,57,99
-outseq	seqoutall	[<sequence>.<format>] sequence set(s) filename and optional format (output USA)

Table 9.8 Standard (mandatory) qualifiers of seqret from EMBOSS

Parameters	Value	Description
-sequence	seqall	(Gapped) sequence(s) filename and optional format, or reference (input USA)
-outseq	seqoutall	[<sequence>.<format>] sequence set(s) filename and optional format (output USA)

Table 9.9 Standard (mandatory) qualifiers of einverted from EMBOSS

Parameters	Value	Description
-sequence	seqall	Nucleotide sequence(s) filename and optional format, or reference (input USA)
-gap	integer	[12] Gap penalty (integer 0 or more)
-threshold	integer	[50] Minimum score threshold (integer 0 or more)
-match	integer	[3] Match score (integer 0 or more)
-mismatch	integer	[-4] Mismatch score (integer up to 0)
-outfile	outfile	[*.einverted] Sanger Centre program inverted output file
-outseq	seqout	[<sequence>.<format>] the sequence of the inverted repeat regions without gap characters

Table 9.10 Standard (mandatory) qualifiers of transeq from EMBOSS

Parameters	Value	Description
-sequence	seqall	Nucleotide sequence(s) filename and optional format, or reference (input USA)
-outseq	seqoutall	[<sequence>.<format>] protein sequence set(s) filename and optional format (output USA)

(4) transeq. This command is under the category "TRANSLATION" and is used to translate nucleic acid sequences. The commonly used parameters for this command are listed in Table 9.10.
(5) distmat. This command is under the category "PHYLOGENY ANALYSIS" and is used to create a distance matrix from a multiple sequence alignment. The commonly used parameters for this command are shown in Table 9.11.

9.3.3 Extracting Transcript Sequences by gffread

The gffread software can be used to generate a FASTA file with the DNA sequences for all transcripts based on a GFF file. For this operation, a FASTA file for the genomic sequences has to be provided. For example, one might want to extract the sequence of all transfrags assembled from a Cufflinks assembly session. This can be accomplished with a command line like this:

Table 9.11 Standard (mandatory) qualifiers of transeq from EMBOSS

Parameters	Value	Description
[-sequence]	seqset	File containing a sequence alignment
-nucmethod	menu	[0] Multiple substitution correction methods for nucleotides. [Values: 0 (uncorrected); 1 (Jukes-Cantor); 2 (Kimura); 3 (Tamura); 4 (Tajima-Nei); 5 (Jin-Nei gamma)]
-protmethod	menu	[0] Multiple substitution correction methods for proteins. [Values: 0 (uncorrected); 1 (Jukes-Cantor); 2 (Kimura protein)]
-outfile	outfile	[*.distmat] PHYLIP distance matrix output file

```
>gffread -w transcripts.fa -g /path/genome.fa transcripts.gtf
```

The file genome.fa in this example would be a multi-FASTA file with the genomic sequences of the target genome. This also requires that every contig or chromosome name found in the first column of the input GFF file (transcript.gtf in this example) must have a corresponding sequence entry in the file chromosomes.fa. This should be the case in our example if genome.fa is the file corresponding to the same genome (index) that was used for mapping the reads with TopHat. Note that the retrieval of the transcript sequences in this way is much faster if a FASTA index file (genome.fa.fai in this example) is found in the same directory with the genomic FASTA file. Such an index file can be created with the SAMtools utility prior to running gffread, like this:

```
>samtools faidx genome.fa
```

Then in subsequent runs using the -g option, gffread will find this FASTA index and use it to speed up the extraction of transcript sequences.

The various usage options of gffread are:

```
gffread <input_gff> [-g <genomic_seqs_fasta> | <dir>][-s <seq_info.fsize>] [-o <outfile.gff>] [-t <tname>] [-r [[<strand>]<chr>:]<start>..<end> [-R]] [-CTVNJMKQAFGUBHZWTOLE] [-w <exons.fa>] [-x <cds.fa>] [-y <tr_cds.fa>] [-i <maxintron>]
```

9.3.4 CLUSTALW2

Clustal is a series of widely used computer programs for multiple sequence alignment (Larkin et al. 2007). The most commonly used parameters for CLUSTALW2 are described in Table 9.12.

9.3.5 Useful Tools from the FASTX-Toolkit

In addition to the tools described above, there are various "utility tools" that can be used to solve many handy problems. Below, we will describe a few of them from the FASTX-Toolkit (http://hannonlab.cshl.edu/fastx_toolkit/).

(1) FASTQ_to_fasta

This tool will convert sequences from FASTQ format to FASTA format. The basic usage is:

```
fastq_to_fasta [-r] [-n] [-z] [-i INFILE] [-o OUTFILE]
```

The commonly used parameters are described in Table 9.13.

Table 9.12 Most important parameters used in CLUSTALW2 commands

Parameters	Description
-INFILE = file.ext	Input sequences
-PROFILE1 = file.ext	Profiles (old alignment)
-PROFILE2 = file.ext	Profiles (old alignment)
-INTERACTIVE	Read command line, and then enter normal interactive menus
-QUICKTREE	Use FAST algorithm for the alignment guide tree
-TYPE=	Protein or DNA sequences
-NEGATIVE	Protein alignment with negative values in matrix
-OUTFILE=	Sequence alignment file name
-OUTPUT=	Clustal (default), GCG, GDE, PHYLIP, PIR, NEXUS and FASTA
-OUTORDER=	input or aligned

Table 9.13 Most important parameters used in fastq_to_fasta command

Parameters	Description
-r	Rename sequence identifiers to numbers
-n	Keep sequences with unknown (N) nucleotides. Default is to discard such sequences
-z	Compress output with GZIP
-i INFILE	FASTA/Q input file. Default is STDIN
-o OUTFILE	FASTA output file. Default is STDOUT

Table 9.14 Most important parameters used in fasta_formatter command

Parameters	Description
-i INFILE	FASTA/Q input file. default is STDIN
-o OUTFILE	FASTA/Q output file. default is STDOUT
-w N	Max. sequence line width for output FASTA file. When zero (the default), sequence lines will not be wrapped—all nucleotides of each sequence will appear on a single line (good for scripting)
-t	Output tabulated format (instead of FASTA format). Sequence identifiers will be on first column, and nucleotides will appear on second column (as single line)
-e	Output empty sequences (default is to discard them) Empty sequences are ones who have only a sequence identifier, but not actual nucleotides

(2) fasta_formatter

This tool can reformat the sequence file in FASTA format to various formats. The usage of the tool is "fasta_formatter [-i INFILE] [-o OUTFILE] [-w N] [-t] [-e]". The most commonly used parameters for this tool are described in Table 9.14.

Below, we will show you the results of several runs of the fasta_formatter command with various parameters.

Given an input sequence in the following format:

>MY-ID
AAAAAGGGGG
CCCCCTTTTT
AGCTN
The output with unlimited line width [-w 0] is as follows:
>MY-ID
AAAAAGGGGGCCCCCTTTTTAGCTN
The output with max. line width = 7 [-w 7] is as follows:
>MY-ID
AAAAAGG
GGGTTTT
TCCCCCA
GCTN
The output with tabular output [-t] is as follows:
MY-ID AAAAAGGGGGCCCCCTTTTTAG
CTN

This is particularly useful, as the file can be directly imported into Excel to do various analyses.

9.4 Software Tools for the Prediction of circRNA

Circular RNA (circRNA) is a type of single-stranded RNA which, unlike the better-known linear RNA, forms a covalently closed continuous loop; i.e., in circular RNA, the 3′ and 5′ ends normally present in an RNA molecule have been joined together (Huang et al. 2018). Below, we describe the usage of several software tools for the discovery of circRNAs as examples to identify novel molecules. Software tools useful for the analyses of other type of molecules, such as miRNA, lncRNA and etc, work in similar manners.

9.4.1 circRNA_finder

The software circRNA_finder (Westholm et al. 2014) can find circular RNAs (circRNAs) from RNA-Seq data, based on the alignment software STAR. It is used for the prediction of circRNA with very proximal splice sites (below 100 bp). It works de novo without knowledge of gene annotations and exon–intron structures. The software can be downloaded from https://github.com/orzechoj/circRNA_finder. The files contained in the software are: filterCirc.awk, filterSpliceSiteCircles.pl, nrForwardSplicedReads.pl, postProcessStarAlignment.pl, runStar.pl, starCirclesToBed.pl. These scripts have been tested on various Linux distributions. Before they can be run, make sure that the following prerequisites are installed: perl, awk, STAR (tested on version 2.6.0c), SAMtools. To run the scripts to identify circular RNAs, first run STAR, once for each data set. Assuming you are already in the directory containing the scripts, for paired-end data, the command is:

>./runStar.pl –inFile1 [R1 FASTQ] –inFile2 [R2 FASTQ] –genomeDir [path to STAR genome] –outPrefix [output directory and prefix]

And for single-end data, simply omit inFile2 to run the following command:

>./runStar.pl –inFile1 [R1 FASTQ] –genomeDir [path to STAR genome] –outPrefix [output directory and prefix]

Next, run the post-processing scripts. If there are STAR outputs for many data sets in the same folder, this command will process each of these in turn:

>./postProcessStarAlignment.pl –starDir [directory with STAR results] –minLen [minimum length of circular RNAs] –outDir [output directory]

For each library, the following output files are produced:

(1) _filteredJunctions.bed: A bed file with all circular junctions found by the pipeline. The score column indicates the number reads spanning each junction.
(2) _s_filteredJunctions.bed: A bed file with those junctions in (1) that are flanked by GT-AG splice sites. The score column indicates the number reads spanning each junction.
(3) _s_filteredJunctions_fw.bed: A bed file with the same circular junctions as in file (2), but here the score column gives the average number of forward-spliced reads at both splice sites around each circular junction.

9.4.2 CIRCexplorer2

CIRCexplorer2 (Zhang et al. 2016) is the successor of CIRCexplorer with plenty of new features to facilitate circular RNA identification and characterization. This software package has several features as it is able to: precisely annotate circular RNAs (annotate); support multiple circular RNA aligners (TopHat2/TopHat-Fusion, STAR, MapSplice, BWA, and segemehl) (align and parse); de novo assemble novel circular RNA transcripts (assemble); characterize various of alternative (back-)splicing events of circular RNAs (de novo); fast identify circular RNAs with STAR or BWA (parse); and support both single-read and paired-end sequencing. The software can be downloaded from https://github.com/gitter-badger/CIRCexplorer2. CIRCexplorer2 contains five modules and offers flexibility for multiple circular RNA analysis tasks. However, it would confuse many people who are not very familiar with CIRCexplorer2 and block people from making good use of it. As a result, the fast_circ.py script has been written to integrate different combinations of modules to complete different tasks.

Assuming you are in the right directory, typical usages of the three tools are listed below:

```
>./fast_circ.py parse -r REF -g GENOME -t ALIGNER [–pe] [-o OUT] <fusion>
>./fast_circ.py annotate -r REF -g GENOME -G GTF [-p THREAD] [-o OUT] -f FQ
>./fast_circ.py denovo -r REF -g GENOME -G GTF [-n PLUS_OUT] [-p THREAD] [-o OUT] -f FQ
```

If you have mapped RNA-Seq reads using one of the listed aligners (TopHat2/TopHat-Fusion, STAR, segemehl, and MapSplice), you should use fast_circ.py parse with gene annotation file (via -r) and reference genome sequence file (via -g). Meanwhile, you should also indicate its aligner (via -t) and whether reads are paired-end or not (via –pe). Last but not least, the fusion junction file (<fusion>) should be correct. This command is just like CIRCexplorer. If you only have raw RNA-Seq reads, you could use fast_circ.py annotate to align RNA-Seq reads with TopHat2/TopHat-Fusion. You need to provide gene annotation file (via -r), gene annotation GTF file (via -G), reference genome sequence file (via -g), and the raw read FASTQ file (via -f).

fast_circ.py de novo would align raw RNA-Seq reads with TopHat2/TopHat-Fusion, de novo assemble circular RNA transcripts with Cufflinks, and last extract alternative (back-)splicing events. Some options are the same as fast_circ.py annotate. If you provide a TopHat mapping directory for p(A)+ RNA-Seq (via -n), fast_circ.py de novo will fetch all the alternative splicing events. Otherwise, it only fetches alternative back-splicing events.

9.4.3 CIRI2

CIRI2 (Gao et al. 2018) is designed to differentiate back-spliced junction (BSJ) reads from non-BSJ reads using efficient maximum likelihood estimation (MLE) based on multiple seed matching. The software is optimized for the key steps in circRNA detection, including inferring original region for sequencing read segments and distinguishing BSJ reads from forward-spliced junction (FSJ) reads. The software is applicable to sequencing data with mixed read lengths and can be run with multiple threads. The software can be downloaded from https://sourceforge.net/projects/ciri/files/CIRI2/.

To use CIRI2, the first step is to prepare the data using the following commands:

```
>bwa index -a bwtsw ref.fa
>bwa mem –T 19 ref.fa reads.fq > aln-se.sam (single-end reads)
>bwa mem –T 19 ref.fa read1.fq read2.fq > aln-pe.sam (paired-end reads)
```

Then the following command can be issued to run the pipeline. > perl CIRI2.pl -I in.sam -O outfile -F ref.fa (-R ref_dir/) The sample results are presented in Table 9.15.

In summary, with the rapid development in bioinformatic tools, what challenges the users is not that there are no software tools, but frequently, there are so many tools and there are so many parameters. It is not only difficult to choose which software tools to use, but it is even harder to set the parameters. And there is no standard way to combine analysis results obtained from multiple analysis results. In our own experience, one can rank the software tools based on the impact factor of the journal it was published and numbers of citations of the tools. In terms of the parameters, using the settings described in the paper where the software tools were originally published is a good choice. For any bioinformatic analysis results, experimental validation is a must. As a general researcher studying *S. miltiorrhiza*, development of new algorithms and implementation of new software tools are generally not needed. Bioinformatic skills are like any other laboratory skills, and repeated practicing is the single most effective method to master various tools and techniques. With the abundant information on the Internet, any of your questions should have been encountered by other researchers already, and one should be able to find the solutions easily.

Table 9.15 Format of the output files for the CIRI2 program

Col	Description
1	circRNA_ID
2	Chromosome of a predicted circRNA
3	circRNA_start
4	circRNA_end
5	Circular junction read count of a predicted circRNA
6	Unique CIGAR types of a predicted circRNA. For example, a circRNAs have three junction reads: read A (80 M20S, 80S20 M), read B (80 M20S, 80S20 M), and read C (40 M60S, 40S30M30S, 70S30M), and then it has two SM types (80S20M, 70S30M), two MS types (80M20S, 70M30S), and one SMS type (40S30M30S). Thus, its SM_MS_SMS should be 2_2_1
7	Non-junction read count of a predicted circRNA that mapped across the circular junction but consistent with linear RNA instead of being back-spliced
8	Ratio of circular junction reads calculated by 2*#junction_reads/(2*#junction_reads + #non_junction_reads). #junction_reads is multiplied by two because a junction read is generated from two ends of circular junction but only counted once while a non-junction read is from one end. It has to be mentioned that the non-junction reads are still possibly from another larger circRNA, so the junction_reads_ratio based on it may be an inaccurate estimation of relative expression of the circRNA
9	Type of a circRNA according to positions of its two ends on chromosome (exon, intron, or intergenic_region; only available when annotation file is provided)
10	ID of the gene(s) where an exonic or intronic circRNA locates
11	Strand info of a predicted circRNAs (new in CIRI2)
12	All of the circular junction read IDs (split by ",")

Acknowledgements This work has been supported by Chinese Academy of Medical Sciences, Innovation Funds for Medical Sciences (CIFMS) (2016-I2M-3-016, 2017-I2M-1-013) and National Natural Science Foundation of China (81872966).

References

Altschul SF, Gish W, Miller W, Myers EW, Lipman JD (1990) Basic local alignment search tool. J Mol Biol 215:403–410

Gao Y, Zhang J, Zhao F (2018) Circular RNA identification based on multiple seed matching. Brief Bioinform 19:803–810

Huang X, Chen Y, Xiao J, Huang Z, Peng J (2018) Identification of differentially expressed circular RNAs during TGF-ß1-induced endothelial-to-mesenchymal transition in rat coronary artery endothelial cells. Anat J Cardiol 19:192

Langmead B, Salzberg SL (2012) Fast gapped-read alignment with Bowtie 2. Nat Meth 9:357–359

Langmead B, Trapnell C, Pop M, Salzberg SL (2009) Ultrafast and memory-efficient alignment of short DNA sequences to the human genome. Genome Biol 10:R25

Larkin MA, Blackshields G, Brown NP, Chenna R, McGettigan PA, McWilliam H, Valentin F, Wallace IM, Wilm A, Lopez R (2007) Clustal W and Clustal X version 2.0. Bioinformatics 23:2947–2948

Li H, Handsaker B, Wysoker A, Fennell T, Ruan J, Homer N, Marth G, Abecasis G, Durbin R, 1000 Genome Project Data Processing Subgroup (2009) The sequence Alignment/Map format and SAMtools. Bioinformatics 25:2078–2079

Rice P, Longden I, Bleasby A (2000) EMBOSS: the European molecular biology open software suite. Trends Genet 16:276–277

Westholm JO, Miura P, Olson S, Shenker S, Joseph B, Sanfilippo P, Celniker SE, Graveley BR, Lai EC (2014) Genome-wide analysis of drosophila circular RNAs reveals their structural and sequence properties and age-dependent neural accumulation. Cell Rep 9:1966–1980

Zhang XO, Dong R, Zhang Y, Zhang JL, Luo Z, Zhang J, Chen LL, Yang L (2016) Diverse alternative back-splicing and alternative splicing landscape of circular RNAs. Genome Res 26:1277–1287

Biosynthetic Pathway of Tanshinones in *Salvia miltiorrhiza*

10

Juan Guo and Ying Ma

Abstract

Tanshinones are one of the active compounds of Danshen. They are a group of abietane-type norditerpenoid quinone natural products. Biosynthetic pathway of tanshinones in *Salvia miltiorrhiza* includes formation of the terpenoids common precursors, structural formation of terpenoid skeleton, and post-structural modification. With the development of omics analysis, the biopathway analysis of tanshinone has been deeply investigated. The upstream pathway genes of biosynthesis of tanshinone are conserved in plant and have been cloned. By comparative omics analysis from different tissues, different development stages, or different treatments, two terpenoid synthases and three P450s specific for biosynthesis of tanshinones have been cloned from *S. miltiorrhiza* and functional characterized both in vivo and in vitro. Further, metabolic engineering and synthetic biology utilization of the biosynthesis pathway genes were illustrated in this section. However, the biosynthetic pathway of tanshinones from the intermediate to the end products such as tanshinone IIA still needs further investigation. The functionally known genes and metabolic engineering yeast strains for production of intermediate provide foundation for further biopathway elucidation.

10.1 Introduction

The chemical constituents of Danshen have been studied since the 1930s. More than 200 chemical constituents have been identified (Wang et al. 2017; Mei et al. 2019). It includes hydrophilic phenolic acids and lipophilic diterpenoid tanshinons, which are used to evaluate the quality of medicinal materials (Chinese Pharmacopoeia 2015; Wang 2010). Modern pharmacological studies indicated that both components contribute to Danshen's pharmacological and therapeutic effects. Phenolic acid compounds are distributed in both the aboveground and underground parts of *S. miltiorrhiza*, while tanshinones mainly accumulate in the roots and rhizomes. More than 40 structurally diverse tanshinones have been isolated and identified (Zhang et al. 2012). Among them, cryptotanshinone, tanshinone IIA, and tanshinone I are the main active ingredients (Zhong et al. 2009). These compounds have been extensively studied for their well-known cardiovascular activities, as well as

J. Guo (✉) · Y. Ma
State Key Laboratory Breeding Base of Dao-Di Herbs, National Resource Center for Chinese Materia Medica, China Academy of Chinese Medical Sciences, 100700 Beijing, China
e-mail: guojuanzy@163.com

Y. Ma
e-mail: xiaoma1110@126.com

S. Lu (ed.), *The Salvia miltiorrhiza Genome*, Compendium of Plant Genomes,
https://doi.org/10.1007/978-3-030-24716-4_10

for their anticancer activities in vitro and in vivo (Zhang et al. 2012). It has been shown that tanshinone IIA functions as antioxidant protection of cardiomyocytes and has been reported to have potential activity for treating human inflammation (Robertson et al. 2014), as well as anti-adipogenic effects on 3T3-L1 cells and in zebrafish (Park et al. 2017).

S. miltiorrhiza Bunge is a commonly used traditional Chinese medicine. Before the 1970s, raw materials of *S. miltiorrhiza* were mainly from wild resources; however, wild resources had become decreasing with the increasing demand. Nowadays, cultivation has become the main resource of *S. miltiorrhiza*. The main productive areas of *S. miltiorrhiza* are Shandong, Henan, Shanxi, Sichuan, in China. Through years of cultivation and breeding, some cultivars with regional adaptability have been selected. In comparison with wild resources, cultivated *S. miltiorrhiza* under good agricultural practice (GAP) cultivation is relatively stable in quality to provide uniformly raw materials for pharmaceutical enterprises, but there also have disadvantages including pest diseases and continuous cropping obstacles.

Studies on the pharmacological effects, chemical synthesis, and compound acquisition of tanshinones have attracted much attention due to its various biological activities. There are some ways to obtain tanshinones (Zhang et al. 2012). Firstly, tanshinone compounds can be obtained by chemical total synthesis (Yang et al. 2001). But the high cost, low yield, and environmental pollution limited the commercial application by chemical synthesis of tanshinones. Secondly, tanshinones could be isolated from the original plant or tissue culture of *S. miltiorrhiza* (Dreger et al. 2010; Shimomura et al. 1991; Wang and Wu 2010). Since tanshinones mainly accumulated in the roots and rhizomes of *S. miltiorrhiza*, the content of tanshinones could be increased by elicited with exogenous inducer in the hairy roots of *S. miltiorrhiza* (Kai et al. 2012). Hairy roots are hormone autotrophy, with genetically operable, rapid growth, stable heredity and relatively simple culture and so on. Therefore, hairy roots displayed the ability to produce secondary metabolites under the culture condition in vitro. Furthermore, the active ingredient content in some hairy roots could be increased by metabolic engineering or by improving the culture conditions and medium composition. However, it is still difficult to produce tanshinones on a large scale by hairy root culture. There are also reports of tanshinones production by suspension cell culture of *S. miltiorrhiza*. However, in suspension cells, the upstream compounds of tanshinone biosynthesis accumulate significantly in suspension cells, while the downstream active compounds such as tanshinone IIA are relatively low (Miyasaka et al. 1985). At present, there are no mature hairy root, suspension cell line, or culture system of *S. miltiorrhiza* for production of tanshinones. Therefore, the production of tanshinone compounds by suspension cells or hairy roots is still in the laboratory stage. However, these tissue cultures have provided stable and controllable materials for scientific research of active ingredient biosynthesis and accumulation regulation in *S. miltiorrhiza*. Therefore, the extraction of tanshinone compounds from the roots and rhizomes of the original plants of *S. miltiorrhiza* is still the main source of tanshinones.

With the development of new drugs and the industrialization of traditional Chinese medicine, the demand for active component monomer increases, forcing us to use modern scientific and technology to continuously improve the yield of tanshinones, to meet the increasing clinical and market drug demand. Analyzing the biosynthetic pathway of tanshinones, and reconstructing the biosynthesis pathway in microorganisms by synthetic biology can provide new options for raw material supply of tanshinones.

10.2 Tanshinone Biosynthetic Pathway Analysis

Tanshinones are secondary metabolites of *S. miltiorrhiza*. The secondary metabolites of plants are derived from primary metabolites. The primary metabolism provides energy and basic stacking modules by photosynthesis, glycolysis, and tricarboxylic acid cycle. Then these modules are catalyzed by various enzymes to form the common skeleton structure. These steps are supposed to be conserved in secondary metabolism pathway in plants. Tanshinones are a kind of diterpenoids. Which precursor skeleton is geranylgeranyl pyrophosphate (GGPP), GGPP is then modified with cyclization, hydroxylation, and oxidation by specific catalytic enzymes of *S. miltiorrhiza* to form tanshinones.

The biosynthetic pathway of tanshinone can be divided into two parts (Fig. 10.1). First is the formation of terpenoid precursors. The upstream synthesis pathway of terpenoids is conservative in plants, starting from either acetyl coenzyme A via the mevalonate pathway (MVA pathway) or from pyruvic acid and glyceraldehyde-3-phosphate via the deoxyxylose-5-phosphoric acid in Methyl erythritol-4-phosphate pathway (MEP pathway) to produce the common precursor of all terpenoids, iso-prentenyl diphosphate (IPP), and its isomer dimethylallyl pyrophosphate (DMAPP). Currently, studies conducted in different species showed relatively clear understanding of the catalytic enzymes and rate-limiting enzymes in the upstream pathway. Therefore, the upstream enzymes could be cloned from *S. miltiorrhiza* by homologous cloning. Second is the structural formation and modification of carbon skeleton of tanshinones. Tanshinones are special metabolites in *S. miltiorrhiza* catalyzed by the specific enzymes of *S. miltiorrhiza*. Most of these enzymes share low homology with those of other species, so molecular biological methods are needed for further exploration and analysis.

10.2.1 Candidate Gene Selection

Candidate genes could be selected by comparative multiomics analysis. The plant genome is complex, and the functional genes involved in plant secondary metabolism often exist as gene families in plants. Screening out effective candidate genes from gene families is a prerequisite for functional research. Systematic integration and application of genomics technologies, including metabolome, transcriptome, proteome, and genome, provide technical support for candidate genes screening. (1) Tanshinones mainly accumulate in roots and rhizomes of *S. miltiorrhiza*, especially in the cortex part of roots and rhizomes. Therefore, by constructing comparative transcriptome and metabolome data at different tissue or different developmental stages, it could effectively narrow down the screening range and obtain candidate genes for further functional characterization. Such as, Yang et al. (2013) obtained 64,139 unigenes by conducting transcriptome sequencing on roots and leaves of *S. miltiorrhiza*, respectively, and predicted functional genes of the downstream tanshinone biosynthesis pathway by tissue-specific gene expression analysis. Xu et al. (2016) reported the draft sequence and analysis of the *S. miltiorrhiza* genome using high-throughput sequencing. Combined with tissue-specific expression of the genes using RNA-Seq data was employed to investigate the candidate genes involving in biosynthesis of tanshinones. In this way, it was found that 32 cytochrome P450s (P450s) exhibited similar expression pattern to the upstream genes. This provides candidates for further biopathway analysis of tanshinones. (2) The secondary metabolites of plants are produced by the interaction between plants and the environment during long-term evolution process. They have the functions of resistance to stress and pests and could be accumulated under the induction of elicitors. The hairy root system of *S. miltiorrhiza* is easily controllable and operable, which provides ideal research materials for candidate genes selection of the tanshinone biosynthesis pathway. Elicitor can induce accumulation of secondary metabolites in *S. miltiorrhiza* hairy roots, while different elicitors show different effects on different types of secondary metabolites (Dreger et al. 2010). Therefore, candidate functional genes can be obtained by screening a suitable elicitor against *S. miltiorrhiza* hairy roots using comparative omics analysis. Based on this, Cui et al. (2011) constructed a DNA chip using hairy root after elicited and obtained 114 differentially expressed unigenes from 4354 clones. It was found that 52 of which were firstly reported

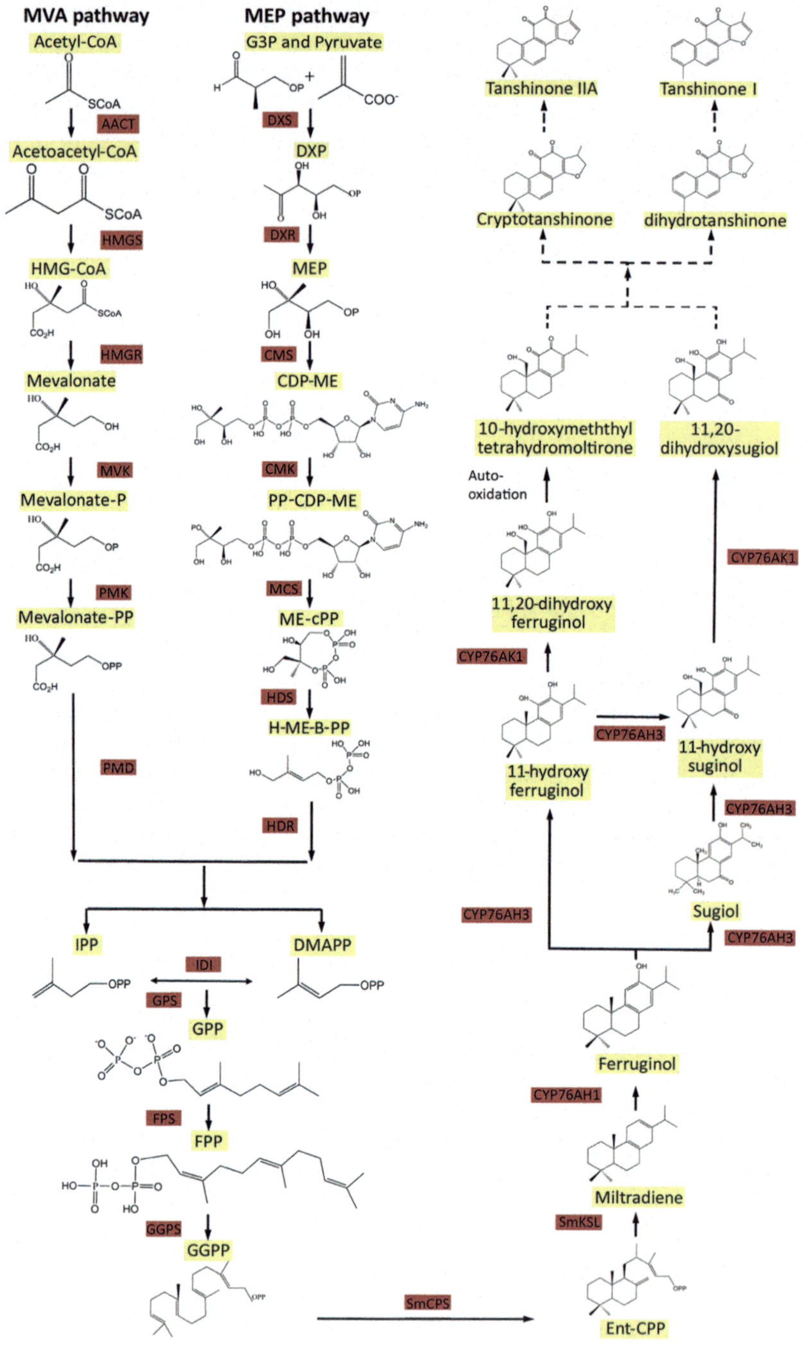

Fig. 10.1 Proposed biosynthetic pathway of tanshinones. Catalyzing enzymes are showed in brown box. Compounds or intermediates involving in biopathway are showed in yellow box

in *S. miltiorrhiza*. Gao et al. (2014) used a combination of yeast extracts and silver ions to induce hairy roots of *S. miltiorrhiza* to study the molecular mechanism of the formation of active tanshinone components. By combining with metabolome, a comprehensive comparative transcription dataset of hairy root of *S. miltiorrhiza* was obtained. This resulted in 6358 differentially expressed genes. In addition, eight candidate P450 genes involved in tanshinone downstream biosynthesis were selected. These studies have greatly accelerated the research process of tanshinone biosynthesis pathway.

10.2.2 Functional Characterization of Candidate Genes in Vivo and in Vitro

It is important to functionally characterize and analyze the function involving in tanshinone biosynthesis pathway after obtaining candidate genes. The research methods mainly include functional studies in vitro and in vivo. (1) In vitro analysis includes heterologous expression and enzyme-catalyzed reaction. Currently, relatively mature heterologous expression systems include *Escherichia coli*, yeast, insect cells, and tobacco transient expression, etc. First, by analyzing the characteristics of the enzyme, an appropriate expression system was selected to perform protein expression and purification. Then substrates were added into a suitable reaction system to conduct enzyme-catalyzed reaction in vitro and then conducted for product detection. This method is suitable for the functional study of specific substrates. (2) Functional characterization of candidate genes in vivo includes overexpression, knockdown (e.g., RNA interference), or knockout (e.g., CRISPR/Cas9) of the candidate genes in *S. miltiorrhiza*. Analysis of the changes of metabolites which regulated by the gene expression abundance helps to confirm their biological functions in planta. The genetic transformation system of *S. miltiorrhiza* is available, which provides technical support for in vivo functional studies of candidate genes. Integrated analysis by in vitro and in vivo could accelerate biopathway analysis of tanshinones.

10.3 Advances in Biosynthetic Pathway Analysis of Tanshinone

The biosynthetic pathway of terpenoids can be divided into three processes (Fig. 10.1). The formation of terpenoid includes formation of common precursors isopentenyl pyrophosphate (IPP) and dimethylallyl pyrophosphate (DMAPP), the structural formation of terpenoid skeleton, and the post-structural modification processes including oxidation, glycosylation, methylation, decarboxylation, acetylation, and cyclization, etc. In biosynthesis of tanshinones, the post-modification processes is oxidation, which is mostly catalyzed by P450s.

10.3.1 Formation of Terpenoid Common Precursors

IPP and DMAPP are common precursors of all terpenoids. In higher plants, IPP and DMAPP are produced by the MVA pathway in the cytoplasm or the MEP pathway in the plastid. It is generally believed that the MEP pathway produces the precursors of monoterpenes, diterpenes, carotenoids, chlorophyll, and side chains of plastid quinones, while the MVA pathway mainly produces the precursors of sesquiterpenes, triterpenes, and steroids.

MEP pathway was discovered successively from bacteria and plants through isotope labeling experiments in the 1990s (Rohmer et al. 1993). In the MEP pathway, glyceraldehyde-3-phosphate (G3P) and pyruvate produced from primary metabolic pathway were used as substrates and could be condensed to form D-xylulose-5-phosphate (DXP) under the catalysis of 1-deoxy-D-xylulose-5-phosphate synthase (DXS). DXP was then catalyzed by 1-deoxy-D-xylulose-5-phosphate reductoisomerase (DXR) to generate 2-C-Methyl-D-erythritol 4-phosphate (MEP). MEP is catalyzed by 4-(Cytidine 5'-diphospho)-2-C-methyl-D-erythrito synthase (CMS), 4-(Cytidine-5'-diphospho)-2-C-methyl-D-erythritol kinase (CMK), 2-C-Methyl-D-erythritol 2,4-cyclodiphosphate synthase (MCS), 4-Hydroxy-3-methylbut-2-enyl diphosphate synthase (HDS), and 4-hydroxy-3-methylbut-2-enyl diphosphate reductase (HDR) to form IPP. MVA pathway has been deeply investigated. First, three acetyl-CoA molecules were condensed into 3-Hydroxy-3-methylglutaryl CoA (HMG-CoA) by acetoacetyl-CoA transferase (AACT) and hydroxymethylglutaryl-CoA synthase (HMGS). HMG-CoA is catalyzed by 3-Hydroxy-3-methylglutaryl CoA reductase (HMGR) to form mevalonate (MVA). MVA is catalyzed by mevalonate kinase (MVK), phosphomevalonate kinase (PMK), and diphosphomevalonate decarboxylase (PMD) to generate IPP. IPP generated by MEP pathway and MVA pathway was catalyzed by IDI to form its isomerization product DMAPP.

The MEP and MVA pathway are conserved in plants and have been thoroughly investigated. By transcriptome sequencing, gene annotation, and homologous gene cloning, all the genes involved in upstream biopathway of tanshinones have been cloned from *S. miltiorrhiza* (Ma et al. 2012). Generally, it was believed that the diterpenoid precursors were mainly derived from the MEP pathway; however, reports also indicated that both MEP and MVA pathway contributed to the biosynthesis of tanshinone precursors in *S. miltiorrhiza* (Laule et al. 2003; Yang et al. 2012). Genome sequence of *S. miltiorrhiza* showed that the gene copies of the upstream enzymes ranged from 1 to 6 (Ma et al. 2012). DXS is considered as the rate-limiting step involving in the MEP pathway. There are five *DXS* genes in *S. miltiorrhiza*. Tissue expression analysis showed that the expression of *SmDXS2* was higher than other *DXS* genes in roots. Overexpression of *SmDXS2* gene in *S. miltiorrhiza* hairy roots can significantly increase the accumulation of tanshinone, indicating that SmDXS2 plays an important role in the biosynthesis pathway of tanshinone (Kai et al. 2011). HMGR is considered as the rate-limiting step involving in the MVA pathway (Bach et al. 1991). There are four *HMGR* genes in *S. miltiorrhiza* genome. Overexpression of *SmHMGR1* and *SmHMGR2* in *S. miltiorrhiza* hairy roots can significantly increase the accumulation of tanshinone, indicating that the two *HMGR* genes have certain effect on improving tanshinone accumulation. It further demonstrates that the MVA pathway also plays a role in the biosynthesis of tanshinone (Dai et al. 2011; Kai et al. 2011). Kai et al. (2011) found that the accumulation of tanshinone in the hairy root of *S. miltiorrhiza* could significantly increase by overexpressing *DXS*, *HMGR*, and *GGPPS* in *S. miltiorrhiza* hairy roots, respectively. Specifically, the accumulation of tanshinone in *S. miltiorrhiza* hairy roots by overexpression of *DXS* was higher than that by overexpression of *HMGR*.

10.3.2 Formation of Terpenoid Carbon Skeleton

The second stage of terpenoid biosynthesis is the condensation of DMAPP with one, or two, or three molecules of IPP to form precursors of C10 monoterpenoid geranyl pyrophosphate (GPP), C15 sesquiterpene precursor farnesyl diphosphate (FPP), and C20 diterpenoid precursor geranyl geranyl pyrophosphate (GGPP), respectively. The condensation processes are catalyzed by prenyltransferases. These involve GPP synthase (GPS), FPP synthase (FPS), and GGPP synthase (GGPPS). Then two FPPs are catalyzed by squalene synthase (SQS) to produce the precursor of triterpenoids and steroid compounds. Two GGPPs are catalyzed by phytoene synthase (PSY) to produce phytoene as the precursor of carotenoids. In the biosynthesis of tanshinone, GPS, FPS, and GGPPS catalyze IPP and DMAPP to generate GGPP. In plants, these prenyltransferases are usually present as dimers. In the genome of *S. miltiorrhiza*, there is one copy of *FPPS*. GPPS has four subunits. GGPPS has three homologous genes. Kai et al. (2011) found that overexpression of *GGPPS* in hairy roots of *S. miltiorrhiza* could significantly increase the accumulation of tanshinone. Further, comparative analysis showed that the effect of accumulation of tanshinones was more obviously in hairy root which overexpression of *GGPPS* than overexpression of *DXS* or *HMGR*. This indicates that GGPPS plays an important role in diterpenoid biosynthesis (Laule et al. 2003).

10.3.3 Post-structural Modification of Tanshinone Carbon Skeleton

Structural post-modification in the terpenoid precursors is important to produce various active compounds and structural diversified compounds. Carbon skeleton structures are firstly formed by cyclization and dephosphorylation of GGPP by terpenoid synthase. Most of terpenoids contain special group modifications and even skeleton structure rearrangement. These post-modification includes hydroxylation, epoxidation, methylation, acetylation, glycosylation, and isomerization, and so on etc.

Biosynthesis of tanshinone is firstly catalyzed by terpenoid synthase, which catalyzes GGPP to form carbon skeleton miltiradiene. The terpenoid synthase gene is a medium-sized gene family in plants. There are totally 82 terpene synthase genes annotated in the genome of *S. miltiorrhiza,* which involved in the biosynthesis of monoterpenes, sesquiterpenes, diterpenoids, and triterpenoids (Ma et al. 2012; Xu et al. 2016). Tanshinone belongs to diterpenoids. In angiosperms, most of the cyclic diterpenoids are formed from diterpenoid substrates (E,E,E)-geranylgeranyl pyrophosphate (GGPP), catalyzed by Class II CPS and class I KSL, which are generally considered to be evolved from the diterpene synthase gene for the synthesis of gibberellin. In the genome of *S. miltiorrhiza*, there are nine genes annotated as copalyl diphosphate synthase (CPS) or its fragments, and three genes annotated as *ent*-kaurene synthase or its fragments (Xu et al. 2016). By analyzing the cDNA microarray of the induced hairy root of *S. miltiorrhiza* with those of controls, two terpenoids synthases, including *S. miltiorrhiza* copalyl diphosphate synthase (SmCPS1) and *S. miltiorrhiza* ent-kaurene synthase (SmKSL1), were detected to be upregulated by Ag^+ and yeast extracts. Functional studies found that these two enzymes could sequentially catalyze GGPP to produce *ent*-CPP and its dephosphorylated product. The new product was first reported and found in *S. miltiorrhiza* and was named miltiradiene (Cui et al. 2011; Gao et al. 2009). Guo et al. (2013) fed the hairy roots of *S. miltiorrhiza* with ^{13}C-labeled miltiradiene. After induction culture, ^{13}C-labeled cryptotanshinone was detected in hairy roots. Therefore, it proved that miltiradiene was the precursor of tanshinone compounds. RNA interference analysis showed that after knocking down the expression of *SmCPS1*, the content of tanshinones was significantly reduced. This further confirmed the role of SmCPS1 in the biosynthesis of tanshinones (Cheng et al. 2014). Furthermore, Cui et al. (2015) systematically studied five *CPS* and two *KSL* genes in *S. miltiorrhiza* genome and found that only *CPS1*, *CPS2,* and *KSL1* in these seven genes were involved in the biosynthesis of tanshinone. Among them, *CPS1* and *KSL1* were specifically expressed in cork layer, while other genes were almost not expressed in cork layer. RNAi analysis of *CPS1* in *S. miltiorrhiza* plants showed that inhibiting the expression of *CPS1* had no significant effect on the expression of other *CPS genes*, and there was no significant change in the growth of transgenic plants, but the root color turned white (Cui et al. 2015). Results showed that the content of tanshinone compounds was significantly decreased, which further confirmed that this gene is the only class II terpenoid synthase gene controlling tanshinone biosynthesis in the cork layer of *S. miltiorrhiza* roots. Comparing with *KSL2*, *KSL1* was specifically expressed in cork layer. But different from *CPS1*, higher expression of *KSL1* could also be detected in the aerial part. Further studies showed that co-expression of *CPS2* and *KSL1* could produce miltiradiene. Because *CPS2* is mainly expressed in the aerial part of *S. miltiorrhiza*, and trace tanshinone compounds can also be detected in the aerial part of *S. miltiorrhiza*. Therefore, CPS2 is speculated to be also involved in the tanshinone synthesis in the aerial part of *S. miltiorrhiza*. However, it needs to be further clarified whether tanshinone of the aerial part is dominated by CPS2 or by CPS1 which is also micro-expressed in the aerial part.

As the carbon skeleton structure of tanshinone compounds, miltiradiene does not have any biological activity. Further, structural modification is required to form the active tanshinone compounds. P450 plays an important role in the structural modification of terpenoids. It was reported that more than 97% of terpenoids required oxidative modification. Most of the oxidation are catalyzed by P450 (Hamberger and Bak 2013). Cytochrome P450 is the largest family of enzyme genes in plants, accounting for about 1% of plant-encoded proteins, and plays an important role in the primary and secondary metabolism of plants (Mizutani and Ohta 2010). Moreover, P450 catalysis has substrate specificity or site-specificity, which increases the difficulty in function screening and prediction of P450. There are total 437 P450 genes in *S. miltiorrhiza* genome, so it is important to narrow the range of candidate genes reasonably for functional studies (Xu et al. 2016). Gao et al. (2014) found the significant accumulation of tanshinone and the significantly upregulated expression of *CPS* and *KSL*, by inducing the hairy roots of *S. miltiorrhiza* by yeast extract and Ag^{+}. Based on this, 39 upregulated expression P450 genes were screened from the P450 genes (Gao et al. 2014). Based on the characteristics of the tanshinone accumulation in the *S. miltiorrhiza* roots, Guo et al. (2013) further analyzed the tissue expression of these 39 P450 genes to narrow the candidates to 14. Protein expression and enzymatic analysis in vitro of these 14 genes were conducted by using the yeast expression system, and the results showed that CYP76AH1 could catalyze the hydroxylation and aromatization of the carbon skeleton precursors miltiradiene to generate ferruginol (Guo et al. 2013). RNA interference of *CYP76AH1* in *S. miltiorrhiza* hairy roots showed that inhibiting the expression of this gene resulted in the accumulation of miltiradiene, while the contents of ferruginol and tanshinone compounds were significantly decreased. This proved the function of the gene in *S. miltiorrhiza* (Ma et al. 2016). Guo et al. (2016) further screened two genes involved in tanshinone structure modification from these 14 genes. CYP76AH3 could catalyze both hydroxylation of C11 and oxidation of C6 on the basis of CYP76AH1 to form 11-hydroxy ferruginol and sugiol and could further catalyze the formation of 11-hydroxytaxol using these two products as substrates. The diversity of substrate types and catalytic sites indicated that the gene is promiscuity. Based on these C11 and C12 hydroxylation products, CYP76AK1 further catalyzed the hydroxylation of C20 to form intermediate metabolites. The intermediate metabolite was unstable and spontaneously oxidized into products of C-ring ketonization products (Guo et al. 2016). These studies have greatly promoted the biosynthesis pathway analysis of tanshinone. Further, structural modification is required from C-ring ketonization products to tanshinone IIA and so on. According to the structure of tanshinones, P450s still play important role in the next processes. Screening and functional characterization of P450 genes are still the focuses and difficulties in the elucidation of the tanshinone biosynthesis pathway.

10.4 Metabolic Engineering and Synthetic Biology of Tanshinone Biosynthetic Pathway

Elucidation of the biosynthetic pathways of active ingredients in medicinal plants not only provides targets for metabolic engineering improvement of medicinal plants, but also provides biological elements for heterologous production of medicinal active compounds using synthetic biology. With biosynthetic pathway analysis of tanshinone, metabolic engineering and synthetic biology were employed to improve production of tanshinones or produce tanshinones in microorganisms.

10.4.1 Metabolic Engineering Improvement

HMGR, *DXS*, and *GGPPS* are the rate-limiting enzyme genes in the upstream of the terpene biosynthetic pathways, overexpression, and co-expression of these genes in *S. miltiorrhiza* hairy roots are efficacious ways to increase the accumulation of tanshinones. Results showed that the tanshinone content in hairy roots with the overexpression of *HMGR* was 1.78–3.19-folds higher than that of the control line, and the highest content of tanshinones in the transgenic line was 1.007–2.32 mg/g dw, while the tanshinone content in *GGPPS* overexpression hairy roots was 4.74-folds higher than that in control line. It was also found that compared to *HMGR*, *DXS* is a better target in the overexpression experiment on the accumulation of tanshinones, while the overexpression of *GGPPS* had a more obvious effect on the accumulation of tanshinone compared to upstream genes (Kai et al. 2011). The highest content of tanshinones reached 2.727 mg/g dw when *HMGR* and *GGPPS* were co-expressed in the hairy roots of *S. miltiorrhiza*, which was 4.74-folds of the control group. *GGPPS* and *DXS II* were cotransformed into the hairy roots of *S. miltiorrhiza*. The content of tanshinones in the transgenic *S. miltiorrhiza* hairy roots was up to 12.93 mg/g dw, which was the highest content reported (Shi et al. 2016). However, current studies on improving the quality of *S. miltiorrhiza* by metabolic engineering modification are limited in the upstream biosynthetic pathway of tanshinones and *S. miltiorrhiza* hairy roots. With the screening of the tanshinones biosynthetic pathway, more effective targets will be operated in the downstream of tanshinone biosynthetic pathway. Regulations in modification of tanshinone pathway and metabolic flux in *S. miltiorrhiza* will be intensively studied in the future.

10.4.2 Synthetic Biology of Tanshinones

Lots of trace metabolic intermediates with potential biological activities were discovered in the process of excavation of new functional genes in the tanshinone biosynthetic pathway. These intermediates are not only medicinal active ingredients, but also important substrates for downstream pathways analysis. However, because of extremely low content in the original plant of *S. miltiorrhiza*, it is difficult to separate and purify. Reconstruction of the biosynthetic pathways of these metabolic intermediates in microorganisms by synthetic biology provides a new strategy for more efficient acquisition.

Miltiradiene is the first important intermediate in the biosynthetic pathway of tanshinones, which is produced from GGPP under the catalysis of two diterpenoid syntheses CPS and KSL. Miltiradiene could be directly produced by transferring *CPS1* and *KSL1* from *S. miltiorrhiza* into *Saccharomyces cerevisiae*, although the yield was very low. Zhou et al. (2012) proved the interaction between CPS1 and KSL1 by immunoprecipitation, and then the fusion protein expression vectors were constructed by fusing the two genes in different ways and were introduced into yeast to produce miltiradiene. Three rate-limiting enzymes genes *FPPS*, *GGPPS*, and *HMGR* in the terpenoids upstream pathway were then overexpressed to improve the supply of precursors which significantly increased the production of miltiradiene in yeast engineering bacteria. These genes were designed and fusion optimized with modular pathway engineering strategy. The optimized functional modules were fermented in autotrophic yeast, auxotrophic yeast, haploid yeast, and diploid yeast, respectively. The yield of miltiradiene in the final optimal engineered yeast was up to 365 mg/L (Zhou et al. 2012).

Dai et al. (2012) increased the production of the precursor compound FPP by overexpressing the upstream *HMGR* gene and *upc2.1* in the engineering yeast transferred with *CPS1* and *KSL1*. Although the yield of miltiradiene was not significant improved, the content of the triterpenoid precursor squalene was increased greatly. Considering that most of the metabolic streams flow to the synthesis of triterpene precursors, fusion protein of FPPS and GGPPS were constructed, *GPPS* was overexpressed, and the auxotrophic marker was replaced with the antibiotic marker. Finally, the yield of miltiradiene was 61.8 mg/L in the shake flask, and the final fermentation yield reached 488 mg/L (Dai et al. 2012). These two strains provide sufficient substrates for the analysis of tanshinone biosynthetic pathway and provide solid foundation for the construction of downstream engineering yeast. On these bases, Guo et al. (2013) further transferred *CYP76AH1*, a miltiradiene modified P450 gene, into the miltiradiene engineering yeast, but the target product ferruginol was not detected. Considering that P450 reductase was required to provide electrons for P450s functions, two P450 reductase genes were then cloned from *S. miltiorrhiza* and transferred into engineering yeast. The results showed that a large amount of ferruginol accumulated in the engineering yeast transferred with *SmCPR1* (Guo et al. 2013). Based on these results, the expression modules of CYP76AH3 and CYP76AK1 were further transferred into the engineering yeast and the newly introduced expression modules could successfully convert the products of the previous engineering yeast into 11,20-dihydroxy ferruginol and its ketonization product 10-hydroxymethyl tetrahydromiltirone, and a small amount of 11,20-dihydroxy sugiol (Guo et al. 2016).

At present, the genetical tanshinone biosynthesis engineering strain has been transferred with three rate-limiting enzyme genes from the upstream of terpenoid synthesis, as well as five structural modification enzyme and one coenzyme from *S. miltiorrhiza*. The construction of tanshinone biosynthesis intermediate metabolites engineering yeast provides sufficient substrates and foundation strains for the research of tanshinone biosynthetic pathway and will greatly accelerate the study of tanshinone biosynthesis.

Acknowledgements This work was supported by grants from the National Natural Science Foundation of China (81573532, 81822046), Key project at central government level: The ability to establish sustainable use of valuable Chinese medicine resources (2060302), Hunan Province Universities 2011 Collaborative Innovation Center for Protection and Utilization of Hu-xiang Chinese Medicine Resources.

References

Bach TJ, Wettstein A, Boronat A, Ferrer A, Enjuto M, Gruissemm W, Narita JO (1991) Properties and molecular cloning of plant HMG-CoA reductase. In: Patterson GW, Nes WD (eds) Physiology and biochemistry of sterols. American Oil Chemists Society, Champaign, IL, USA, pp 29–49

Cheng Q, Su P, Hu Y, He Y, Gao W, Huang L (2014) RNA interference-mediated repression of *SmCPS* (*copalyldiphosphate synthase*) expression in hairy roots of *Salvia miltiorrhiza* causes a decrease of tanshinones and sheds light on the functional role of SmCPS. Biotechnol Lett 36(2):363–369

Cui G, Duan L, Jin B, Qian J, Xue Z, Shen G, Snyder JH, Song J, Chen S, Huang L, Peters RJ, Qi X (2015) Functional divergence of diterpene syntheses in the medicinal plant *Salvia miltiorrhiza*. Plant Physiol 169 (3):1607–1618

Cui G, Huang L, Tang X, Zhao J (2011) Candidate genes involved in tanshinone biosynthesis in hairy roots of *Salvia miltiorrhiza* revealed by cDNA microarray. Mol Biol Rep 38(4):2471–2478

Dai Z, Cui G, Zhou S-F, Zhang X, Huang L (2011) Cloning and characterization of a novel 3-hydroxy-3-methy lglutaryl coenzyme A reductase gene from Salvia miltiorrhiza involved in diterpenoid tanshinone accumulation. J Plant Physiol 168(2):148–157

Dai Z, Liu Y, Huang L, Zhang X (2012) Production of miltiradiene by metabolically engineered *Saccharomyces cerevisiae*. Biotechnol Bioeng 109(11):2845–2853

Dreger M, Krajewska-Patan A, Gorska-Paukszta M, Pieszak M, Buchwald W, Mikolajczak P (2010) Production of the secondary metabolites in *Salvia miltiorrhiza* in vitro cultures. Herba Polonica 56:78–90

Gao W, Hillwig ML, Huang L, Cui G, Wang X, Kong J, Yang B, Peters RJ (2009) A functional genomics approach to tanshinone biosynthesis provides stereochemical insights. Org Lett 11(22):5170–5173

Gao W, Sun H-X, Xiao H, Cui G, Hillwig M, Jackson A, Wang X, Shen Y, Zhao N, Zhang L, Wang X-J, Peters R, Huang L (2014) Combining metabolomics and transcriptomics to characterize tanshinone biosynthesis in *Salvia miltiorrhiza*. BMC Genom 15:73

Guo J, Ma X, Cai Y, Ma Y, Zhan Z, Zhou YJ, Liu W, Guan M, Yang J, Cui G, Kang L, Yang L, Shen Y, Tang J, Lin H, Ma X, Jin B, Liu Z, Peters RJ, Zhao ZK, Huang L (2016) Cytochrome P450 promiscuity leads to a bifurcating biosynthetic pathway for tanshinones. New Phytol 210(2):525–534

Guo J, Zhou YJ, Hillwig ML, Shen Y, Yang L, Wang Y, Zhang X, Liu W, Peters RJ, Chen X, Zhao ZK, Huang L (2013) CYP76AH1 catalyzes turnover of miltiradiene in tanshinones biosynthesis and enables heterologous production of ferruginol in yeasts. Proc Natl Acad Sci USA 110(29):12108–12113

Hamberger B, Bak S (2013) Plant P450s as versatile drivers for evolution of species-specific chemical diversity. Philos Trans R Soc Lond B Biol Sci 368 (1612):20120426

Kai G, Liao P, Xu H, Wang J, Zhou C, Zhou W, Qi Y, Xiao J, Wang Y, Zhang L (2012) Molecular mechanism of elicitor-induced tanshinone accumulation in *Salvia miltiorrhiza* hairy root cultures. Acta Physiol Plant 34(4):1421–1433

Kai G, Xu H, Zhou C, Liao P, Xiao J, Luo X, You L, Zhang L (2011) Metabolic engineering tanshinone biosynthetic pathway in *Salvia miltiorrhiza* hairy root cultures. Metab Eng 13(3):319–327

Laule O, Furholz A, Chang H, Zhu T, Wang X, Heifetz P, Gruissem W, Lange M (2003) Crosstalk between cytosolic and plastidial pathways of isoprenoid biosynthesis in *Arabidopsis thaliana*. Proc Natl Acad Sci USA 100(11):6866–6871

Ma Y, Ma XH, Meng FY, Zhan ZL, Guo J, Huang LQ (2016) RNA interference targeting CYP76AH1 in hairy roots of *Salvia miltiorrhiza* reveals its key role in the biosynthetic pathway of tanshinones. Biochem Biophys Res Commun 477(2):155–160

Ma Y, Yuan L, Wu B, Xe Li, Chen S, Lu S (2012) Genome-wide identification and characterization of novel genes involved in terpenoid biosynthesis in *Salvia miltiorrhiza*. J Exp Bot 63(7):2809–2830

Mei XD, Cao YF, Che YY, Li J, Shang ZP, Zhao WJ, Qiao YJ, Zhang JY (2019) Danshen: a phytochemical and pharmacological overview. Chin J Nat Med 17 (1):59–80

Miyasaka H, Nasu M, Yamamoto T, Yoneda K (1985) Production of ferruginol by cell suspension cultures of *Salvia miltiorrhiza*. Phytochemistry 24(9):1931–1933

Mizutani M, Ohta D (2010) Diversification of P450 genes during land plant evolution. Annu Rev Plant Biol 61 (1):291–315

Park YK, Obiang-Obounou BW, Lee J, Lee TY, Bae MA, Hwang KS, Lee KB, Choi JS, Jang BC (2017) Anti-adipogenic effects on 3T3-L1 cells and zebrafish by tanshinone IIA. Int J Mol Sci 18(10):2065

Chinese Pharmacopoeia (2015) Pharmacopoeia Commission of People's Republic of China, 2015 edn, vol 1. China Medical Science Press, Beijing

Robertson AL, Holmes GR, Bojarczuk AN, Burgon J, Loynes CA, Chimen M, Sawtell AK, Hamza B, Willson J, Walmsley SR, Anderson SR, Coles MC, Farrow SN, Solari R, Jones S, Prince LR, Irimia D, Rainger GE, Kadirkamanathan V, Whyte MKB, Renshaw SA (2014) A zebrafish compound screen reveals modulation of neutrophil reverse migration as an anti-inflammatory mechanism. Sci Transl Med 6 (225):225–229

Rohmer M, Knani M, Simonin P, Sutter B, Sahm H (1993) Isoprenoid biosynthesis in bacteria: a novel pathway for the early steps leading to isopentenyl diphosphate. Biochem J 295(2):517–524

Shi M, Luo X, Ju G, Li L, Huang S, Zhang T, Wang H, Kai G (2016) Enhanced diterpene tanshinone accumulation and bioactivity of transgenic *Salvia miltiorrhiza* hairy roots by pathway engineering. J Agric Food Chem 64(12):2523–2530

Shimomura K, Kitazawa T, Okamura N, Yagi A (1991) Tanshinone production in adventitious roots and regenerates of *Salvia miltiorrhiza*. J Nat Prod 54 (6):1583–1587

Wang BQ (2010) *Salvia miltiorrhiza*: Chemical and pharmacological review of a medicinal plant. J Med Plants Res 4(25):2813–2820

Wang L, Ma R, Liu C, Liu H, Zhu R, Guo S, Tang M, Li Y, Niu J, Fu M, Gao S, Zhang D (2017) Salvia miltiorrhiza: a potential red light to the development of cardiovascular diseases. Curr Pharm Des 23(7):1077–1097

Wang JW, Wu JY (2010) Tanshinone biosynthesis in *Salvia miltiorrhiza* and production in plant tissue cultures. Appl Microbiol Biotechnol 88(2):437–449

Xu H, Song J, Luo H, Zhang Y, Li Q, Zhu Y, Xu J, Li Y, Song C, Wang B, Sun W, Shen G, Zhang X, Qian J, Ji A, Xu Z, Luo X, He L, Li C, Sun C, Yan H, Cui G, Li X, Xe Li, Wei J, Liu J, Wang Y, Hayward A, Nelson D, Ning Z, Peters Reuben J, Qi X, Chen S (2016) Analysis of the genome sequence of the medicinal plant *Salvia miltiorrhiza*. Mol Plant 9 (6):949–952

Yang D, Du X, Liang X, Han R, Liang Z, Liu Y, Liu F, Zhao J (2012) Different roles of the mevalonate and methylerythritol phosphate pathways in cell growth and tanshinone production of *Salvia miltiorrhiza* hairy roots. PLoS ONE 7(11):e46797

Yang L, Ding G, Lin H, Cheng H, Kong Y, Wei Y, Fang X, Liu R, Wang L, Chen X, Yang C (2013) Transcriptome analysis of medicinal plant *Salvia miltiorrhiza* and identification of genes related to tanshinone biosynthesis. PLoS ONE 8(11):e80464

Yang Z, Kitano Y, Chiba K, Shibata N, Kurokawa H, Doi Y, Arakawa Y, Tada M (2001) Synthesis of variously oxidized abietane diterpenes and their antibacterial activities against MRSA and VRE. Biorg Med Chem 9(2):347−356

Zhang Y, Jiang P, Ye M, Kim SH, Jiang C, Lu J (2012) Tanshinones: sources, pharmacokinetics and anti-cancer activities. Int J Mol Sci 13(10):13621–13666

Zhong GX, Li P, Zeng LJ, Guan J, Li DQ, Li SP (2009) Chemical characteristics of *Salvia miltiorrhiza* (Danshen) collected from different locations in China. J Agric Food Chem 57(15):6879–6887

Zhou YJ, Gao W, Rong Q, Jin G, Chu H, Liu W, Yang W, Zhu Z, Li G, Zhu G, Huang L, Zhao ZK (2012) Modular pathway engineering of diterpenoid synthases and the mevalonic acid Pathway for miltiradiene production. J Am Chem Soc 134(6):3234–3241

Biochemistry, Biosynthesis, and Medicinal Properties of Phenolic Acids in *Salvia miltiorrhiza*

11

Guoyin Kai, Shucan Liu, Min Shi, Bing Han, Xiaolong Hao and Zhixiang Liu

Abstract

As one group of important bioactive compounds in *Salvia miltiorrhiza*, water-soluble phenolic acids own a variety of bioactivities including anti-oxidation, anti-inflammatory, and anti-cancer. Due to the degradation of genetic resources and low content of phenolic acids in traditionally cultured *S. miltiorrhiza*, limited phenolic acid production cannot meet the increasing market demand. It is extremely important to use modern biotechnology methods to increase the yield of phenolic acids. Here, we summarize pharmacological activities of phenolic acids in *S. miltiorrhiza*, as well as various biological methods including culturing hairy roots, callus, suspension cells, and endophytic fungi for producing phenolic acids and using elicitors treatment, metabolic engineering and transcriptional regulation for increasing the production of phenolic acids.

11.1 Introduction

Phenolic acids possessing various pharmacological activities such as anti-cancer, anti-oxidant, anti-bacterial, and anti-inflammatory activities are widely distributed in nature, especially in some commonly used traditional Chinese medicines such as *Salvia miltiorrhiza* (Zhou et al. 2011, 2012). *S. miltiorrhiza* has the functions of relieving pain, promoting blood circulation, regulating menstruation, and nourishing the heart so that it is widely used in the treatment of cardiovascular diseases (Wang and Cao 2016). There are more than 20 phenolic acids in *S. miltiorrhiza* including rosmarinic acid (RA), salvianolic acid B (SAB), salvianolic acid A (SAA), danshensu (DSU), caffeic acid, cinnamic acid, ferulic acid, and lithospermic acid (Fig. 11.1) (Xing et al. 2018a, b). SAB and RA in crude *S. miltiorrhiza* account for the largest content (Sun et al. 2016).

In recent years, RA, an ester of caffeic acid and 3,4-dihydroxyphenyllactic acid, has been proved to be used for prophylaxis and treatment of neuropathic pain for its anti-apoptotic and

G. Kai (✉) · S. Liu · M. Shi · B. Han · X. Hao · Z. Liu
Laboratory of Medicinal Plant Biotechnology, College of Pharmacy, Zhejiang Chinese Medical University, Hangzhou 310053, Zhejiang, People's Republic of China
e-mail: kaiguoyin@zcmu.edu.cn

S. Liu
e-mail: liushucan210@163.com

M. Shi
e-mail: shimin6047@126.com

B. Han
e-mail: hanbing7525@163.com

X. Hao
e-mail: xiaolong19890217@126.com

Z. Liu
e-mail: zhixiangliu88@163.com

S. Lu (ed.), *The Salvia miltiorrhiza Genome*, Compendium of Plant Genomes,
https://doi.org/10.1007/978-3-030-24716-4_11

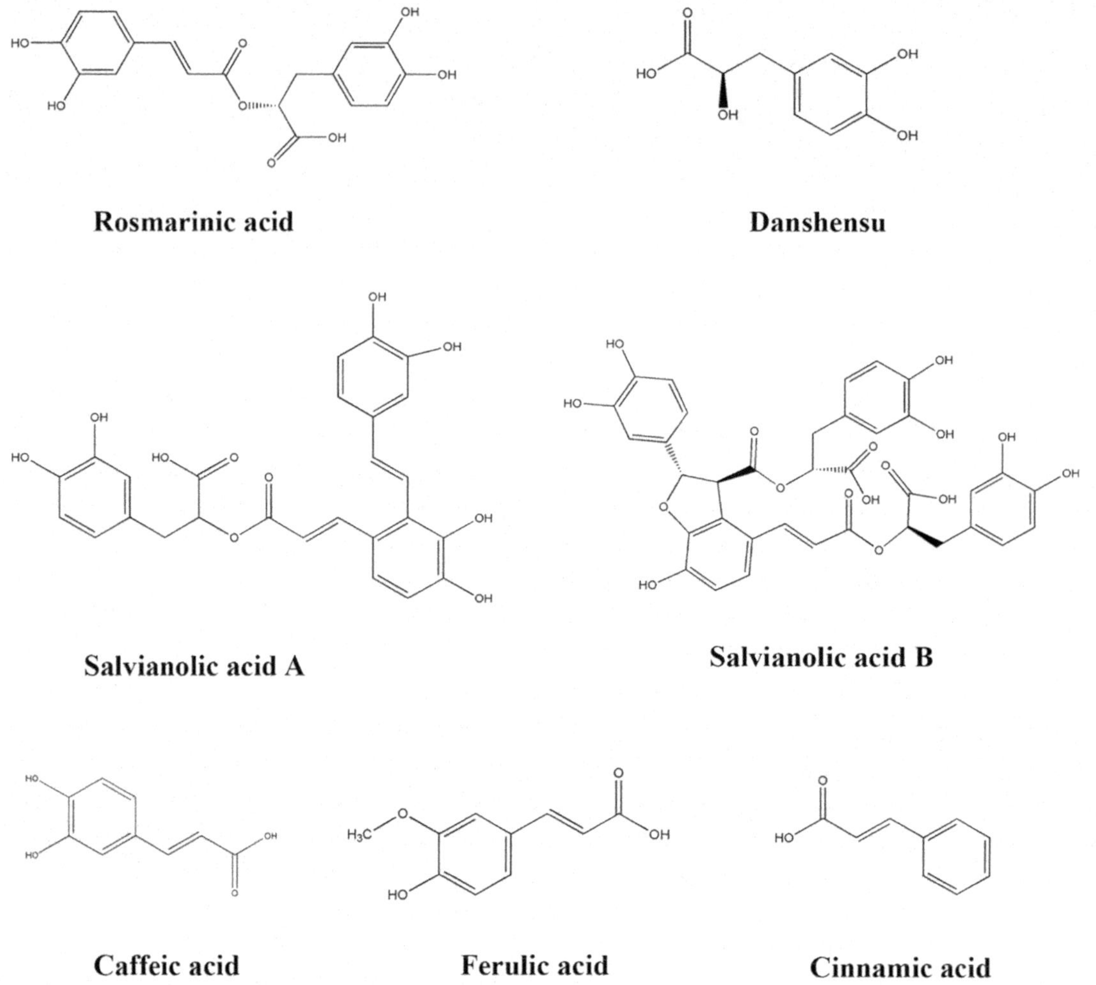

Fig. 11.1 Chemical structure of main phenolic acids including rosmarinic acid (RA), danshensu (DSU), salvianolic acid A (SAA), salvianolic acid B (SAB), caffeic acid, ferulic acid, and cinnamic acid in *S. miltiorrhiza*

anti-inflammatory effects (Pezeshki and Petersen 2018; Rahbardar et al. 2018). SAB and DSU isolated from *S. miltiorrhiza* are extremely effective anti-oxidants and have stronger oxygen free radical scavenging activity than vitamin C. Moreover, SAB shows better anti-oxidant activity than DSU (Zhao et al. 2008). It has been reported that SAB has a protective effect on emphysema-like lung cell death and protect against ischemia/reperfusion-induced cerebral injury (Dhapare and Sakagami 2018; Fan et al. 2018). SAB can effectively inhibit the growth of cultured MDA-MB-231 cells and tumor xenografts via a ceramide-mediated pathway. SAB also enhances apoptosis by regulating ceramide glycosylase and reduces TNBC cell proliferation (Sha et al. 2018). SAA possesses extensive pharmacological activities like treating liver disease, which prevents chronic ethanol-induced liver damage via SIRT1-mediated autophagosome–lysosome fusion recovery (Shi et al. 2018).

With the rapid development of biotechnology, the biosynthetic pathway of phenolic acids in *S. miltiorrhiza* has been gradually revealed (Petersen and Simmonds 2003; Di et al. 2013; Ma et al. 2015). Due to its wide medicinal value, the depletion of wild resources and the low yield of phenolic acid, genetic engineering and metabolic engineering have been used to enhance the production of phenolic acids in *S. miltiorrhiza*, which has become a research hotspot. Here we summarize the research progress of the main substances, pharmacological activities, biosynthetic pathways, and in vitro synthesis of phenolic acids in *S. miltiorrhiza*, in order to more efficiently produce pharmacological phenolic acids and discuss their prospects.

11.2 Medicinal Properties of Phenolic Acids

Phenolic acids are the major active components in *S. miltiorrhiza*, which exerts a variety of pharmacological activities including anti-oxidant, cardio-protective, neuro-protective, anti-platelet, anti-cancer, anti-inflammatory, reno-protective, anti-diabetic properties (summarized in Fig. 11.2). The pharmacological activities of SAA and SAB are listed in Table 11.1.

11.2.1 Anti-oxidant Activity

Phenolic acids have been found to possess potent anti-oxidative ability due to their polyphenolic structure. SAA could enhance the activities of superoxide dismutase (SOD), glutathione peroxidase (GPx), and catalase in an established 5/6 nephrectomized (5/6Nx) rat model, and decreased reactive oxygen species (ROS) in H_2O_2-induced HK-2 cells. The protective effects of SAA on oxidative stress were suggested to be related to the modulation of Akt/GSK-3β/Nrf2 and the NF-κB signaling pathway (Zhang et al. 2019a). In addition, SAA decreased malondialdehyde (MDA) but increased SOD level in angiotensin II-incubated macrophages (Li et al. 2016a). SAB protected against subarachnoid hemorrhage-induced oxidative damage in vivo through activating the SIRT1 and Nrf2 signaling pathway (Zhang et al. 2018a). The levels of antioxidase SOD and GPx were decreased in patients with cysteine stone, which were associated with oxidative stress. Inversely, SAB could

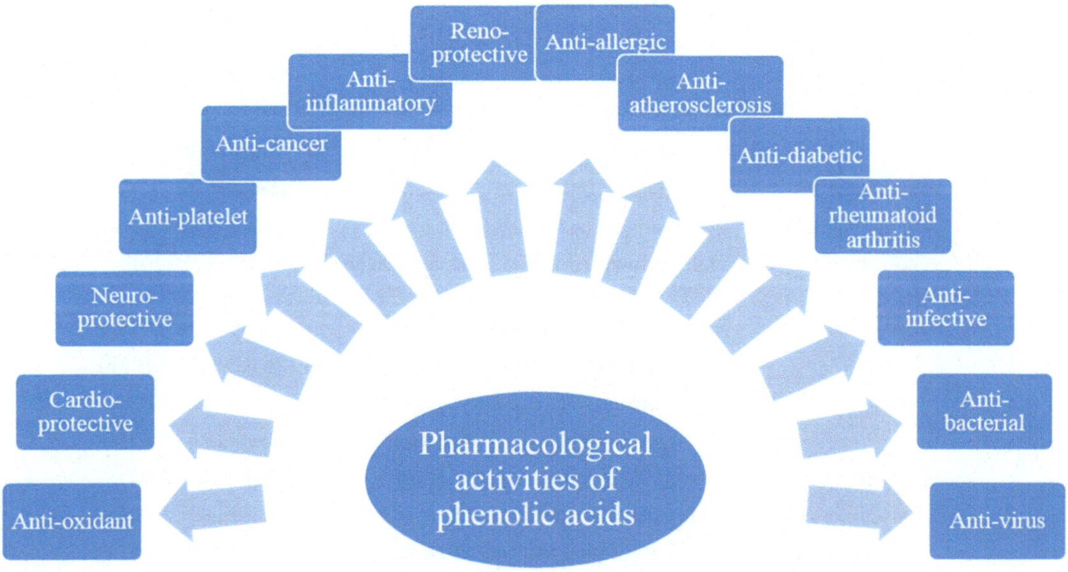

Fig. 11.2 Pharmacological activities of phenolic acids in *S. miltiorrhiza*

Table 11.1 Pharmaceutical effects of SAA and SAB

Phenolic acids	Bioactivities	Targets	References
SAA	Anti-oxidant	5/6 nephrectomized rat	Zhang et al. (2019a, b)
		Angiotensin II-induced murine peritoneal macrophages	Li et al. (2016a)
	Cardio-protective	Myocardial ischemia/reperfusion injury	Qian et al. (2019)
		Arsenic trioxide-induced cardiac injury	Zhang et al. (2018b, c)
	Neuro-protective	Spinal cord injury rats	Yu et al. (2017)
		Middle cerebral artery occlusion mice	Mahmood et al. (2017)
		ROS-induced neuronal damage	Zhang et al. (2012)
	Anti-platelet	Human platelet and $LDLR^{-/-}$ mice	Huang et al. (2010)
		Rat and human platelet, arterio-venous shunt thrombosis rats	Fan et al. (2010)
	Anti-cancer	Acute myeloid leukemia	Pei et al. (2018)
		Oral squamous cell cancer	Fang et al. (2018)
		Lung cancer	Bi et al. (2013), Tang et al. (2017)
		Angiogenesis	Yang et al. (2019)
		Breast cancer	Wang et al. (2015a, b), Cai et al. (2014), Zheng et al. (2015)
	Anti-inflammatory	Ischemia reperfusion-induced rat brain damage	Zhang et al. (2018d)
		LPS-induced macrophages	Liu et al. (2018)
	Reno-protective	Doxorubicin-induced nephropathy	Fan et al. (2015)
		5/6 nephrectomized rats	Zhang et al. (2018e)
		Diabetic nephropathy	Hou et al. (2017)
		Minimal change disease rats	Wang et al. (2019)
	Anti-allergic	Ovalbumin-induced allergic asthma mice	Heo and Im (2019)
	Anti-atherosclerosis	Macrophages and ApoE KO mice	Zhang et al. (2014c)
	Anti-diabetic	Alloxan-induced type 1 diabetic mice and streptozotocin-induced type 2 diabetic rats	Qiang et al. (2015)
SAB	Anti-oxidant	Subarachnoid hemorrhage rats	Zhang et al. (2018a)
		HK-2 cells, *Slc7a9* knockout mice	Zhang et al. (2019b)
		γ-irradiation-radiated mice	Zhou et al. (2019)
		HO^-, O_2^-, DPPH and ABTS radicals	Zhao et al. (2008)
	Cardio-protective	Doxorubicin-induced cardiac injury	Chen et al. (2017)
	Neuro-protective	Subarachnoid hemorrhage rat	Zhang et al. (2018a)
		Aβ25-35 peptide-induced Alzheimer's disease mice	Lee et al. (2013)
		Cerebral small vessel disease rat	Wang and Hu (2018)
		Stroke rat	Lv et al. (2015)
		Traumatic brain-injured mice	Chen et al. (2011)
		Parkinson's disease	Zhou et al. (2014)

(continued)

Table 11.1 (continued)

Phenolic acids	Bioactivities	Targets	References
	Anti-platelet	Human platelet	Liu et al. (2014)
		Rat mesentery	Wang et al. (2009)
	Anti-cancer	Colon cancer	Guo et al. (2018), Jing et al. (2016)
		Breast cancer	Sha et al. (2018)
		Hepatocellular cancer	Gong et al. (2016)
	Anti-inflammatory	Rheumatoid arthritis rats	Xia et al. (2018)
		LPS-induced cell injury	Meng et al. (2019)
		Liver injury	Zhao et al. (2019)
	Reno-protective	Ischemic reperfusion-induced renal injury	Ma et al. (2017)
	Anti-atherosclerosis	Macrophages and ApoE KO mice	Bao et al. (2012)
	Anti-diabetic	Streptozotocin-induced diabetic rats	Raoufi et al. (2015)
	Anti-rheumatoid arthritis	Collagen-induced rheumatoid arthritis rat	Xia et al. (2018)
	Anti-infective	*Neisseria meningitidis*	Huttunen et al. (2016)
	Anti-bacterial	*Pseudomonas aeruginosa*	Kong et al. (2017)

prevent cysteine stone formation, protect against oxidative injury (Zhang et al. 2019b). Moreover, SAB and DSU showed potent scavenging activities against HO^{-}, O^{2-}, DPPH, and ABTS radicals than vitamin C (Zhao et al. 2008). SAB also protected the mice from radiation injury through nuclear factor (erythroid-derived 2)-like 2 protein/BTB-mediated anti-oxidant effect (Zhou et al. 2019).

11.2.2 Cardio-protective Activity

Myocardial reperfusion during infarction causes damage in cardiomyocytes. Previous reports indicated that total salvianolic acid injection (TSI) of *S. miltiorrhiza* improved ischemia/reperfusion (I/R)-induced myocardial injury, decreased apoptosis, and reduced infarct size in Sprague-Dawley rats (Huang et al. 2019). Recent studies demonstrated that SAA could attenuate apoptosis and prevent I/R injury in cardiomyocytes through the PI3K/Akt, GSK-3β, JNK, and ERK1/2 pathways, and probably via the JNK-ERK1/2 crosstalk (Qian et al. 2019). Mitochondrial dysfunction contributes to the heart diseases such as coronary heart disease and heart failure. However, pretreatment with SAA could maintain normal mitochondrial function and biogenesis, alleviate the damage, and protect against arsenic trioxide-induced cardiac injury in vivo (Zhang et al. 2018c). Furthermore, treatment with SAB showed protective effect against doxorubicin-induced cardiac injury, which is a common clinical syndrome that causes severe pain to cancer patients (Chen et al. 2017).

11.2.3 Neuro-protective Activity

Previously, the TSI has been approved by Chinese State Food and Drug Administration (SFDA) for the treatment of ischemic stroke (Han et al. 2017). Besides, SAA was confirmed to recover the neurological function, improve the motor ability, and inhibit apoptosis-related proteins in spinal cord injury rats (Yu et al. 2017). In another middle cerebral artery occlusion mice model, SAA administration ameliorated neuronal damage and decreased infarcted volume via the inhibition of eNOS uncoupling and peroxynitrite formation (Mahmood et al. 2017). Elevated ROS level involves in stroke and other neurodegenerative diseases. However, pretreatment with SAA increased cell survival against ROS-induced neuronal damage (Zhang et al. 2012). Treatment with SAB inhibited oxidative damage, prevented neurologic impairment, and improved cell viability in a rat subarachnoid hemorrhage model and cultured neurons, which was associated with the activation of Nrf2 and SIRT1 signaling pathway (Zhang et al. 2018e). Subchronic SAB administration (10 mg/kg) ameliorated the memory impairment by decreasing the expression of nitric oxide synthase and cyclooxygenase-2 in Aβ25-35 peptide-induced Alzheimer's disease mouse (Lee et al. 2013). In addition, SAB also affected vasculature and cognitive function. Studies revealed that SAB could recover the angiogenesis and cognitive deficits in cerebral small vessel disease rat though modulation of STAT3/VEGF pathway (Wang and Hu 2018). Other studies have indicated that SAB has beneficial effects on stroke, brain injury, and Parkinson's disease (Lv et al. 2015; Chen et al. 2011; Zhou et al. 2014).

11.2.4 Anti-platelet Activity

Salvianolate, the bioactive constituents of *S. miltiorrhiza*, has been approved by Chinese SFDA for the treatment of coronary artery disease since 2005. Clinical research found salvianolate enhanced the anti-platelet activity of standard aspirin plus clopidogrel therapy in acute coronary syndrome patients. Further in vitro studies demonstrated that SAB, the main component of salvianolate (>85%), suppressed thrombin, arachidonic acid, collagen, and U46619-induced platelet aggregation via inhibiting phosphodiesterase and antagonizing P2Y12 receptor (Liu et al. 2014). The beneficial effect of SAB on thrombosis was related to direct peroxide scavenge or indirect adhesion molecules suppression (Wang et al. 2009). SAA inhibited platelet aggregation and attenuates arterial thrombus formation by inhibiting the PI3K expression (Huang et al. 2010). Additionally, intravenous administration of SAA (2.5–10 mg/kg) showed anti-thrombotic activity in vivo. It modulated the hemorheology without influence on the coagulation function and was presumed to be related to the cAMP induction (Fan et al. 2010).

11.2.5 Anti-cancer Activity

SAA exhibited anti-cancer activities against various carcinomas such as acute myeloid leukemia, oral squamous cell, and lung cancer (Pei et al. 2018; Fang et al. 2018; Bi et al. 2013). Moreover, SAA inhibited GRP78 secretion and angiogenesis in tumor microenvironment (Yang et al. 2019). Chemotherapy resistance is a major challenge in cancer treatment. Studies revealed that SAA enhanced the efficacy of cisplatin in lung cancer A549 cells by inhibiting the AKT/mTOR signaling pathway (Tang et al. 2017). Besides, SAA treatment selectively attenuated the growth of multidrug-resistant MCF-7 breast cancer cells, which was correlated with ROS production (Wang et al. 2015b). Overexpression of transgelin 2 increased the resistance of cancer cells to paclitaxel therapy. However, SAA treatment could reverse resistance, induce apoptosis, and inhibit invasion in paclitaxel-resistant breast cancer cells (Cai et al. 2014; Zheng et al. 2015). It was indicated that SAB could reverse the multidrug resistance in colon cancer cells and promote apoptosis in triple-negative breast cancer (Sha et al. 2018; Guo et al. 2018). Autophagy functions as a death executioner that induces autophagic cell death. SAB was demonstrated to be a novel autophagy inducer that mediated colorectal cancer cells autophagy through modulation of the AKT/mTOR signaling pathway (Jing et al. 2016). Evidence also showed that autophagy as well as apoptosis was involved in SAB-induced hepatocellular carcinoma cell death. While pretreatment with autophagy inhibitors attenuates the effects induced by SAB (Gong et al. 2016).

11.2.6 Anti-inflammatory Activity

SAB decreased the levels of inflammatory cytokines (IL-1β, IL-6, and TNF-α) and increased anti-oxidant enzyme (SOD, CAT, and GSH) activities in collagen-induced rheumatoid arthritis rat (Xia et al. 2018). Treatment with SAB markedly ameliorated LPS-induced injury on MH7 A cells via the modulation of NF-κB and JNK pathways, suggesting the potential anti-inflammatory capacity (Meng et al. 2019). Also, SAB decreased the level of pro-inflammatory cytokines in liver injury (Zhao et al. 2019). SAA protected blood-brain barrier by reducing inflammation response and NF-κB inactivation (Zhang et al. 2018d). Liu and colleagues demonstrated that all the phenolic acids isolated from *S. miltiorrhiza* exhibited anti-inflammatory activity in LPS-stimulated THP-1 cells, among which the inhibitory effect of lithospermic acid was the strongest and similar to that of SAB (Liu et al. 2018).

11.2.7 Reno-protective Activity

Nephrotic syndrome is a common nephrology disorder accompanied by heavy proteinuria, hypoalbuminemia, and hyperlipidaemia. Recent studies demonstrated that SAA administration ameliorated histological damages, podocyte injury, and improved hemorheology in doxorubicin-induced nephropathy (Fan et al. 2015). Intraperitoneal with SAA at 10 mg/kg per day relieved urinary proteins and TNF-α level, alleviated pathological lesions in the kidney of 5/6Nx rats (Zhang et al. 2018e). SAA could also protect against early stage diabetic nephropathy, restored glomerular endothelial permeability through suppressing AGE-RAGE pathway (Hou et al. 2017). Furthermore, combination of SAA with prednisone relieved urinary proteins, improved renal function indices including blood urea nitrogen and serum creatinine level in rats (Wang et al. 2019). SAB showed reno-protective activity in a renal I/R rat model by attenuating inflammatory process and oxidative stress through activating the PI3K/Akt signaling pathway (Ma et al. 2017).

11.2.8 Anti-atherosclerosis Activity

SAB acted as a CD36 antagonist that inhibited lipid uptake in macrophages, leading to the inhibition of atherosclerotic lesions formation in ApoE knockout mice (Bao et al. 2012). SAA attenuated TNF-α-induced CC chemokine ligand-20 (CCL-20) secretion, which plays a crucial role in atherogenesis (Zhang et al. 2014c). In addition, DSU prevented atherosclerosis through inhibiting the expression of adhesion molecules in arterial endothelia (Yang et al. 2010).

11.2.9 Anti-diabetic Activity

Diabetic rats treated with SAB (40 mg/kg) for three weeks showed decreased serum glucose and MDA levels as well as increased serum insulin level. Meanwhile, SAB could protect the pancreatic islet cells against cytotoxicity partly through the inhibition of apoptosis and oxidative stress (Raoufi et al. 2015). In alloxan-induced type 1 and high-fat diet with low-dose streptozotocin-induced type 2 diabetic animal models, SAA administration improved mitochondrial function in liver and skeletal muscle, increased ATP production through activating AMPK/CaMKKβ signaling pathway (Qiang et al. 2015).

11.2.10 Other Activities

SAA was potential anti-allergic therapy that reduced the number of eosinophils and secretion of inflammatory IL-4 and IL-13 in the lung tissue of ovalbumin-induced allergic asthma mice (Heo and Im 2019). SAB showed anti-rheumatoid arthritis activity on collagen-induced rat model (Xia et al. 2018). In addition, SAB exhibited anti-infective property against human pathogen *Neisseria meningitidis* by inhibiting meningococcal binding (Huttunen et al. 2016). Protocatechualdehyde and SAB were major components in *S. miltiorrhiza* that possess anti-bacterial properties (Kong et al. 2017). Besides, the combination of salvianolic acid V, a new salvianolic acid, with Colistin sulfate or Levofloxacin showed effects on MRSA or *Acinetobacter baumannii* (Zhang et al. 2018b). Protocatechuic aldehyde markedly inhibited hepatitis B virus replication in vitro and reduced viremia in virus-infected ducks (Zhou et al. 2007).

11.3 Biosynthesis of Phenolic Acids in *S. miltiorrhiza*

11.3.1 Biosynthetic Pathway of Phenolic Acids

Phenolic acids are biosynthesized via two pathways including the phenylpropanoid pathway and the tyrosine-derived pathway in *S. miltiorrhiza* (Xiong et al. 2010; Ma et al. 2015). In the phenylpropanoid pathway, L-phenylalanine produces 4-coumaroyl-CoA under the catalysis of phenylalanine ammonia-lyase (PAL), cinnamic acid 4-hydroxylase (C4H), hydroxycinnamate coenzyme A ligase (4CL) (Di et al. 2013; Shi et al. 2019). In tyrosine-derived pathway, L-tyrosine produces 4-hydroxyphenylpyruvic acid under oxidative deamination of tyrosine aminotransferase (TAT), following 4-hydroxyphenylpyruvate reductase (HPPR) and forming 3,4-dihydroxy phenyllactic acid at last (Di et al. 2013; Shi et al. 2019).

Finally, 3,4-dihydroxyphenyllactic acid and 4-coumaroyl-CoA are catalyzed by the RAS to form the precursor substance 4-coumaroyl-3′,4′-dihydroxyphenyllactic acid (4C-DHPL), and finally, the CYP98A14 enzyme synthesizes salvianolic acid (Di et al. 2013). In *S. miltiorrhiza*, 4C-DHPL is regarded as the major intermediate for RA biosynthesis involved in this pathway (Fig. 11.3). Meanwhile, RA is also considered to be a precursor of biosynthesis of SAB which is a structural dimer of RA, but the enzyme involved in catalysis is still unclear (Zhang et al. 2014a, b; Ma et al. 2015).

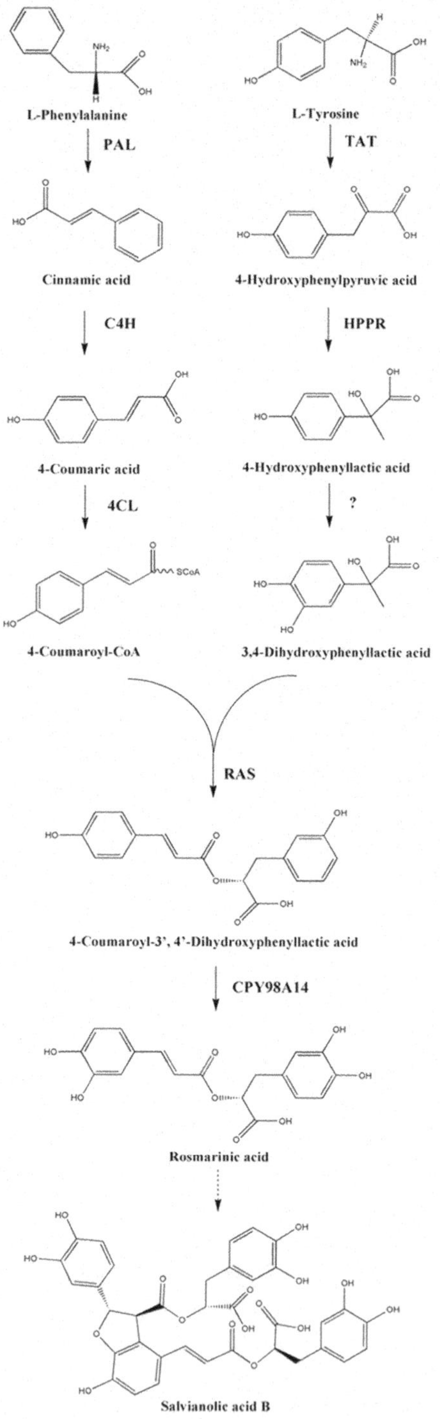

Fig. 11.3 Putative phenolic acid synthesis pathway. PAL, phenylalanine ammonia-lyase; C4H, cinnamic acid 4-hydroxylase; 4CL, 4-coumarate: CoA ligase; TAT, tyrosine aminotransferase; HPPR, 4-hydroxyphenylpyruvate reductase; RAS, rosmarinic acid synthase; CYP98A14, a cytochrome P450-dependent monooxygenase

11.3.2 Related Genes Involved in Phenolic Acid Biosynthesis

With the extensive study of secondary metabolic pathways in *S. miltiorrhiza*, several key enzyme genes involved in these two pathways of phenolic acids biosynthesis have been identified, including *SmPAL*, *SmC4H*, *Sm4CL*, *SmTAT*, *SmHPPR*, *SmRAS*, and *CYP98A14* (Table 11.2; Fig. 11.4) (Ma et al. 2015; Shi et al. 2019).

11.3.2.1 Genes in the Phenylpropanoid Pathway

Phenylalanine ammonia-lyase (PAL), the first point enzyme to catalyze in the phenylpropanoid pathway, plays a significant role in primary metabolism and secondary metabolism regulation. Three members in *SmPALs* have been revealed from *S. miltiorrhiza*, including *SmPAL1*, *SmPAL2*, and *SmPAL3*. Gene expression pattern shows that *SmPAL1* and *SmPAL3* highly express in both roots and leaves of

Table 11.2 Key enzymes of phenolic acids biosynthesis pathway

Key enzymes	Accession no.	Functions	Highest expressed tissues	References
SmPAL1	EF462460	Catalyzing the L-phenylalanine to produce trans-cinnamic acid	Root and leaf	Wang et al. (2009, 2015a), Hou et al. (2013)
SmPAL2	GQ249111		Stem and flower	Hou et al. (2013), Wang et al. (2015a), Ma et al. (2015)
SmPAL3	KF220569		Root and leaf	Hou et al. (2013), Wang et al. (2015a), Ma et al. (2015)
SmC4H1	EF377337 DQ355979	Catalyzing the hydroxylation of cinnamate to 4-coumarate	Root and stem	Wang et al. (2015a), Ma et al. (2015)
SmC4H2	KF220564		Root and stem	Wang et al. (2015a), Ma et al. (2015)
Sm4CL1	AY237163	Catalyzing a series of aromatic substrates to form their corresponding hydroxycinnamoyl-CoA esters	Leaf	Wang et al. (2015a), Ma et al. (2015)
Sm4CL2	AY237164		Root	Wang et al. (2015a), Ma et al. (2015)
Sm4CL3	KF220556		Root	Wang et al. (2015a), Ma et al. (2015)
Sm4CL-like1	KF220557		Root	Wang et al. (2015a), Ma et al. (2015)
Sm4CL-like2	KF220558			Wang et al. (2015a), Ma et al. (2015)
Sm4CL-like3	KF220559			Wang et al. (2015a), Ma et al. 2015
Sm4CL-like4	KF220560		Root	Wang et al. (2015a), Ma et al. 2015
Sm4CL-like5	KF220561			Wang et al. (2015a), Ma et al. (2015)
Sm4CL-like6	KF220562			Wang et al. (2015a), Ma et al. (2015)
Sm4CL-like7	KF220563			Wang et al. (2015a), Ma et al. (2015)

(continued)

Table 11.2 (continued)

Key enzymes	Accession no.	Functions	Highest expressed tissues	References
SmTAT1	DQ334606 EF192320	Catalyzing the transamination from L-tyrosine to 4-hydroxyphenylpyruvate	Stem	Wang et al. (2015a), Ma et al. (2015)
SmTAT2	KF220575		Flower	Wang et al. (2015a), Ma et al. (2015)
SmTAT3	KF220555		Root and stem	Wang et al. (2015a), Ma et al. (2015)
SmHPPR1	DQ099741 DQ266514 EF458148	Participate in the synthesis of 4-hydroxyphenyl lactic acid	Stem and flower	Wang et al. (2015a), Ma et al. (2015)
SmHPPR2	KF220565		Stem and leaf	Wang et al. (2015a), Ma et al. (2015)
SmHPPR3	KF220566		Stem	Wang et al. (2015a), Ma et al. (2015)
SmHPPR4	KF220567			Wang et al. (2015a), Ma et al. (2015)
SmRAS1	FJ906696	Catalyzing 4-coumaroyl-CoA and 3, 4-hydroxyphenyllactate to form an ester (4-coumaroyl-3′,4′-dihydroxyphenyllactic acid)	Root	Wang et al. (2015a), Ma et al. (2015), Di et al. (2013)
SmRAS1-like	GU647199		Root	Wang et al. (2015a), Ma et al. (2015), Song (2010)
SmHCT1	KF220570		Root	Wang et al. (2015a), Ma et al. 2015
SmHCT2	KF220571		Stem	Wang et al. (2015a), Ma et al. (2015)
SmHCT3	KF220572		Stem	Wang et al. (2015a), Ma et al. (2015)
SmHCT4	KF220573		Stem	Wang et al. (2015a), Ma et al. (2015)
SmHCT5	KF220574		Stem	Wang et al. (2015a), Ma et al. (2015)
SmCPY98A14	HQ316179	Catalyzing rosmarinic acid synthesis	Root	Wang et al. (2015a), Ma et al. (2015), Di et al. (2013)

S. miltiorrhiza, while *SmPAL2* is mainly expressed in stems and flowers. The full-length ORF of *SmPAL1* is 2827 bp, encoding a 711-amino-acid peptide. Meanwhile, SmPAL1 has been revealed to be induced by abscisic acid (ABA), wounding, polyethylene glycol (PEG), methyl jasmonate (MJ), salicylic acid (SA), Ca^{2+}, gibberellin (GA), and ethylene (Ma et al. 2015; Hou et al. 2013). *SmPAL2* contains 2127 bp ORF encoding a 683-amino-acid peptide, and *SmPAL3* has 2283 bp ORF encoding 760 amino acid, which are induced by PEG and MJ (Ma et al. 2015; Song and Wang 2009). Interestingly, all three *SmPALs* are regulated by MeJA, and the expression of *SmPAL*1 and *SmPAL3* is drastically raised at 6 h after MeJA treatment.

Fig. 11.4 Schematic diagram of key enzyme genes highly expressed in different tissues

Meanwhile, RNAi of *SmPAL1* has been found to cause a significant decrease in the content of RA and SAB in *S. miltiorrhiza*, indicating that *SmPAL1* plays a more significant role in phenylpropanoid pathway (Wang et al. 2015a; Song and Wang 2011).

Cinnamate 4-hydroxylase (C4H) is a cytochrome P450 enzyme involved in the catalysis of the production of *p*-coumaric acid by trans-cinnamic acid produced by PAL (Huang et al. 2008a). The full length of *SmC4H1* is 1512 bp encoding a 504-amino-acid protein. *SmC4H1* is highly expressed in roots and stems; meanwhile, it is induced by elicitor such as MJ, ABA, Ag^{+}, and UV-B radiation. *SmC4H2* encoding a 397-amino-acid protein is absent ER-targeting peptide, probably causing by N-terminal deletion of gene (Shi et al. 2019; Wang et al. 2015a, b). The tissue expression of *SmC4H2* is the same to *SmC4H1*, expressed highly in stem and root (Ma et al. 2015). However, *SmC4H2* does not seem to be involved in MJ regulation because of their insensitive to MJ, but its promoter contains some elements that respond to other stresses including fungal attack and salicylic acid (Wang et al. 2015a, b).

4-Coumarate: CoA ligase (4CL) catalyzing 4-coumaroyl acid to form 4-coumaroyl-CoA in phenylpropanoid pathway has ten members such as *Sm4CL1*, *Sm4CL2*, *Sm4CL3*, *Sm4CL-like 1*, *Sm4CL-like 2*, *Sm4CL-like 3*, *Sm4CL-like 4*, *Sm4CL-like 5*, *Sm4CL-like 6*, and *Sm4CL-like 7*. *Sm4CL1* has been found highly expressed in stem, low in leaves and rear in roots, while the expression levels of *Sm4CL2*, *Sm4CL3*, *Sm4CL-like1*, and *Sm4CL-like4* are high in roots. Furthermore, the expression of *Sm4CL1* is affected by elicitors like MJ and YE, and *Sm4CL2* is induced by MJ, YE and Ag^{+}, implying that *Sm4CL1* may be involved in the biosynthesis of phenolic acids in the stems and leaves, while *Sm4CL2* may participate in the biosynthesis of phenolic acids in roots of *S. miltiorrhiza* (Jin et al. 2012; Shi et al. 2019; Wang et al. 2015a; Zhao et al. 2006).

11.3.2.2 Genes in the Tyrosine-Derived Pathway

Tyrosine aminotransferase (TAT) and 4-hydroxyphenylpyruvate reductase (HPPR) are two main key enzymes which are involved in forming 3,4-dihydroxyphenyllactic acid (DHPL) in tyrosine-derived pathway. Tyrosine aminotransferase (TAT) is the first enzyme in the tyrosine-derived pathway of phenolic acids biosynthesis. *SmTAT1* contains an ORF of 1233 bp encoding 411 amino acid (Huang et al. 2008b). The expression patterns show that *SmTAT1* is expressed higher in stem compared to root and leaf. Meanwhile, *SmTAT1* responds to MJ, ABA, SA, UV-B, GA, ethylene, Ag^{+}, YE (Xing et al. 2015; Yan et al. 2006; Liang et al. 2013). *SmTAT2* and *SmTAT3* are found to express in flower, stem, and root, respectively, also have been cloned from *S. miltiorrhiza*. *SmTAT1* and *SmTAT3* are grouped in the same branch in the phylogenetic tree analysis, but *SmTAT2* is divided into another branch, indicating that *SmTAT1* and *SmTAT3* may play a similar role in the phenolic acid synthesis pathway (Wang et al. 2015a).

4-Hydroxyphenylpyruvate reductase (HPPR) is the second enzyme in tyrosine-derived pathway, catalyzing 4-hydroxyphenylpyruvic acid to form 4-hydroxyphenyllactic acid. Three HPPRs are designed as *SmHPPR1*, *SmHPPR2*, and *SmHPPR3*. *SmHPPR1* and *SmHPPR2* encode 313 amino acid, and *SmHPPR3* encodes a 319-amino-acid protein. Promoter region of *SmHPPR1* expressed highest in stem which possesses many stress-responsive elements. At the same time, *SmHPPR1* has been found that it can be induced by MJ, SA, GA3, ABA, Ag^{+}, and UV-B radiation (Xing et al. 2015; Wang et al. 2015a; Xiao et al. 2011).

11.3.2.3 Genes Involved in Rosmarinic Acid (RA) Biosynthesis

Rosmarinic acid synthase (RAS) is regard as a rate-limiting enzyme catalyzing 4-hydroxyphenyllactic acid to form precursor 4-coumaroyl-3′,4′-dihydroxyphenyllactic acid (4C-DHPL). Seven *SmRASs* including *SmRAS1*, *SmRAS1-like,* and five *SmHCTs* have been revealed so far (Ma et al. 2015; Shi et al. 2019). *SmRAS1* containing 1284 bp ORF and encoding a 426-amino-acid polypeptide has been proved expressed predominantly in roots and stems, and induced by MJ, Ag^{+} (Di et al. 2013; Ma et al. 2015). *SmRAS-like* is expressed higher in stem than other tissues and sensitive to *Pseudomonas lachrymans*, MJ, light, and SA (Song. 2010; Ma et al. 2015). *SmHCTs* include *SmHCT1*, *SmHCT2*, *SmHCT3*, *SmHCT4,* and *SmHCT5*, encoding 341, 425, 426, 439, 427 amino acid, respectively. *SmHCT2*, *SmHCT3*, *SmHCT4*, and *SmHCT5* are highly expressed in stem, while *SmHCT1* is highly expressed in root. Meanwhile, *SmHCT3*, *SmHCT4*, and *SmHCT5* were responsive to MJ elicitation (Wang et al. 2015a; Ma et al. 2015).

Cytochrome P450-dependent monooxygenase (CYP98A14) in *S. miltiorrhiza* participates the synthesis of RA by catalyzing 4C-DHPL. *SmCYP98A14* has 1525 bp ORF encoding a 508-amino-acid protein, and its expression is higher in roots than that in stems and leaves; moreover, *SmCYP98A14* is sensitive to MJ and Ag^{+} (Di et al. 2013; Wang et al. 2015a; Ma et al. 2015).

11.4 Biotechnological Approaches to Improve the Production of Phenolic Acids

Due to the serious degradation of the quality of traditionally cultivated *S. miltiorrhiza*, slow growth cycle and low yield of active ingredients, the yield and quality of *S. miltiorrhiza* cannot meet the growing market demand. Phenolic acids are active ingredients with important economic and medicinal properties; its synthetic route has been basically clear, how to improve the amount of phenolic acid compounds in *S. miltiorrhiza* has become a research hotspot and difficulty. The application of plant biotechnology in improving biological activity and the required ingredients is more attractive and efficient than traditional methods. Here, we will introduce several methods which were reported to enhance the production of PAs including elicitors, metabolic engineering, and transcriptional regulation.

11.4.1 Elicitation Treatment to Increase the Production of Phenolic Acids

Elicitors are usually an agent that stimulates a plant defense response. Simulating biotic and abiotic stresses, elicitors classified into biotic elicitor and abiotic elicitor can stimulate plants and plant cultures to respond to them and lead to the accumulation of secondary metabolites in plants and plant cultures, which is important to produce valuable pharmaceutical ingredients (Wang and Wu 2013). Some elicitors have been utilized to enhance the production of secondary metabolites such as phenolic acids in the hairy roots or cell culture of *S. miltiorrhiza*, including methyl jasmonate (MeJA), yeast extract (YE), abscisic acid (ABA), silver ions (Ag^+), gibberellic acid (GA), salicylic acid (SA), polyamines, and ethylene. For example, treating with MeJA (0.1 mM), RA and lithospermic acid B accumulation were significantly increased. Meanwhile, several RA biosynthesis genes were induced by 0.1 mM MeJA, including phenylalanine ammonia-lyase, cinnamic acid 4-hydroxylase, tyrosine aminotransferase, 4-hydroxyphenylpyruvate reductase, and 4-hydroxyphenylpyruvate dioxygenase (Xiao et al. 2009). SA was utilized to treat suspension cultures of *S. miltiorrhiza*, which lead to a significant increase in the RA content. Furthermore, increase of the PAL activity was measured with SA treatment (Jiao et al. 2012). As the most studied heavy metal ion elicitor, low concentration of Ag^+ can stimulate 3-*O*-glucosylresveratrol production, which do not affect cell growth at the same time (Cai et al. 2013). With Ag^+ treatment, the total phenolic acids of *S. miltiorrhiza* hairy roots were increased and the tyrosine aminotransferase (TAT) activity showed a remarkably rise (Yan et al. 2006). It was found that accumulation of RA on day 6 after treatment of *S. miltiorrhiza* hairy roots with Ag^+ treatment was significantly increased to 1.3 times than that of the control and productions of caffeic acid and ferulic acid were also increased with Ag^+ treatment. However, productions of DSU, cinnamic acid, and lithospermic acid B (LAB) were significantly dropped by Ag^+ treatment. The results of qRT-PCR showed that expression of five key enzymes genes in RA biosynthesis pathways was significantly up-regulated by Ag^+ (Xing et al. 2015). YE was reported much more effective than Ag^+ to enhance the content of RA and accumulation of total phenolic acids. The activity of PAL showed a notable repression with YE treatment, while the activity of TAT was enhanced by YE (Yan et al. 2006). It was reported that exogenous ABA and polyamines increased the production of salvianolic acids in hairy root cultures of *S. miltiorrhiza*. The results of HPLC showed that contents of SAB and SAA enhanced 2.0-fold and 3.3-fold, respectively, after 80 $\mu mol\ L^{-1}$ ABA treatment in *S. miltiorrhiza* hairy roots. Meanwhile, PAL activity also was detected, which increased 1.8-fold after ABA treatment (Hao et al. 2012). Liang et al. (2013) found that three significant phytohormones including abscisic acid (ABA), gibberellin (GA), and ethylene (Eth) could enhance the production of phenolic acids and activities of PAL and TAT in *S. miltiorrhiza* hairy roots.

11.4.2 Metabolic Engineering for Production of Phenolic Acids in *S. miltiorrhiza*

The use of modern biotechnology and molecular biology methods regulates the content of phenolic acids in *S. miltiorrhiza*. The first is the regulation of key enzyme genes in its synthetic pathway. Hairy roots have the advantages of high genetic stability, rapid growth, etc., which is considered as a promising system to generate pharmacological active ingredient of traditional Chinese medicine plant (Guillon et al. 2006; Kai et al. 2012; Shi et al. 2019, 2014). In order to improve the synthesis of pharmacologically active ingredients such as phenolic acids in *S. miltiorrhiza*, researchers attempted to overexpress one or more key enzymes of the phenolic acid synthesis pathway in hairy roots of *S. miltiorrhiza*. Previous research revealed that the content of rosmarinic acid (RA) was ~3.6-fold more than control in *SmC4H* transgenic lines. Furthermore, the concentration of LAB enhanced 11.1-fold of that in control (EV) when overexpressed *SmC4H* in hairy roots. Overexpression of *SmTAT* and *SmHPPR* in hairy root also showed an increase in RA and LAB levels. It is worth noting that the *SmTAT-SmHPPR* co-transformed lines had the highest level of metabolites that the content more than 16.1 and 18.8 times that in EV line (Xiao et al. 2011). Overexpressed genes such as *Sm4CL, SmRAS,* and Sm*CYP98A14* in the biosynthesis pathway of phenolic acids are also a strategy to be used for enhancing the yields of phenolic acids in the future. Not only endogenous genes, but also foreign genes can be involved in the regulation of phenolic acid synthesis. It was reported that the accumulation of phenolic acids in *S. miltiorrhiza,* especially SAB, was affected by overexpressing the foreign gene *AtPAP1* (Zhang et al. 2010).

11.4.3 Transcriptional Regulation of Phenolic Acids Biosynthesis in S. *miltiorrhiza*

The effects of overexpressing one or more critical enzyme genes of synthetic pathway to enhance the target product are limited, and the increase in the yield of the target product is often limited. The regulatory functions and mechanisms of transcription factors for plant secondary metabolism have become research hotspots. As research continues, it has been found that the use of upstream transcription factors to activate the entire metabolic regulatory network is often more efficient than simply transferring to one or several rate-limiting enzyme genes. JA is involved in plant growth and development regulation as a plant hormone, as well as response to stress and leads to accumulation of secondary metabolites (Wasternack and Hause 2013; Namdeo. 2007). With the continuous advancement of sequencing technology, transcription factors mainly including ERF, WRKY, bHLH, MYB, and other transcription factors in responsive to JAs have been discovered and cloned for secondary metabolic regulation in *S. miltiorrhiza* (Zhou and Memelink 2016; Yu et al. 2018; Cao et al. 2018; Du et al. 2018; Ding et al. 2017).

Members of the AP2/ERF TF family responding to JA are significant biosynthesis of secondary metabolites in plant (Zhou and Memelink 2016). *SmERF115*, most sensitive to MeJA, has been isolated and characterized. The phenolic acids production of hairy roots in *S. miltiorrhiza* is enhanced when *SmERF115* overexpressed, while silencing of *SmERF115* leads to the decreased of phenolic acids. Meanwhile, *SmERF115* is binding directly the promoter of *SmRAS* and up-regulating the expression of *SmRAS* to mediate the yield of phenolic acids (Sun et al. 2019).

Basic helix-loop-helix (bHLH) transcription factor, one of the largest families of transcription factors in plants, plays an important role in plant growth and development and secondary metabolism. SmbHLH37, SmbHLH51, and SmbHLH148 in *S. miltiorrhiza* were revealed to participate the regulation of phenolic acids. SmbHLH37 was reported binding the promoter regions of *SmTAT* and *SmPAL* to repress their expression, inhibiting the SAB synthesis pathway and resulting in decreased SAB production. In contrast, SmbHLH51 and SmbHLH148 played a positive role in mediating the pathway of phenolic acid biosynthesis. The content of *SmbHLH51*-OE lines increased 2.19-fold and 1.59-fold, respectively, and overexpressed SmbHLH148 significantly improved three salvianolic acid levels including caffeic acid, rosmarinic acid, and salvianolic acid B (Du et al. 2018; Wu et al. 2018; Xing et al. 2018a, b). As a special member of the bHLH family, the MYC2 transcription factor has been widely studied. MYC2 is not only the core of the response MJ in plants, but also plays a significant role in secondary metabolism and various growth and development processes (Kazan and Manners 2013; Gangappa et al. 2010). For example, it was reported that SmMYC2 regulated the generation of phenolic acids by activating both primary and secondary metabolic pathways in *S. miltiorrhiza.* Overexpressed *SmMYC2* transgenic plants showed higher SAB content which was 1.88-fold higher than in control lines (Yang et al. 2017).

As the largest transcription factor family, MYB TFs are widely found in plant. The MYB transcription factors are divided into four subfamilies, called 1R-MYB, 2R-MYB, 3R-MYB, and 4R-MYB, respectively, depending on the number of incomplete repeats (one, two, three, or four) in the DNA-binding domain (Katiyar et al. 2012; Zhang et al. 2013; Li and Lu 2014). MYB transcription factors have been shown to be involved in primary and secondary metabolism, cell fate and traits, developmental processes, and responses to biotic and abiotic stresses (Dubos et al. 2010; Katiyar et al. 2012). As a R2R3-MYB transcription factor, SmMYB36 was in involved in tanshinones and phenolic acids biosynthesis regulation. In the overexpressing SmMYB36 hairy roots, the content of SAB, RA, and total phenolic acid decreased significantly compared with the control lines, while the content of tanshinone increased in overexpressing lines (Ding et al. 2017). It also was reported that SmMYB111 promoted accumulation of SAB and RA. The concentrations of RA in *SmMYB111*-OE lines were 3.05 and 3.10 times higher than that in control lines. Meanwhile, the content of SAB in *SmMYB111*-OE lines was about 3.54- and 2.50-fold higher than control (Li et al. 2018).

JASMONATE ZIM-DOMAIN (JAZ) transcriptional repressor plays an important role in the JA signaling pathway. Meanwhile, JAZs can interact with MYC2 transcription factor and repress it function when MJ absent (Thines et al. 2007). SmJAZ8 was found involved in biosynthesis of phenolic acids and tanshinones in *S. miltiorrhiza* hairy roots. The production of phenolic acids was decreased when overexpressed *SmJAZ8*, but increased in RNAi transgenic lines. Meanwhile, the content of tanshinones also declined or enhanced in *SmJAZ8*-OE lines or RNAi lines, respectively (Pei et al. 2017).

The above results indicate that it is feasible to increase phenolic acid production at the level of transcription factor regulation. Altering the expression of a transcription factor can directly or indirectly result in a change in the expression of one or more key enzyme genes in the phenolic acid biosynthetic pathway to accumulate or reduce phenolic acid. The production of phenolic acids can be increased to meet needs of markets by constructing single plants or hairy roots that overexpress transcription factors that positively regulate phenolic acid synthesis.

11.4.4 Callus Cultures of *S. miltiorrhiza* for Enhancing Production of Phenolic Acids

Hairy roots have been previously introduced for the mass production of valuable secondary metabolites, especially some slow-growing medicinal plants. However, studies on using callus and cell suspension cultures in *S. miltiorrhiza* to generate secondary metabolite, especially important pharmacological components, are limited. In 1996, *Taxus* cell cultures were reported that could be an alternative source of paclitaxel and related taxane production. A significant increase in paclitaxel and baccatin III levels was observed in cultured cells of *Taxus* species after treatment with MeJA (Yukimune et al. 1996). In recent years, callus cultures of *S. miltiorrhiza* were reported to be used for producing RA and SAB. Callus cultures were cultivated in MS medium using stem and leaf explants; subsequently, the active ingredients of the extracts from callus cultures were analyzed by high-performance liquid chromatography coupled to DAD and MS (HPLC-DAD-MS). The results showed that extraction of callus cultures from stem produced higher amounts of RA and SAB than callus leaves. The content of RA in stem was $1.27 \pm 0.38\%$ but $0.28 \pm 0.02\%$ in leaves, meanwhile, the SAB production in stem and leaves was $0.87 \pm 0.20\%$ and $0.07 \pm 0.03\%$, respectively (Wu et al. 2016).

Plant callus and cell suspension cultures possessing high-value secondary metabolites are a promising potential alternative source for industrial production of medicinal ingredients like phenolic acids. Cell cultures are insensitive to the external environment and rapidly produce metabolites with pharmacological active ingredients (Shi et al. 2019). How to use the suspension cells and callus of *S. miltiorrhiza* to carry out the production of secondary metabolites of pharmaceutical ingredients needs further investigation.

11.4.5 Making Use of Endophytic Fungus in *S. miltiorrhiza* to Produce Phenolic Acids

Making use of endophytic fungus to produce plant secondary metabolites is a novel technology. *Phoma glomerata* D14, an endophytic fungus isolated from *S. miltiorrhiza*, was also found to produce salvianolic acid C (SAC). However, HPLC analysis found that the production of salvianolic acid C in extract of mycelium and broth was very low, 47.67 ± 0.04 μg/g and 0.054 μg/mL, respectively (Li et al. 2016b). Furthermore, *Fusarium proliferatum* SaR-2 and *Alternaria alternata* SaF-2 were found to exhibit higher levels of phenolics than plant roots. Results of total phenol production detection showed that total phenolic content of *F. proliferatum* SaR-2 and *A. alternata* SaF-2 was 21.75 ± 0.11 mg/g and 20.53 ± 0.08 mg/g, respectively (Li et al. 2015). *Trichoderma atroviride* D16, an endophytic fungus isolated from roots of *S. miltiorrhiza*, has been reported to produce tanshinone I and tanshinone IIA, which was a potential source for industrially production of tanshinone I and tanshinone IIA to meet pharmaceutical needs (Lou et al. 2013). *Chaetomium globosum* D38 isolated from *S. miltiorrhiza* roots could enhance the production of tanshinones, especially for dihydrotanshinone I and cryptotanshinone. Furthermore, this endophytic fungus could co-exist with the root of *S. miltiorrhiza* without toxicity. Both live fungus and its mycelia extract were revealed that could increase the production of tanshinones (Zhai et al. 2018). Due to the low variety and low content of endophytic fungus-producing phenolic acids, more endophytic fungus that produce phenolic acid should be explored and culture conditions optimized to increase phenolic acids production in the future.

11.5 Conclusions and Prospects

S. miltiorrhiza is an important traditional Chinese herbal medicine, which has been widely used in the treatment of cardiovascular and cerebrovascular diseases. Phenolic acid is one group of biologically active compounds in *S. miltiorrhiza*. It has great curative effect and pharmacological activity. However, due to the market demand is increasing, the germplasm resources of *S. miltiorrhiza* are degraded, and the content of phenolic acids in traditionally cultured *S. miltiorrhiza* is low. How to improve the production of phenolic acid has become an urgent problem to be solved.

It was a strategy that the use of modern biological means to regulate the synthesis of phenolic acids, through the expression of genes leading to the accumulation of secondary metabolites. However, its mechanism of action and regulation mode needs further exploration due to its complex secondary metabolic network. Combined with molecular biology, transcriptomics, and metabolomics, the expression of secondary metabolite-related genes and their mechanisms are important for understanding the synthesis and regulation mechanisms of phenolic acids. Currently, the upstream pathway of RA synthesis and the key enzymes involved in regulation have been basically understood, including SmPAL, SmC4H, Sm4CL, SmTAT, SmHPPR, SmRAS, and SmCYP98A14. However, the synthetic pathways of other phenolic acids such as SAB downstream of RA are still unclear. Therefore, exploring the synthetic pathways downstream of RA to other phenolic acids, discovering the key enzymes involved in catalysis and further researching the metabolic mechanism of metabolites, will be beneficial to promote the production of phenolic acids and drug development in *S. miltiorrhiza*.

To date, several types of transcription factors such as bHLH, MYB, JAZ, and AP2/EARF have been reported to regulate phenolic acid biosynthesis. Through the genome and transcriptome data of *S. miltiorrhiza*, it is possible to dig deeper and isolate more genes involved in the regulation of phenolic acids. The strategy of constructing a "transcription factor-biosynthetic pathway critical gene-metabolite" network will contribute to the improved synthesis of phenolic acids in *S. miltiorrhiza*. Meanwhile, comprehensive utilization of synthetic biology, elicitors, hairy roots and transgenic plants, endophytic fungus, genetic engineering and transcriptional regulation to enhance phenolic acid production in *S. miltiorrhiza* is also the direction of future research.

Acknowledgements This work was supported by National Natural Science Fund of China (81522049, 31571735, 31270007), the "Dawn" Program of Shanghai Education Commission (16SG38), Shanghai Science and Technology Committee Project (17JC1404300), Zhejiang Provincial Ten Thousands Program for Leading Talents of Science and Technology Innovation, Zhejiang Provincial Program for the Cultivation of High-level Innovative Health talents.

References

Bao Y, Wang L, Xu Y, Yang Y, Wang L, Si S, Cho S, Hong B (2012) Salvianolic acid B inhibits macrophage uptake of modified low density lipoprotein (mLDL) in a scavenger receptor CD36-dependent manner. Atherosclerosis 223(1):152–159

Bi L, Chen J, Yuan X, Jiang Z, Chen W (2013) Salvianolic acid A positively regulates PTEN protein level and inhibits growth of A549 lung cancer cells. Biomed Rep 1(2):213–217

Cai Z, Kastell A, Speiser C, Smetanska I (2013) Enhanced resveratrol production in *Vitis vinifera* cell suspension cultures by heavy metals without loss of cell viability. Appl Biochem Biotech 171(2):330–340

Cai J, Chen S, Zhang W, Zheng X, Hu S, Pang C, Lu J, Xing J, Dong Y (2014) Salvianolic acid A reverses paclitaxel resistance in human breast cancer MCF-7 cells via targeting the expression of transgelin 2 and attenuating PI3K/Akt pathway. Phytomedicine 21 (12):1725–1732

Cao WZ, Wang Y, Shi M, Hao XL, Wang Y, Zhao WW, Wang Y, Ren J, Kai GY (2018) Transcription factor SmWRKY1 positively promotes the biosynthesis of tanshinones in *Salvia miltiorrhiza*. Front Plant Sci 9:554

Chen T, Liu W, Chao X, Zhang L, Qu Y, Huo J, Fei Z (2011) Salvianolic acid B attenuates brain damage and inflammation after traumatic brain injury in mice. Brain Res Bull 84(2):163–168

Chen RC, Sun GB, Ye JX, Wang J, Zhang MD, Sun XB (2017) Salvianolic acid B attenuates doxorubicin-induced ER stress by inhibiting TRPC3 and TRPC6 mediated Ca^{2+} overload in rat cardiomyocytes. Toxicol Lett 276:21–30

Dhapare S, Sakagami M (2018) Salvianolic acid B as an anti-emphysema agent I: in vitro stimulation of lung cell proliferation and migration, and protection against lung cell death, and in vivo lung STAT3 activation and VEGF elevation. Pulm Pharmacol Ther 53:107–115

Di P, Zhang L, Chen JF, Tan H, Xiao Y, Dong X, Zhou X, Chen WS (2013) ^{13}C tracer reveals phenolic acids biosynthesis in hairy root cultures of *Salvia miltiorrhiza*. ACS Chem Biol 8(7):1537–1548

Ding K, Pei TL, Bai ZQ, Jia YY, Ma PD, Liang ZS (2017) SmMYB36, a novel R2R3-MYB transcription factor, enhances tanshinone accumulation and decreases phenolic acid content in *Salvia miltiorrhiza* hairy roots. Sci Rep 7(1):5104

Du TZ, Niu JF, Su J, Li SS, Guo XR, Li L, Cao XY, Kang JF (2018) SmbHLH37 functions antagonistically with SmMYC2 in regulating jasmonate-mediated biosynthesis of phenolic acids in *Salvia miltiorrhiza*. Front Plant Sci 9:1720

Dubos C, Stracke R, Grotewold E, Weisshaar B, Martin C, Lepiniec L (2010) MYB transcription factors in *Arabidopsis*. Trends Plant Sci 15(10):573–581

Fan HY, Fu FH, Yang MY, Xu H, Zhang AH, Liu K (2010) Antiplatelet and antithrombotic activities of salvianolic acid A. Thromb Res 126(1):e17–e22

Fan HY, Yang MY, Qi D, Zhang ZK, Zhu L, Shang-Guan XX, Liu K, Xu H, Che X (2015) Salvianolic acid A as a multifunctional agent ameliorates doxorubicin-induced nephropathy in rats. Sci Rep 5:12273

Fan Y, Luo QP, Wei JJ, Lin RH, Lin LL, Li YK, Chen ZR, Chen Q, Lin W (2018) Mechanism of salvianolic acid B neuroprotection against ischemia/reperfusion induced cerebral injury. Brain Res 1679:125–133

Fang CY, Wu CZ, Chen PN, Chang YC, Chuang CY, Lai CT, Yang SF, Tsai LL (2018) Antimetastatic potentials of salvianolic acid A on oral squamous cell carcinoma by targeting MMP-2 and the c-Raf/MEK/ERK pathway. Environ Toxicol 33 (5):545–554

Gangappa SN, Prasad VBR, Chattopadhyay S (2010) Functional interconnection of MYC2 and SPA1 in the photomorphogenic seedling development of *Arabidopsis*. Plant Physiol 154(3):1210–1219

Gong L, Di C, Xia X, Wang J, Chen G, Shi J, Chen P, Xu H, Zhang W (2016) AKT/mTOR signaling

pathway is involved in salvianolic acid B-induced autophagy and apoptosis in hepatocellular carcinoma cells. Int J Oncol 49(6):2538–2548
Guillon S, Trémouillaux-Guiller J, Pati PK, Rideau M, Gantet P (2006) Hairy root research: recent scenario and exciting prospects. Curr Opin Plant Biol 9 (3):341–346
Guo P, Wang J, Gao W, Liu X, Wu S, Wan B, Xu L, Li Y (2018) Salvianolic acid B reverses multidrug resistance in nude mice bearing human colon cancer stem cells. Mol Med Rep 18(2):1323–1334
Han JY, Li Q, Ma ZZ, Fan JY (2017) Effects and mechanisms of compound Chinese medicine and major ingredients on microcirculatory dysfunction and organ injury induced by ischemia/reperfusion. Pharm Ther 177:146–173
Hao GP, Ji HW, Li YL, Shi RJ, Wang JM, Feng L, Huang LQ (2012) Exogenous ABA and polyamines enhanced salvianolic acids contents in hairy root cultures of *Salvia miltiorrhiza* Bge. *f. alba*. Plant Omics 5(5):446
Heo JY, Im DS (2019) Anti-allergic effects of salvianolic acid A and tanshinone IIA from *Salvia miltiorrhiza* determined using in vivo and in vitro experiments. Int Immunopharmacol 67:69–77
Hou X, Shao F, Ma Y, Lu S (2013) The phenylalanine ammonia-lyase gene family in *Salvia miltiorrhiza*: genome-wide characterization, molecular cloning and expression analysis. Mol Biol Rep 40(7):4301–4310
Hou BY, Qiang GF, Zhao YR, Yang XY, Chen X, Yan Y, Wang XB, Liu C, Zhang L, Du GH (2017) Salvianolic acid A protects against diabetic nephropathy through ameliorating glomerular endothelial dysfunction via inhibiting AGE-RAGE signaling. Cell Physiol Biochem 44(6):2378–2394
Huang BB, Duan YB, Yi B, Sun LN, Lu B, Yu XH, Sun H, Zhang W, Chen WS (2008a) Characterization and expression profiling of cinnamate 4-hydroxylase gene from *Salvia miltiorrhiza* in rosmarinic acid biosynthesis pathway. Russ J Plant Physiol 55(3):390
Huang BB, Yi B, Duan YB, Sun LN, Yu XH, Guo J, Chen WS (2008b) Characterization and expression profiling of tyrosine aminotransferase gene from *Salvia miltiorrhiza* (Dan-shen) in rosmarinic acid biosynthesis pathway. Mol Biol Rep 35(4):601–612
Huang ZS, Zeng CL, Zhu LJ, Jiang L, Li N, Hu H (2010) Salvianolic acid A inhibits platelet activation and arterial thrombosis via inhibition of phosphoinositide 3-kinase. J Thromb Haemost 8(6):1383–1393
Huang D, Wei X, Mu H, Pan C, Li Q, Hu B, Chang X, Yan L, Fan J, Liu Y, Luo J, Han J (2019) Total salvianolic acid injection prevents ischemia/reperfusion-induced myocardial injury via antioxidant mechanism involving mitochondrial respiratory chain through the upregulation of sirtuin1 and sirtuin3. Shock 51(6):745–756
Huttunen S, Toivanen M, Liu C, Tikkanen-Kaukanen C (2016) Novel anti-infective potential of salvianolic acid B against human serious pathogen *Neisseria meningitidis*. BMC Res Notes 9(1):25
Jiao M, Cao R, Chen H, Hao W, Dong JE (2012) Effects of salicylic acid on synthesis of rosmarinic acid and related enzymes in the suspension cultures of *Salvia miltiorrhiza*. Chin J Biotechnol 28(3):320–328
Jin XQ, Chen ZW, Tan RH, Zhao SJ, Hu ZB (2012) Isolation and functional analysis of 4-coumarate: coenzyme A ligase gene promoters from *Salvia miltiorrhiza*. Biol Plant 56(2):261–268
Jing Z, Fei W, Zhou J, Zhang L, Chen L, Zhang X, Liang X, Xie J, Fang Y, Sui X, Han W, Pan H (2016) Salvianolic acid B, a novel autophagy inducer, exerts antitumor activity as a single agent in colorectal cancer cells. Oncotarget 7(38):61509–61519
Kai GY, Zhang A, Guo YY, Li L, Cui LJ, Luo XQ, Liu C, Xiao JB (2012) Enhancing the production of tropane alkaloids in transgenic *Anisodus acutangulus* hairy root cultures by over-expressing tropinone reductase I and hyoscyamine-6β-hydroxylase. Mol BioSyst 8 (11):2883–2890
Katiyar A, Smita S, Lenka SK, Rajwanshi R, Chinnusamy V, Bansal KC (2012) Genome-wide classification and expression analysis of MYB transcription factor families in rice and *Arabidopsis*. BMC Genom 13(1):544
Kazan K, Manners JM (2013) MYC2: the master in action. Mol Plant 6(3):686–703
Kong WJ, Zhang SS, Zhao YL, Wu MQ, Chen P, Wu XR, Ma XP, Guo WY, Yang MH (2017) Combination of chemical fingerprint and bioactivity evaluation to explore the antibacterial components of *Salvia miltiorrhiza*. Sci Rep 7(1):8112
Lee YW, Kim DH, Jeon SJ, Park SJ, Kim JM, Jung JM, Lee HE, Bae SG, Oh HK, Son KH, Ryu JH (2013) Neuroprotective effects of salvianolic acid B on an Abeta25-35 peptide-induced mouse model of Alzheimer's disease. Eur J Pharmacol 704(1–3):70–77
Li CL, Lu SF (2014) Genome-wide characterization and comparative analysis of R2R3-MYB transcription factors shows the complexity of MYB-associated regulatory networks in *Salvia miltiorrhiza*. BMC Genom 15(1):277
Li YL, Xin XM, Chang ZY, Shi RJ, Miao ZM, Ding J, Hao GP (2015) The endophytic fungi of *Salvia miltiorrhiza* Bge. *f. alba* are a potential source of natural antioxidants. Bot Stud 56(1):5
Li L, Xu T, Du Y, Pan D, Wu W, Zhu H, Zhang Y, Li D (2016a) Salvianolic acid A attenuates cell apoptosis, oxidative stress, Akt and NF-κB activation in angiotensin-II induced murine peritoneal macrophages. Curr Pharm Biotechnol 17(3):283–290
Li XQ, Zhai X, Shu ZH, Dong RF, Ming QL, Qin LP, Zheng CJ (2016b) Phoma glomerata D14: an endophytic fungus from *Salvia miltiorrhiza* that produces salvianolic acid C. Curr Microbiol 73(1):31–37
Li SS, Wu YC, Kuang J, Wang HQ, Du TZ, Huang YY, Zhang Y, Cao XY, Wang ZZ (2018) SmMYB111 is a key factor to phenolic acid biosynthesis and interacts with both SmTTG1 and SmbHLH51 in *Salvia miltiorrhiza*. J Agric Food Chem 66(30):8069–8078

Liang ZS, Ma YN, Xu T, Cui BM, Liu Y, Guo ZX, Yang DF (2013) Effects of abscisic acid, gibberellin, ethylene and their interactions on production of phenolic acids in *Salvia miltiorrhiza* Bunge hairy roots. PLoS ONE 8(9):e72806

Liu L, Li J, Zhang Y, Zhang S, Ye J, Wen Z, Ding J, Kunapuli SP, Luo X, Ding Z (2014) Salvianolic acid B inhibits platelets as a P2Y12 antagonist and PDE inhibitor: evidence from clinic to laboratory. Thromb Res 134(4):866–876

Liu H, Ma S, Xia H, Lou H, Zhu F, Sun L (2018) Anti-inflammatory activities and potential mechanisms of phenolic acids isolated from *Salvia miltiorrhiza f. alba* roots in THP-1 macrophages. J Ethnopharmacol 222:201–207

Lou JF, Fu LY, Luo RY, Wang XH, Luo HY, Zhou LG (2013) Endophytic fungi from medicinal herb *Salvia miltiorrhiza* Bunge and their antimicrobial activity. Afr J microbiol Res 7(47):5343–5349

Lv HD, Wang L, Shen JC, Hao SJ, Ming AM, Wang XD, Su F, Zhang ZC (2015) Salvianolic acid B attenuates apoptosis and inflammation via SIRT1 activation in experimental stroke rats. Brain Res Bull 115:30–36

Ma XH, Ma Y, Tang JF, He YL, Liu YC, Ma XJ, Shen Y, Cui GH, Lin HX, Guo J, Huang LQ (2015) The biosynthetic pathways of tanshinones and phenolic acids in *Salvia miltiorrhiza*. Molecules 20(9):16235–16254

Ma ZG, Xia HQ, Cui SL, Yu J (2017) Attenuation of renal ischemic reperfusion injury by salvianolic acid B via suppressing oxidative stress and inflammation through PI3K/Akt signaling pathway. Braz J Med Biol Res 50(6):e5954

Mahmood Q, Wang GF, Wu G, Wang H, Zhou CX, Yang HY, Liu ZR, Han F, Zhao K (2017) Salvianolic acid A inhibits calpain activation and eNOS uncoupling during focal cerebral ischemia in mice. Phytomedicine 25:8–14

Meng D, Li J, Li H, Wang K (2019) Salvianolic acid B remits LPS-induced injury by up-regulating miR-142-3p in MH7A cells. Biomed Pharmacother 115:108876

Namdeo AG (2007) Plant cell elicitation for production of secondary metabolites: a review. Pharmacogn Rev 1 (1):69–79

Pei T, Ma P, Ding K, Liu S, Jia Y, Ru M, Dong J, Liang ZS (2017) SmJAZ8 acts as a core repressor regulating JA-induced biosynthesis of salvianolic acids and tanshinones in *Salvia miltiorrhiza* hairy roots. J Exp Bot 69(7):1663–1678

Pei R, Si T, Lu Y, Zhou JX, Jiang L (2018) Salvianolic acid A, a novel PI3K/Akt inhibitor, induces cell apoptosis and suppresses tumor growth in acute myeloid leukemia. Leuk Lymphoma 59(8):1959–1967

Petersen M, Simmonds MS (2003) Rosmarinic acid. Phytochemistry 62(2):121–125

Pezeshki S, Petersen M (2018) Rosmarinic acid and related metabolites. In: Biotechnology of natural products. Springer, Cham, pp 25–60

Qian W, Wang Z, Xu T, Li D (2019) Anti-apoptotic effects and mechanisms of salvianolic acid A on cardiomyocytes in ischemia-reperfusion injury. Histol Histopathol 34(3):223–231

Qiang G, Yang X, Shi L, Zhang H, Chen B, Zhao Y, Zu M, Zhou D, Guo J, Yang H, Zhang L, Du G (2015) Antidiabetic effect of salvianolic acid A on diabetic animal models via AMPK activation and mitochondrial regulation. Cell Physiol Biochem 36(1):395–408

Rahbardar MG, Amin B, Mehri S, Mirnajafi-Zadeh SJ, Hosseinzadeh H (2018) Rosmarinic acid attenuates development and existing pain in a rat model of neuropathic pain: an evidence of anti-oxidative and anti-inflammatory effects. Phytomedicine 40:59–67

Raoufi S, Baluchnejadmojarad T, Roghani M, Ghazanfari T, Khojasteh F, Mansouri M (2015) Antidiabetic potential of salvianolic acid B in multiple low-dose streptozotocin-induced diabetes. Pharm Biol 53 (12):1803–1809

Sha W, Zhou YF, Ling ZQ, Xie GQ, Pang XW, Wang P, Gu XB (2018) Antitumor properties of salvianolic acid B against triple-negative and hormone receptor-positive breast cancer cells via ceramide-mediated apoptosis. Oncotarget 9(91):36331

Shi M, Luo XQ, Ju GH, Yu XH, Hao XL, Huang Q, Xiao JB, Cui LJ, Kai GY (2014) Increased accumulation of the cardio-cerebrovascular disease treatment drug tanshinone in *Salvia miltiorrhiza* hairy roots by the enzymes 3-hydroxy-3-methylglutaryl CoA reductase and 1-deoxy-D-xylulose 5-phosphate reductoisomerase. Func Integr Genomics 14(3):603–615

Shi X, Sun RM, Zhao Y, Fu R, Wang RW, Zhao HY, Wang ZC, Tang F, Zhang N, Tian XF, Yao JH (2018) Promotion of autophagosome–lysosome fusion via salvianolic acid A-mediated SIRT1 up-regulation ameliorates alcoholic liver disease. RSC Adv 8 (36):20411–20422

Shi M, Huang FF, Deng CP, Wang Y, Kai GY (2019) Bioactivities, biosynthesis and biotechnological production of phenolic acids in *Salvia miltiorrhiza*. Crit Rev Food Sci 59(6):953–964

Song J (2010) Function analysis of the genes involved in rosmarinic acid biosynthesis pathway in *Salvia miltiorrhiza* Bunge. Unpublished Ph.D. thesis, Shaanxi Normal University, Xian, China

Song J, Wang ZZ (2009) Molecular cloning, expression and characterization of a phenylalanine ammonia-lyase gene (SmPAL1) from *Salvia miltiorrhiza*. Mol Biol Rep 36(5):939

Song J, Wang ZZ (2011) RNAi-mediated suppression of the phenylalanine ammonia-lyase gene in *Salvia miltiorrhiza* causes abnormal phenotypes and a reduction in rosmarinic acid biosynthesis. J Plant Res 124 (1):183–192

Sun WY, Tong L, Miao JZ, Huang JY, Li DX, Li YF, Xiao HT, Sun H, Bi KS (2016) Separation and analysis of phenolic acids from *Salvia miltiorrhiza* and its related preparations by off-line two-dimensional hydrophilic interaction chromatography × reversed-phase liquid chromatography coupled with ion trap time-of-flight mass spectrometry. J Chromatogr A 1431:79–88

Sun MH, Wang Y, Huang Q, Yuan TP, Wang Q, Wang C, Zhou W, Kai GY (2019) The biosynthesis of phenolic acids is positively regulated by the JA-responsive transcription factor ERF115 in *Salvia miltiorrhiza*. J Exp Bot 70(1):243–254

Tang XL, Yan L, Zhu L, Jiao DM, Chen J, Chen QY (2017) Salvianolic acid A reverses cisplatin resistance in lung cancer A549 cells by targeting c-met and attenuating Akt/mTOR pathway. J Pharmacol Sci 135 (1):1–7

Thines B, Katsir L, Melotto M, Niu Y, Mandaokar A, Liu G, Nomura K, He S, Howe G, Browse J (2007) JAZ repressor proteins are targets of the SCF COI1 complex during jasmonate signalling. Nature 448 (7154):661

Wang YM, Cao JL (2016) Advances in the chemical and pharmacological studies of phenolic acids in *Salvia miltiorrhiza*. World Chin Med 11(6):1126–1130

Wang W, Hu W (2018) Salvianolic acid B recovers cognitive deficits and angiogenesis in a cerebral small vessel disease rat model via the STAT3/VEGF signaling pathway. Mol Med Rep 17(2):3146–3151

Wang JW, Wu JY (2013) Effective elicitors and process strategies for enhancement of secondary metabolite production in hairy root cultures. Adv Biochem Eng Biotechnol 134:55–89

Wang F, Liu YY, Liu LY, Zeng QJ, Wang CS, Sun K, Yang JY, Guo J, Fan JY, Han JY (2009) The attenuation effect of 3,4-dihydroxy-phenyl lactic acid and salvianolic acid B on venular thrombosis induced in rat mesentery by photochemical reaction. Clin Hemorheol Microcirc 42(1):7–18

Wang B, Sun W, Li QS, Li Y, Luo HM, Song JY, Sun C, Qian J, Zhu YJ, Hayward A, Xu HB (2015a) Genome-wide identification of phenolic acid biosynthetic genes in *Salvia miltiorrhiza*. Planta 241(3):711–725

Wang X, Wang CY, Zhang LJ, Li YJ, Wang SJ, Wang JD, Yuan CY, Niu J, Wang CS, Lu GM (2015b) Salvianolic acid A shows selective cytotoxicity against multidrug-resistant MCF-7 breast cancer cells. Anticancer Drugs 26(2):210–223

Wang XK, Qi D, Fu FH, LiX Liu Y, Ji K, Gao ZF, Kong LL, Yu C, Xie H, Yue G, Zhu H, Liu K, Fan HY (2019) Therapeutic and antiproteinuric effects of salvianolic acid A in combined with low-dose prednisone in minimal change disease rats: Involvement of PPARγ/Angptl4 and Nrf2/HO-1 pathways. Eur J Pharmacol. https://doi.org/10.1016/j.ejphar.2019.04.023

Wasternack C, Hause B (2013) Jasmonates: biosynthesis, perception, signal transduction and action in plant stress response, growth and development. An update to the 2007 review in Annals of Botany. Ann Bot 111 (6):1021–1058

Wu CF, Karioti A, Rohr D, Bilia AR, Efferth T (2016) Production of rosmarinic acid and salvianolic acid B from callus culture of *Salvia miltiorrhiza* with cytotoxicity towards acute lymphoblastic leukemia cells. Food Chem 201:292–297

Wu YC, Zhang Y, Li L, Guo XR, Wang B, Cao XY, Wang ZZ (2018) AtPAP1 interacts with and activates SmbHLH51, a positive regulator to phenolic acids biosynthesis in *Salvia miltiorrhiza*. Front Plant Sci 9:1687

Xia ZB, Yuan YJ, Zhang QH, Li H, Dai JL, Min JK (2018) Salvianolic acid B suppresses inflammatory mediator levels by downregulating NF-κB in a rat model of rheumatoid arthritis. Med Sci Monit 24:2524–2532

Xiao Y, Gao SH, Di P, Chen JF, Chen WS, Zhang L (2009) Methyl jasmonate dramatically enhances the accumulation of phenolic acids in *Salvia miltiorrhiza* hairy root cultures. Physiol Plant 137(1):1–9

Xiao Y, Zhang L, Gao SH, Saechao S, Di P, Chen JF, Chen WS (2011) The *c4h*, *tat*, *hppr* and *hppd* genes prompted engineering of rosmarinic acid biosynthetic pathway in *Salvia miltiorrhiza* hairy root cultures. PLoS ONE 6(12):e29713

Xing BC, Yang DF, Guo WL, Liang ZS, Yan XJ, Zhu YH, Liu Y (2015) Ag^+ as a more effective elicitor for production of tanshinones than phenolic acids in *Salvia miltiorrhiza* hairy roots. Molecules 20(1):309–324

Xing BC, Liang LJ, Liu L, Hou ZN, Yang DF, Yan KJ, Zhang XM, Liang ZS (2018a) Overexpression of SmbHLH148 induced biosynthesis of tanshinones as well as phenolic acids in *Salvia miltiorrhiza* hairy roots. Plant Cell Rep 37(12):1681–1692

Xing BC, Yang DF, Liu L, Han RL, Sun YF, Liang ZS (2018b) Phenolic acid production is more effectively enhanced than tanshinone production by methyl jasmonate in *Salvia miltiorrhiza* hairy roots. PCTOC 134(1):119–129

Xiong W, Liu LX, Wu C, Yang C, Wu QY (2010) ^{13}C-tracer and gas chromatography-mass spectrometry analyses reveal metabolic flux distribution in the oleaginous microalga *Chlorella protothecoides*. Plant Physiol 154(2):1001–1011

Yan Q, Shi M, Ng J, Wu JY (2006) Elicitor-induced rosmarinic acid accumulation and secondary metabolism enzyme activities in *Salvia miltiorrhiza* hairy roots. Plant Sci 170(4):853–858

Yang RX, Huang SY, Yan FF, Lu XT, Xing YF, Liu Y, Liu YF, Zhao YX (2010) Danshensu protects vascular endothelia in a rat model of hyperhomocysteinemia. Acta Pharmacol Sin 31(10):1395–1400

Yang N, Zhou WB, Su J, Wang XF, Li L, Wang LR, Cao XY, Wang ZZ (2017) Overexpression of SmMYC2 increases the production of phenolic acids in *Salvia miltiorrhiza*. Front Plant Sci 8:1804

Yang YF, Zhang LC, La XQ, Li ZY, Li HQ, Guo SJ (2019) Salvianolic acid A inhibits tumor-associated angiogenesis by blocking GRP78 secretion. Naunyn Schmiedebergs Arch Pharmacol 392(4):467–480

Yu DS, Wang YS, Bi YL, Guo ZP, Yuan YJ, Tong SM, Su RC, Ge LH, Wang J, Pan YL, Guan TT, Cao Y (2017) Salvianolic acid A ameliorates the integrity of blood-spinal cord barrier via miR-101/Cul3/Nrf2/HO-1 signaling pathway. Brain Res 1657:279–287

Yu HZ, Guo WL, Yang DF, Hou ZN, Liang ZS (2018) Transcriptional profiles of *SmWRKY* family genes and their putative roles in the biosynthesis of tanshinone and phenolic acids in *Salvia miltiorrhiza*. Int J Mol Sci 19(6):1593

Yukimune Y, Tabata H, Higashi Y, Hara Y (1996) Methyl jasmonate-induced overproduction of paclitaxel and baccatin III in Taxus cell suspension cultures. Nat Biotechnol 14(9):1129–1132

Zhai X, Luo D, Li X, Han T, Jia M, Kong ZY, Ji JC, Rahman K, Qin LP, Zheng CJ (2018) Endophyte *Chaetomium globosum* D38 promotes bioactive constituents accumulation and root production in *Salvia miltiorrhiza*. Front Microbiol 8:2694

Zhang Y, Yan YP, Wang ZZ (2010) The Arabidopsis PAP1 transcription factor plays an important role in the enrichment of phenolic acids in *Salvia miltiorrhiza*. J Agric Food Chem 58(23):12168–12175

Zhang HA, Gao M, Zhang L, Zhao Y, Shi LL, Chen BN, Wang YH, Wang SB, Du GH (2012) Salvianolic acid A protects human SH-SY5Y neuroblastoma cells against H_2O_2-induced injury by increasing stress tolerance ability. Biochem Biophys Res Commun 421(3):479–483

Zhang SC, Ma PD, Yang DF, Li WJ, Liang ZS, Liu Y, Liu FH (2013) Cloning and characterization of a putative R2R3 MYB transcriptional repressor of the rosmarinic acid biosynthetic pathway from *Salvia miltiorrhiza*. PLoS ONE 8(9):e73259

Zhang SC, Yan Y, Wang BQ, Liang ZS, Liu Y, Liu FH, Qi ZH (2014a) Selective responses of enzymes in the two parallel pathways of rosmarinic acid biosynthetic pathway to elicitors in *Salvia miltiorrhiza* hairy root cultures. J Biosci Bioeng 117(5):645–651

Zhang XC, Chen JQ, Li B (2014b) Salvianolic acid A suppresses CCL-20 expression in TNF-α-treated macrophages and ApoE-deficient mice. J Cardiovasc Pharmacol 64(4):318–325

Zhang Y, Yan YP, Wu YC, Hua WP, Chen C, Ge Q, Wang ZZ (2014c) Pathway engineering for phenolic acid accumulations in *Salvia miltiorrhiza* by combinational genetic manipulation. Metab Eng 21:71–80

Zhang X, Wu Q, Lu Y, Wan J, Dai H, Zhou X, Lv S, Chen X, Zhang X, Hang C, Wang J (2018a) Cerebroprotection by salvianolic acid B after experimental subarachnoid hemorrhage occurs via Nrf2- and SIRT1-dependent pathways. Free Radical Biol Med 124:504–516

Zhang JL, Zhang X, Zhang JB, Li MY, Chen DJ, Wu T (2018b) Minor compounds of the high purity salvianolic acid B freeze-dried powder from *Salvia miltiorrhiza* and antibacterial activity assessment. Nat Prod Res 32(10):1198–1202

Zhang JY, Wang M, Wang RY, Sun X, Du YY, Ye JX, Sun GB, Sun XB (2018c) Salvianolic acid A ameliorates arsenic trioxide-induced cardiotoxicity through decreasing cardiac mitochondrial injury and promotes its anticancer activity. Front Pharmacol 9:487

Zhang W, Song JK, Zhang X, Zhou QM, He GR, Xu XN, Yan R, Zhou WX, Du GH (2018d) Salvianolic acid A attenuates ischemia reperfusion induced rat brain damage by protecting the blood brain barrier through MMP-9 inhibition and anti-inflammation. Chin J Nat Med 16(3):184–193

Zhang H, Wang Y, Gao C, Gu Y, Huang J, Wang J, Wang J, Zhang Z (2018e) Salvianolic acid A attenuates kidney injury and inflammation by inhibiting NF-κB and p38 MAPK signaling pathways in 5/6 nephrectomized rats. Acta Pharmacol Sin 39(12):1855–1864

Zhang HF, Wang JH, Wang YL, Gao C, Gu YT, Huang J, Wang JH, Zhang Z (2019a) Salvianolic acid A protects the kidney against oxidative stress by activating the Akt/GSK-3β/Nrf2 signaling pathway and inhibiting the NF-κB signaling pathway in 5/6 nephrectomized rats. Oxid Med Cell Longev 2019:2853534

Zhang YF, Xu LW, Liang K, Zhou LH, Ge YZ, Jia RP (2019b) Protective effect of salvianolic acid B against oxidative injury associated with cystine stone formation. Urolithiasis. https://doi.org/10.1007/s00240-019-01114-4

Zhao SJ, Hu ZB, Liu D, Leung FC (2006) Two divergent members of 4-coumarate: coenzyme A ligase from *Salvia miltiorrhiza* Bunge: cDNA cloning and functional study. J Integr Plant Biol 48(11):1355–1364

Zhao GR, Zhang HM, Ye TX, Xiang ZJ, Yuan YJ, Guo ZX, Zhao LB (2008) Characterization of the radical scavenging and antioxidant activities of danshensu and salvianolic acid B. Food Chem Toxicol 46(1):73–81

Zhao J, Yang XC, Fujino M, Ichimaru N, Que W, Li XK, Takahara S (2019) Salvianolic acid B ameliorates liver injury in a murine aGvHD model by decreasing inflammatory responses via upregulation of HO-1. Transpl Immunol. https://doi.org/10.1016/j.trim.2019.03.002

Zheng X, Chen S, Yang Q, Cai J, Zhang W, You H, Xing J, Dong Y (2015) Salvianolic acid A reverses the paclitaxel resistance and inhibits the migration and invasion abilities of human breast cancer cells by inactivating transgelin 2. Cancer Biol Ther 16(9):1407–1414

Zhou ML, Memelink J (2016) Jasmonate-responsive transcription factors regulating plant secondary metabolism. Biotechnol Adv 34(4):441–449

Zhou Z, Zhang Y, Ding XR, Chen SH, Yang J, Wang XJ, Jia GL, Chen HS, Bo XC, Wang SQ (2007) Protocatechuic aldehyde inhibits hepatitis B virus replication both in vitro and in vivo. Antiviral Res 74 (1):59–64

Zhou YQ, Li WZ, Xu L, Chen L (2011) In *Salvia miltiorrhiza*, phenolic acids possess protective properties against amyloid β-induced cytotoxicity, and tanshinones act as acetylcholinesterase inhibitors. Environ Toxicol Pharmacol 31(3):443–452

Zhou XL, Chan SW, Tseng HL, Deng Y, Hoi PM, Choi PS, Or PM, Yang JM, Lam FF, Lee SM, Leung GP, Kong SK, Ho HP, Kwan YW, Yeung JHK (2012) Danshensu is the major marker for the antioxidant and vasorelaxation effects of Danshen (*Salvia miltiorrhiza*) water-extracts produced by different heat water-extractions. Phytomedicine 19 (14):1263–1269

Zhou J, Qu XD, Li ZY, Wei J, Liu Q, Ma YH, He JJ (2014) Salvianolic acid B attenuates toxin-induced neuronal damage via Nrf2-dependent glial cells-mediated protective activity in Parkinson's disease models. PLoS ONE 9(7):e101668

Zhou R, Long H, Zhang B, Lao Z, Zheng Q, Wang T, Zhang Y, Wu Q, Lai X, Li G, Lin L (2019) Salvianolic acid B, an antioxidant derived from Salvia militarize, protects mice against γ-radiation-induced damage through Nrf2/Bach1. Mol Med Rep 19(2): 1309–1317

Tissue Culture and Hairy Root Induction of *Salvia miltiorrhiza*

12

Fenjuan Shao and Deyou Qiu

Abstract

Salvia miltiorrhiza is a well-known medicinal plant with great economic and medicinal value. Its root contains two major groups of bioactive compounds, including lipophilic diterpenoid tanshinones and hydrophilic phenolic acids. *S. miltiorrhiza* has been widely used in traditional Chinese medicine (TCM) for treating various human diseases for thousands of years, such as dysmenorrhea, amenorrhea, and cardiovascular disease. The wild *S. miltiorrhiza* resources are far from being able to meet the market demand. It is necessary to develop more efficient methods to produce reliable and sustainable resources. The in vitro culture is a high-quality method to improve the quality and the proliferation of *S. miltiorrhiza* seedlings. It can directly produce the active constituents of *S. miltiorrhiza* on a large scale, and it will solve the problems in the current cultivation process and the quality control of *S. miltiorrhiza*. In this chapter, we review recent progress in vitro cultures of *S. miltiorrhiza*, including cell culture, tissue culture, and hairy root culture, as well as possible factors affecting the production of active constituents in the culture.

F. Shao (✉) · D. Qiu
State Key Laboratory of Tree Genetics and Breeding, Key Laboratory of Tree Breeding and Cultivation of State Forestry Administration, Research Institute of Forestry, Chinese Academy of Forestry, Beijing, China
e-mail: sfjstar@126.com

D. Qiu
e-mail: qiudy@caf.ac.cn

12.1 Introduction

Salvia miltiorrhiza, known as Danshen in Chinese and belonging to the Lamiaceae family, is a well-known medicinal plant with great economic and medicinal value (Wagstaff et al. 1995). The root of *S. miltiorrhiza* has been widely used in traditional Chinese medicine (TCM). *S. miltiorrhiza* contains two major groups of bioactive compounds, lipophilic diterpenoid tanshinones and hydrophilic phenolic acids (Cheng 2006a). They were used for treating various human diseases, such as dysmenorrhea, amenorrhea, and cardiovascular disease for thousands of years (Cheng 2006b; Li et al. 2009; Wang et al. 2007). Furthermore, *S. miltiorrhiza* is an emerging model plant for TCM studies because of its relatively small genome size, short life cycle, undemanding growth requirements, and significant medicinal value (Xu et al. 2016).

Due to the medicinal and healthcare function, *S. miltiorrhiza* has a broad international market prospect and development potential. It has been estimated that 20 million kg of *S. miltiorrhiza* is needed in China each year. With the expansion

S. Lu (ed.), *The Salvia miltiorrhiza Genome*, Compendium of Plant Genomes,
https://doi.org/10.1007/978-3-030-24716-4_12

of *S. miltiorrhiza* market and the dramatic increase of demand, the wild *S. miltiorrhiza* resources are far from being able to meet the market demand. Tissue culture and hairy root culture are high-quality methods for improvement of the quality and yield of *S. miltiorrhiza*, and for direct production of effective components of *S. miltiorrhiza*.

This chapter summarizes the achievements in the field of tissue culture and hairy root culture of *S. miltiorrhiza*.

12.2 Cell Culture

There are a large number of secondary metabolites in plants. These secondary metabolites can be utilized as food, cosmetic, and drug, which are closely related to human life and have great values in use. However, there are many disadvantages in the production of these metabolites from plants. So, plant cell culture is developed into a potential technology for the production of active constituents. The approach of plant cell culture has many advantages in the production of active constituents. For example, it is more sustainable and well-controlled, and is easy to produce the constituents with complex structures (Ochoa-Villarreal et al. 2016). Cell culture systems of *S. miltiorrhiza* were first established in 1981 (Chen et al. 1981; Huang et al. 1981; Qiu and Song 2015). The callus was induced from roots on the MS medium with 1 mg/L 2,4-Dichlorophenoxyacetic acid (2,4-D) and 0.1 mg/L kinetin (KT). The callus contained active constituents, including cryptotanshinone, protocatechualdehyde, and diterpenoid quinone compounds. After that, cell cultures of *S. miltiorrhiza* attracted many scientists' attention. Many laboratories have established cell culture methods using different explants (Shan et al. 2007). In 1983, six cell lines were induced from seedlings of *S. miltiorrhiza* on MS medium 1 mg/L 2,4-D and 0.1 mg/L KT. Suspension cell culture of these lines in a liquid MS medium showed that line A contained two terpenoids, cryptotanshinone and ferruginol (Nakanish et al. 1983).

In order to realize large-scale industrial production of plant cell secondary metabolites, researchers began to study the possible factors affecting active constituent production. They included the conditions of cell cultures and the addition of elicitors. Miyasaka et al. (1987) studied the effects of the components of MS medium and the light on the production of the diterpenes cryptotanshinone and ferruginol in cell suspension cultures of *S. miltiorrhiza*. They showed that the components of MS medium, including sucrose, nitrogen source, and thiamine, affected the production of cryptotanshinone and ferruginol. Phosphate, $MnSO_4$, and KT showed slight beneficial effects. Light had little effects on cell growth, but inhibited the production of ferruginol. Based on these results, Miyasaka et al. (1987) designed a revised medium for the production of cryptotanshinone.

Phytohormone plays an important role in plant cell culture. It not only affects the growth of cells but also affects the biosynthesis of secondary metabolites in cells (Putalun et al. 2004). Huang et al. (2000) reported the effects of phytohormone in secondary metabolite biosynthesis. They investigated the influences of 2,4-D, Gibberellin A3 (GA3) and Benzylaminopurine (BA) in the growth of cells and the yields of secondary metabolites. The results showed that 2,4-D promoted cell growth, but inhibited lithospermic acid B production. GA3 inhibited cell growth, but promoted the production of rosmarinic acid and lithospermic acid B. BA was intermediate between 2,4-D and GA3.

Elicitors are the biotic and abiotic factors supplemented in the medium. They can promote the production of secondary metabolites in plant tissue cultures. The most common biotic elicitors include jasmonic acid (JA), methyl jasmonate (MeJA), and salicylic acid (SA), which are signaling molecules associated with plant defense. The abiotic elicitors contain ultraviolet (UV) radiation, osmotic stress, and heavy metal salts. The dosage and time of elicitor addition to the medium are very important for the production of secondary metabolites. In addition, the elicitors can be used alone or in combination to promote the production of secondary metabolites

(Park et al. 2017; Zhou and Wu 2006). The effect of elicitors on the secondary metabolites in the suspension cells of *S. miltiorrhiza* has been widely studied. Table 12.1 shows the elicitors used to induce the production of secondary metabolites in cell and hairy root cultures.

Yu et al. (2018) investigated the effect of different elicitors on the primary and secondary metabolites in the suspension cell culture and tissue culture of *S. miltiorrhiza* Bunge. The results showed that SA could significantly promote glucose metabolism and increase the production of ten secondary metabolites, including caffeic acid, rosmarinic acid, salvianolic acid B, cryptotanshinone, shikimic acid, salvianolic acid A, dihydrotanshinone, *p*-coumaric acid, tanshinol, and tanshinone IIA, in suspension cells. However, SA had no significant effects on the production of protocatechuic acid and tanshinone I. In addition, the content of the ten secondary metabolites was also promoted by NaCl and $AgNO_3$ in the suspension cells.

Dong et al. (2010) reported that SA promoted the production of phenolic compounds. Meanwhile, the activities of phenylalanine ammonia lyase (PAL), tyrosine aminotransferase (TAT), superoxide Dismutase (SOD), catalase (CAT), and peroxidase (POD) enzymes were increased in *S. miltiorrhiza* cell culture (Dong et al. 2010).

Zhao et al. (2010a, b) studied the effects of biotic and abiotic elicitors on the production of diterpenoid tanshinones in *S. miltiorrhiza* cell culture. Heavy metal ions (Co^{2+}, Ag^+, Cd^{2+}), polysaccharides (yeast extract and chitosan), plant response-signaling compound (SA and MeJA), and hyperosmotic stress with sorbitol were used as elicitors. The results showed that the production of tanshinone was significantly

Table 12.1 Elicitors used to induce the production of secondary metabolites in cell culture

Elicitor	Tissue culture type	Medium	Metabolites	References
SA (200 μmol/L), NaCl (50 mmol/L), $AgNO_3$ (60 μmol/L)	Suspension cells	MS with 0.5 mg/L 2,4-D, 1 mg/L kinetin, 30 g/L sucrose, 6 g/L gelrite and 1 g/L acid hydrolyzed casein	Shikimic acid, tanshinol, protocatechuic acid, caffeic acid, *p*-coumaric acid, rosmarinic acid, salvianolic acid B, salvianolic acid A, dihydrotanshinone, cryptotanshinone, tanshinone I and tanshinone II A	Yu et al. (2018)
SA	Cell culture	MS with 100 mg/L myoinositol, 30 g/L sucrose and 4.2 g/L agar	Phenolic compounds	Dong et al. (2010)
Heavy metal ions (Co^{2+}, Ag^+, Cd^{2+}), polysaccharides (yeast extract and chitosan), plant response-signaling compounds (SA and MeJA) and hyperosmotic stress with sorbitol	Cell culture	MS with 8 g/L agar and 30 g/L sucrose	Diterpenoid tanshinones	Zhao et al. (2010a, b)
H_2O_2 and NO	SA-induced cells	MS with 100 mg/L myoinositol, 30 g/L sucrose and 4.2 g/L agar	Salvianolic acid B	Guo et al. (2014)

stimulated by silver nitrate, cadmium chloride, and polysaccharide from yeast extract. The total tanshinone content of cells was increased more than tenfold (2.3 mg/g vs. 0.2 mg/g in control) (Zhao et al. 2010a, b). The stimulating effect was related to the concentration. The most significant effect was achieved at 25 μM of Ag and Cd as well as 100 mg/L (carbohydrate content) of yeast extract. Among the three tanshinones detected, cryptotanshinone was stimulated most dramatically to about 30-fold, and tanshinones I and IIA to no more than fivefold. On the other hand, most of the elicitors suppressed cell growth. The yield of biomass was decreased to about 50% (5.1–5.5 g/L vs. 8.9 g/L in control). In addition, the elicitors could promote the activity of phenylalanine ammonia lyase and transiently increase pH and conductivity of the medium. It suggests that the elicitor-stimulated tanshinone accumulation is a stress response of cells. Guo et al. (2014) investigated the effects of H_2O_2 and NO on the production of salvianolic acid B in SA-induced cells. H_2O_2 and NO were significantly elicited by SA, even if inhibitor or quencher was used to inhibit them. The elicited H_2O_2 significantly increased NO production, the induced NO could also significantly enhance H_2O_2 production, and both of them could induce salvianolic acid B accumulation. H_2O_2 or NO was induced by SA independently, and the elicited H_2O_2 or NO could induce salvianolic acid B accumulation in SA-elicited cells independently or synergistically (Guo et al. 2014).

12.3 Tissue Culture

Plant tissue culture is a useful and rapid technique for plant propagation and secondary metabolite production. It includes cell and organ cultures (Saito 1993; Shasmita et al. 2018; Espinosa-Leal et al. 2018). Here, we concentrate on organ cultures of *S. miltiorrhiza*. Field cultivation of *S. miltiorrhiza* is influenced by geography, season and invasion of microbe pathogens (Wang and Wu 2010). The commercial supply of tanshinones is mainly based on the extraction from the cultivated *S. miltiorrhiza* roots, but the resources of *S. miltiorrhiza* are limited. It is necessary to develop more reliable and sustainable resources. Plant tissue culture of *S. miltiorrhiza* is an effective method to improve the quality and the proliferation of *S. miltiorrhiza* seedlings. It can directly produce the active constituents of *S. miltiorrhiza* on a large scale and can solve the problems in the cultivation process and the quality control of *S. miltiorrhiza* (Wang and Wu 2010).

From the 1990s to present, many studies have been carried out on tissue culture and regeneration of *S. miltiorrhiza*. The effects of different explants and hormone ratios on callus formation and regeneration bud differentiation have been investigated. In 1987, Wang and Liu reported the first study on micropropagation of *S. miltiorrhiza*. They showed that buds and shoots could be induced from the leaf and petiole on MS medium supplemented with 1 mg/L 2,4-D or 0.2 mg/L 6-benzylaminopurine (6-BA) and 1 mg/L 2,4-D or 1 mg/L 6-BA and 0.2 mg/L 2,4-D. Root could be induced and developed into an entire plantlet on phytohormone-free 1/2 MS medium. In 1991, Cai et al. studied tissue culture conditions of *S. miltiorrhiza* in detail. It included the effects of explants, basic media, and phytohormones on shoot induced and rooting. The results showed that the best conditions for inducing shoots from leaf explant were MS medium supplemented with 0.5–1.0 mg/L 6-BA; The best conditions for rooting were 1/2 MS medium supplemented with 0.2 mg/L indole-3-butyric acid (IBA) (Cai et al. 1991). In the same year, Shimomura and Kitazawa (1991) studied the regeneration of adventitious shoots from nodal segments of *S. miltiorrhiza*. They showed that adventitious roots could be induced from petiole segments of shoots on B5 medium. When the roots were cultured in liquid medium with 0.5 mg/L IBA for eight weeks, the content of tanshinones was reached to the highest (Shimomura and Kitazawa 1991). Liang et al. (2009) studied tissue culture and plant regeneration of *S. miltiorrhiza* Bunge f. *alba*. They established a regeneration system of *S. miltiorrhiza* Bunge f. *alba* and found that the clustered shoots could be induced on MS medium with 2.0 mg/L 6-BA and 1.0 or 0.5 mg/L naphthaleneacetic acid (NAA). In order to strengthen shoots and eliminate vitrified shoots, the shoots were transferred to MS medium with 1.5 mg/L KT and 0.2 mg/L NAA, and then roots were induced on 1/2 MS medium with 0.2 mg/L IBA. Zhao et al. (2010a, b) reported tissue culture and plantlet regeneration of *S. miltiorrhiza*. They studied callus induction and germination rate of three explants, including young stem, leaf, and alabastrum. The results showed that young stem was the best explants for callus formation and shoot regeneration. MS medium supplemented with 6-BA 0.5 mg/L and 2,4-D 0.5 mg/L was suitable for callus induction, whereas MS medium supplemented with 6-BA 1.0 mg/L and NAA 0.5 mg/L was good for shoot propagation. For rooting of regenerated shoots, 1/2 MS supplemented with IBA 0.5 mg/L was best among the three hormones (IAA, IBA, and NAA) tested.

12.4 Hairy Root Culture

In the 1980s, it had been reported that *Agrobacterium rhizogenes* contained Ri plasmid. It could infect wounded higher plants and transfer a DNA segment (T-DNA) from the Ri plasmid into the genome of the infected plant (Moyano et al. 2002; Shanks and Morgan 1999). Since T-DNA carries a class of genes that encode enzymes involved in auxin and cytokinin biosynthesis, the infected plant cultivated on a hormone-free medium can produce massive roots, called hairy roots (Guillon et al. 2006). The active secondary metabolites of many plants accumulated in the roots. Plant cultivation requires a long time. Harvesting roots from plants are destructive to plant growth. Therefore, hairy root culture was developed to produce valuable metabolites. It has the advantages of genetic stability, grow fast, no need for exogenous hormones, lack of geotropism, lateral branching, and high content of secondary metabolites (Sevon and Oksman-Caldentey 2002; Srivastava and Srivastava 2007). So far, hairy roots are used to produce secondary metabolites for more and more plants (Guillon et al. 2006; Srivastava and Srivastava 2007).

S. miltiorrhiza mainly contains two classes of bioactive constituents, including diterpenoid tanshinones and phenolic acids. Tanshinone I (T-I), tanshinone IIA (T-IIA), cryptotanshinone (CT), and dihydrotanshinone I (DT-I) are the representatives of tanshinones, whereas rosmarinic acid, caffeic acid, and salvianolic acid B are the major phenolic acids in *S. miltiorrhiza* (Yan et al. 2014). In 1993, Hu and Alfermann used different *A. rhizogenes* strains, including LBA 9402, ATCC 15834, TR 105, R 1601, and A4 1027, to induce hairy roots. As a result, seven tanshinones, including tanshinone I, tanshinone IIA, tanshinone IIB, tanshinone V, dihydrotanshinone I, cryptotanshinone, tanshinone VI, and ferruginol were found in hairy roots and in the medium (Hu and Alfermann 1993). After that, a growing number of researchers take efforts to improve the production of tanshinone and phenolic acids from hairy roots of *S. miltiorrhiza* by increasing the yield of hairy roots and the ratio of active secondary metabolites (Kai et al. 2011; Khalili et al. 2010). The results showed that the production of hairy roots and active secondary metabolites was affected by many factors, including infected strains, explants, culture conditions, medium contents, and elicitors.

Liu et al. (2009) investigated the induction rate of different explants infected with *A. rhizogenes* strain R1601. The results showed that the induction rate of leaves is the highest, followed by stems. The petiole could not induce hairy roots in the culture periods. Liu et al. (2009) also studied the effects of culture time on medicinal active ingredients in the hairy root of *S. miltiorrhiza*. The results showed that the hairy root increased by about 15 times on the 56th day after culture. The yields of hairy roots in different culture stages were positively correlated with the culture time, and the contents of active ingredients were increased significantly at the 28th day, 42nd day and 56th day of the culture time, respectively (Liu et al. 2009).

Tan et al. (2014) investigated the effects of different agrobacteria and culture media on the yields of salvianolic acids in hairy roots of *S. miltiorrhiza*. *A. rhizogenes* A4, LBA9402 and 15834 were used to infect *S. miltiorrhiza* leaves. Hairy roots of *S. miltiorrhiza* induced by *A. rhizogenes* were inoculated into culture media, including MSOH, MS, B5, and 6,7-V liquid media. All *A. rhizogenes* strains tested could induce the generation of hairy roots. Hairy roots induced by *A. rhizogenes* LBA9402 and A4 produced significantly more salvianolic acids in MSOH liquid medium with a pH of 4.81 than other media (Tan et al. 2014).

Various biotic and abiotic elicitors, such as yeast extracts, Ag^+, MeJA, SA and abscisic acid, are able to improve the production of secondary metabolites in *S. miltiorrhiza* hairy roots (Cheng et al. 2013; Ming et al. 2013; Hao et al. 2015). In hairy root culture, the yield of the active ingredients could be improved with an optimum concentration of the elicitors. Using a combination of different elicitors is more effective than one elicitor. Table 12.2 shows the elicitors used to induce the production of secondary metabolites in *S. miltiorrhiza* hairy root culture (Cheng et al. 2013; Ming et al. 2013; Hao et al. 2015). Chen et al. (2001) reported that the hairy roots of *S. miltiorrhiza* infected with *A. rhizogenes* ATCC 15834 could produce tanshinones and phenolic acids under normal growth conditions. When hairy roots were treated with yeast elicitor, the production of both phenolic acids and tanshinones was enhanced, and the growth of hairy roots was stimulated (Chen et al. 2001).

Fang et al. (2011) studied the effects of phytohormones on hairy root growth and tanshinone biosynthesis. The results showed that the growth of *S. miltiorrhiza* hairy roots was inhibited by different concentrations of GA3 and 6-BA but obviously promoted by the combination of 1.0 mg/L KT and 1.0 mg/L IBA. Total content of tanshinone IIA was the highest under the combination of 0.2 mg/L NAA and 3.0 mg/L 6-BA. Yang et al. (2018) studied the responses of tanshinone biosynthesis to yeast extract (YE) and Ag^+ treatment in hairy roots of *S. miltiorrhiza* and *Salvia castanea* Diels f. *tomentosa*. They found that YE could enhance the growth and tanshinone biosynthesis of two hairy roots and contributed more to tanshinone accumulation in *S. castanea* than *S. miltiorrhiza*. In *S. miltiorrhiza*, dihydrotanshinone I and cryptotanshinone were more responsive to YE, while tanshinone IIA was more responsive to Ag^+. In *S. castanea*, accumulation of other four tanshinones and related genes was more responsive to YE than Ag^+ (Yang et al. 2018). Liang et al. (2013) investigated the effects of three phytohormones, including abscisic acid, gibberellin and ethylene,

Table 12.2 Elicitors used to induce the production of secondary metabolites in *S. miltiorrhiza* hairy root cultures

Elicitor	Tissue culture type	Medium	Metabolites	References
Yeast	Hairy roots	1/2 MS with 0.5 g/L sodium cefotaxime	Phenolic acids and tanshinones	Chen et al. (2001)
Yeast extract (YE) and Ag^+	Hairy roots	6,7-V liquid medium	Tanshinones	Yang et al. (2018)
Abscisic acid, gibberellin, and ethylene	Hairy roots	6,7-V medium with 20 g/L sucrose	Phenolic acids	Liang et al. (2013)
β-aminobutyric acid	Hairy roots	MS with 8 g/L agar and 30 g/L sucrose but without ammonium nitrate	Tanshinones	Ge and Wu (2005)
Methyl jasmonate	Hairy roots	B5 medium with 8 g/L agar and 30 g/L sucrose	Rosmarinic acid and lithospermic acid B	Xiao et al. (2009)
Rhizosphere bacteria	Hairy roots	6,7-V medium	Phenolic acids	You et al. (2018)

and their interactions on phenolic production in *S. miltiorrhiza* hairy roots. They showed that abscisic acid (ABA), GA, and ethylene could induce the production of phenolic acids and increase the activities of PAL and TAT in *S. miltiorrhiza* hairy roots (Liang et al. 2013).

Ge and Wu (2005) studied the effects of β-aminobutyric acid on the production of diterpenoid tanshinones in *S. miltiorrhiza* hairy root cultures. They found that β-aminobutyric acid could strongly induce secondary metabolites. Compared with MeJA, β-aminobutyric acid was more effective in improving the production of tanshinones. The combination of β-aminobutyric acid and YE elicitor could further improve the production of tanshinones (Ge and Wu 2005). Xiao et al. (2009) investigated the effects of MeJA on the accumulation of rosmarinic acid and its derivative lithospermic acid B as well as related gene transcript and metabolite profiling in *S. miltiorrhiza* Bunge hairy root cultures. The results showed that MeJA could dramatically improve the production of rosmarinic acid and lithospermic acid B (Xiao et al. 2009). You et al. (2018) investigated the effects of *rhizosphere* bacteria on the accumulation of phenolic compounds in *S. miltiorrhiza* hairy roots. The results showed that six *Pseudomonas* and *Pantoea* bacteria, including HYR1, HYR26, SCR22, 14DSR23, DS6, and LNHR13, could significantly promote the production of RA and SAB. Among the six bacteria, LNHR13 was the most effective (You et al. 2018).

So far, the production of secondary metabolites by the hairy root culture has moved from small-scale laboratory to large-scale production (Shasmita et al. 2018). Qiu et al. (2004) first reported the large-scale production of *S. miltiorrhiza* hairy roots using ball-type airlift bioreactor. They found that the growth rate and tanshinone content of the 50-day-old hairy roots obtained from ten-liter ball-type airlift bioreactor were 241.71 and 0.11% of dry weight, respectively, whereas the growth rate and tanshinone content of the hairy roots obtained from the flask was 203.33 and 0.16% of dry weight, respectively. Thus, it is feasible to cultivate hairy roots of *S. miltiorrhiza* in large scale by airlift bioreactor, providing a new way for the sustainable utilization of hairy root resources of *S. miltiorrhiza* (Qiu et al. 2004).

In addition, numerous studies have been carried out on the regulatory mechanism of secondary metabolic pathways. Overexpression or RNA-interfering one or two of the genes involved in tanshinone biosynthesis could significantly improve the production of tanshinones. Xing et al. (2018) overexpressed *SmbHLH10* in *S. miltiorrhiza* hairy roots. The production of tanshinones was increased in *SmbHLH10*-overexpressing line 6. The expression level of genes involved in tanshinones biosynthetic pathway was consistent with tanshinone content. *SmbHLH10* could regulate tanshinones biosynthesis via binding to G-box in the promoter of tanshinone biosynthetic pathway genes (Xing et al. 2018). Zhao et al. (2015) reported that maize transcription factor C1 could improve the accumulation of tanshinones in the hairy roots through up-regulating *SmMDC* and *SmPMK*. Xiao et al. (2009) showed that overexpression of single *C4H*, *TAT*, or *HPPR*, or both the *TAT* and *HPPR* genes could induce the production of RA and SAB in the hairy roots, suggesting that these genes play an important role in the biosynthesis of salvianolic acids.

References

Cai Z, Gao S, Xu D (1991) Study on the rapid propagation of tissue culture of *Salvia miltiorrhiza*. J China Pharm Univ 22:65–68

Chen SQ, Luo HW, Yan SQ (1981) Preliminary investigation of chemical constituents in callus tissue of Danshen. J China Pharm Univ 16:6–8

Chen H, Chen F, Chiu F, Lo C (2001) The effect of yeast elicitor on the growth and secondary metabolism of hairy root cultures of *Salvia miltiorrhiza*. Enzyme Microb Technol 28:100–105

Cheng TO (2006a) Danshen: a popular Chinese cardiac herbal drug. J Am Coll Cardiol 47:1498

Cheng TO (2006b) Cardiovascular effects of Danshen. Int J Cardiol 121:9–22

Cheng Q, He Y, Li G, Liu Y, Gao W, Huang L (2013) Effects of combined elicitors on tanshinone metabolic profiling and *SmCPS* expression in *Salvia miltiorrhiza* hairy root cultures. Molecules 18:7473–7485

Dong J, Wan G, Liang Z (2010) Accumulation of salicylic acid-induced phenolic compounds and raised activities of secondary metabolic and antioxidative enzymes in *Salvia miltiorrhiza* cell culture. J Biotechnol 148:99–104

Espinosa-Leal CA, Puente-Garza CA, García-Lara S (2018) In vitro plant tissue culture: means for production of biological active compounds. Planta 248:1–18

Fang CP, Wang WT, Wang ZF, Shan CG, Ni DP, Zhang YQ (2011) Effects of phytohormones on hairy root growth and tanshinone biosynthesis of *Salvia miltiorrhiza*. Zhong Yao Cai 34:661–664

Ge X, Wu J (2005) Induction and potentiation of diterpenoid tanshinone accumulation in *Salvia miltiorrhiza* hairy roots by beta-aminobutyric acid. Appl Microbiol Biotechnol 68:183–188

Guillon S, Trémouillaux-Guiller J, Pati PK, Rideau M, Gantet P (2006) Hairy root research: recent scenario and exciting prospects. Curr Opin Plant Biol 9: 341–346

Guo H, Dang X, Dong J (2014) Hydrogen peroxide and nitric oxide are involved in salicylic acid-induced salvianolic acid B production in *Salvia miltiorrhiza* cell cultures. Molecules 19:5913–5924

Hao XL, Shi M, Cui L, Xu C, Zhang Y, Kai G (2015) Effects of methyl jasmonate and salicylic acid on tanshinone production and biosynthetic gene expression in transgenic *Salvia miltiorrhiza* hairy roots. Biotechnol Appl Biochem 62:24–31

Hu ZB, Alfermann AW (1993) Diterpenoid production in hairy root cultures of *Salvia miltiorrhiza*. Phytochemistry 32:699–703

Huang X, Hu Z, Yang B, Zhu X (1981) Study on diterpenoid quinones in callus of *Salvia miltiorrhiza*. Chin Pharm Bull 16:22–23

Huang D, Liu D, Hu Z (2000) Effects of plant hormones on cell growth and phenolic acid content in suspension culture of *Salvia miltiorrhiza*. Zhong Yao Cai 23:1–4

Kai G, Hui X, Zhou C, Liao P, Xiao J, Luo X, You L, Zhang L (2011) Metabolic engineering tanshinone biosynthetic pathway in *Salvia miltiorrhiza* hairy root cultures. Metab Eng 13:319–327

Khalili M, Hasanloo T, Tabar S (2010) Ag^+ enhanced silymarin production in hairy root cultures of *Silybum marianum* (L.) Gaertn. Plant Omics 3:109–114

Li YG, Song L, Liu M, Hu ZB, Wang ZT (2009) Advancement in analysis of *Salvia miltiorrhiza* Radix et Rhizoma (Danshen). J Chromatogr A 1216:1941–1953

Liang H et al (2009) Tissue culture and plant regeneration of *Salvia miltiorrhiza* Bge. f. *alba*. J Anhui Agric Sci 37:9876–9904

Liang Z, Ma Y, Xu T, Cui B, Liu Y, Guo Z, Yang D (2013) Effects of abscisic acid, gibberellin, ethylene and their interactions on production of phenolic acids in *Salvia miltiorrhiza* Bunge hairy roots. PLoS ONE 8:e72806

Liu X, Wang W, Li X (2009) Establishment of hairy roots culture system of *Salvia*. Acta Agric Boreali-occidentalis Sin 18:183–186

Ming Q, Su C, Zheng C et al (2013) Elicitors from the endophytic fungus Trichoderma atroviride promote *Salvia miltiorrhiza* hairy root growth and tanshinone biosynthesis. J Exp Bot 64:5687–5694

Miyasaka H, Nasu M, Yamamoto T, Shiomi Y, Ohno H, Endo Y, Yoneda K (1987) Effect of nutritional factors on cryptotanshinone and ferruginol production by cell suspension cultures of *Salvia miltiorrhiza*. Phytochemistry 26:1421–1424

Moyano E, Fornale S, Palazon J, Cusidó RM, Bagni N, Piñol MT (2002) Alkaloid production in Duboisia hybrid hairy root cultures over expressing the Pmt gene. Phytochemistry 59:697–702

Nakanishi T, Miyasaka H, Nasum Hashimoto H, Yoneda K (1983) Production of cryptotanshinone and ferruginol in cultured cells of *Salvia miltiorrhiza*. Phytochemistry 22:721–722

Ochoa-Villarreal M, Howat S, Hong S, Jang M, Jin Y, Lee E, Loake G (2016) Plant cell culture strategies for the production of natural products. BMB Rep 49:149–158

Park IS, Kim HJ, Jeong YS, Kim WK, Kim JS (2017) Differential abilities of Korean soybean varieties to biosynthesize glyceollins by biotic and abiotic elicitors. Food Sci Biotechnol 26:255–261

Putalun W, Prasamsiwamai P, Tanaka H, Shoyama Y (2004) Solasodine glycoside production by hairy root cultures of *Physalis minima* Linn. Biotechnol Lett 26:545–548

Qiu D, Song J (2015) Danshen's in vitro culture. In: Yan X (ed) Dan Shen (*Salvia miltiorrhiza*) in medicine, vol 1. Biology and chemistry. People's Medical Publishing House, Beijing, China, and Springer, Dordrecht, Netherlands, pp 69–87

Qiu D, Song J, Ma X, Qi J, Zhang Y (2004) Studies on the mass culture of the hairy roots of *Salvia miltiorrhiza* with bioreactor. Mol Plant Breed 2:699–703

Saito K (1993) Genetic engineering in tissue culture of medicinal plants. Plant Tissue Cult Lett 10:1–8

Sevon N, Oksman-Caldentey KM (2002) *Agrobacterium rhizogenes*-mediated transformation: root cultures as a source of alkaloids. Planta 68:859–868

Shan C, Wang Z, Su X, Yan S, Sun H (2007) Study advance in *Salvia miltiorrhiza* tissue culture. Res Pract Chin Med 22:4–57

Shanks JV, Morgan J (1999) Plant 'hairy root' culture. Curr Opin Biotechnol 10:151–155

Shasmita, Rai MK, Naik SK (2018) Exploring plant tissue culture in *Withania somnifera* (L.) Dunal: in vitro propagation and secondary metabolite production. Crit Rev Biotechnol 38:836–850

Shimomura K, Kitazawa T (1991) Tanshinone production in adventitious roots and regenerates of *Salvia miltiorrhiza*. J Nat Prod 54:1583–1587

Srivastava S, Srivastava AK (2007) Hairy root culture for mass production of high-value secondary metabolites. Crit Rev Biotechnol 27:29–43

Tan R, Zhang J, Zhao S (2014) Optimization of induction and culture conditions for hairy roots of *Salvia miltiorrhiza*. Zhongguo Zhong Yao Za Zhi 39:3048–3053

Wagstaff SJ, Olmstead RG, Cantino PD (1995) Parsimony analysis of cpDNA restriction site variation in subfamily Nepetoideae (Labiatae). Am J Bot 82:886–892

Wang JY, Liu D (1987) Organogenesis of *Salvia miltiorrhiza*. Plant Physiol Commun 6:46

Wang JW, Wu JY (2010) Tanshinone biosynthesis in *Salvia miltiorrhiza* and production in plant tissue cultures. Appl Microbiol Biotechnol 88:437–449

Wang XH, Morris-Natschke SL, Lee KH (2007) Developments in the chemistry and biology of the bioactive constituents of Tanshen. Med Res Rev 27:133–148

Xiao Y, Gao S, Di P, Chen J, Chen W, Zhang L (2009) Methyl jasmonate dramatically enhances the accumulation of phenolic acids in *Salvia miltiorrhiza* hairy root cultures. Physiol Plant 137:1–9

Xing B, Yang D, Yu H, Zhang B, Yan K, Zhang X, Han R, Liang Z (2018) Overexpression of SmbHLH10 enhances tanshinones biosynthesis in *Salvia miltiorrhiza* hairy roots. Plant Sci 276:229–238

Xu H, Song J, Luo H, Zhang Y, Li Q, Zhu Y, Xu J, Li Y, Song C, Wang B, Sun W, Shen G, Zhang X, Qian J, Ji A, Xu Z, Luo X, He L, Li C, Sun C, Yan H, Cui G, Li X, Li X, Wei J, Liu J, Wang Y, Hayward A, Nelson D, Ning Z, Peters RJ, Qi X, Chen S (2016) Analysis of the genome sequence of the medicinal plant *Salvia miltiorrhiza*. Mol Plant 9:949–952

Yan Y, Zhang SC, Zhang JY, Ma PD, Duan JL, Liang ZS (2014) Effect and mechanism of endophytic bacteria on growth and secondary metabolite synthesis in *Salvia miltiorrhiza* hairy roots. Acta Physiol Plant 36:1095–1105

Yang D, Fang Y, Xia P, Zhang X, Liang Z (2018) Diverse responses of tanshinone biosynthesis to biotic and abiotic elicitors in hairy root cultures of *Salvia miltiorrhiza* and *Salvia castanea* Diels f. *tomentosa*. Gene 643:61–67

You H, Yang S, Zhang L, Hu X, Li O (2018) Promotion of phenolic compounds production in *Salvia miltiorrhiza* hairy roots by six strains of rhizosphere bacteria. Eng Life Sci 18:160–168

Yu Y, Wang T, Wu Y, Zhou Y, Jiang Y, Zhang L (2018) Effect of elicitors on the metabolites in the suspension cell culture of *Salvia miltiorrhiza* Bunge. Physiol Mol Biol Plants 25:229–242

Zhao D, Cao X, Qiu G, Liu B, Lu J, Pan Y (2010a) Tissue culture and plantlets regeneration of *Salvia miltiorrhiza* Bunge. J Beijing Univ Agric 25:5–7

Zhao JL, Zhou LG, Wu JY (2010b) Effects of biotic and abiotic elicitors on cell growth and tanshinone accumulation in *Salvia miltiorrhiza* cell cultures. Appl Microbiol Biotechnol 87:137–144

Zhao S, Zhang J, Tan R, Yang L, Zheng X (2015) Enhancing diterpenoid concentration in *Salvia miltiorrhiza* hairy roots through pathway engineering with maize C1 transcription factor. J Exp Bot 66:7211–7226

Zhou LG, Wu JY (2006) Development and application of medicinal plant tissue cultures for production of drugs and herbal medicinals in China. Nat Prod Rep 23:789–810

13 Genetic Transformation of *Salvia miltiorrhiza*

Meizhen Wang and Shanfa Lu

Abstract

Salvia miltiorrhiza is a valuable traditional Chinese medicinal herb. Due to the great importance of tanshinones and phenolic acids in *S. miltiorrhiza*, it is necessary to increase their yields to satisfy the rapidly growing demand. Many efforts have been made to elucidate the biosynthetic pathways of tanshinones and phenolic acids. This chapter describes two *Agrobacterium*-mediated *S. miltiorrhiza* transformation systems, named *Agrobacterium rhizogenes*-mediated hairy root transformation, and *Agrobacterium tumefaciens*-mediated 'composite' plant transformation, respectively. These systems provide a solid foundation for genetic engineering of *S. miltiorrhiza* to enhance these bioactive compounds. The achievements of *S. miltiorrhiza* metabolic engineering obtained through transformation are summarized. In addition, recent advances on the application of activation tagging mutagenesis (ATM) and CRISPR/Cas9 for mutation of genes in *S. miltiorrhiza* are discussed.

M. Wang · S. Lu (✉)
Institute of Medicinal Plant Development Chinese Academy of Medical Sciences and Peking Union Medical College, No. 151 Malianwa North Road, Haidian District, Beijing 100193, China
e-mail: sflu@implad.ac.cn

M. Wang
e-mail: mzwang@implad.ac.cn

13.1 Introduction

Salvia miltiorrhiza Bunge, also known as Danshen in China, is a well-known medicinal plant belonging to the Lamiaceae family. Dried *S. miltiorrhiza* roots have been widely used in traditional Chinese medicine (TCM) for treating various human diseases, such as cardiovascular and cerebrovascular diseases, Alzheimer's diseases, and cancer (Su et al. 2015). The Compound Danshen Dripping Pill is the first Chinese proprietary medicine that has completed phase-III clinical trial for the treatment of stable angina in USA (NCT01659580, https://clinicaltrials.gov/). Lipophilic diterpenoid tanshinones and hydrophilic phenolic acids are two major types of bioactive compounds in *S. miltiorrhiza* (Zhou et al. 2005; Su et al. 2015). By far, more than 90 diterpenoid compounds and 40 hydrophilic phenolic acids have been identified from *S. miltiorrhiza*, including tanshinone I, tanshinone IIA, tanshinone IIB, cryptotanshinone, salvianolic acid A, salvianolic acid B, and so on (Su et al. 2015; Wei et al. 2017; Mei et al. 2019; Ngo et al. 2019). In vitro cell and tissue cultures of *S. miltiorrhiza*, particularly Ti-transformed hairy root, have been widely used to produce these bioactive components (Tsutomu et al. 1983; Shimomura et al. 1991; Song et al. 1997; Zhang et al. 2004, 2013a; Xiao et al. 2009, 2010; Wu et al. 2016). With the release of *S. miltiorrhiza* genome sequence (Zhang et al. 2015a; Xu et al. 2016), studies have attempted to

S. Lu (ed.), *The Salvia miltiorrhiza Genome*, Compendium of Plant Genomes,
https://doi.org/10.1007/978-3-030-24716-4_13

elucidate the biosynthetic pathways of the active compounds (Guo et al. 2013; Luo et al. 2014; Gao et al. 2014; Xu et al. 2015). Understanding of the biosynthetic pathways and regulatory mechanisms paves the way for metabolic engineering of *S. miltiorrhiza*, which relies on simple and efficient gene transfer approaches. Here, we summarize the procedures of two *Agrobacterium*-mediated *S. miltiorrhiza* transformation systems, together with their applications on improving the major bioactive constituents or other phenotypic traits. Moreover, recent progress on metabolic production by gene mutations is also present.

13.2 *Agrobacterium*-Mediated *S. miltiorrhiza* Transformation

Genetic transformation is the foundation for gene function study and genetic engineering of plants. Biolistic transformation and *Agrobacterium*-mediated system are two successfully used methods for gene delivery. A major disadvantage of biolistic transformation is the multiple-copy insertion of transgenes into plant genome, which leads to considerable variation in their expression and stability (Kohli et al. 1999). *Agrobacterium*-mediated transformation, on the other hand, offers single or lower copy numbers and more stability for the transgene (Dai et al. 2001). It has become the first choice of transformation method for many plants.

So far, *Agrobacterium* is the only known organism that can transfer DNA between kingdoms. The genus *Agrobacterium* contains a number of species, such as *Agrobacterium tumefaciens*, *Agrobacterium rhizogenes*, *Agrobacterium radiobacter,* and *Agrobacterium rubi*. For plant genetic engineering, *A. tumefaciens* and *A. rhizogenes* are two most successfully used species. Their ability to transform plant cells is associated with a region of tumorigenic (Ti) or rhizogenic (Ri) plasmid present in *Agrobacterium*, called transfer DNA (T-DNA). T-DNA regions are approximately 10–30 kb in size, accounting for less than 10% of native Ti or Ri plasmids (Zambryski et al. 1980; Lemmers et al. 1980; Byrne et al. 1983; Barker et al. 1983; Suzuki et al. 2000). Due to large size and lack of unique restriction endonuclease sites, it is complex to clone a gene of interest into the T-DNA region of native Ti or Ri plasmids (Gelvin 2003). However, the finding of binary vector system has revolutionized *Agrobacterium*-mediated gene transformation (Hoekema et al. 1983; de Frammond et al. 1983). In this system, Ti or Ri plasmids are disarmed by complete or partial deletion of the T-region but contain *vir* genes, called the *vir* helper plasmid (Hoekema et al. 1983; de Frammond et al. 1983). Another vector contains two 25-bp-long T-DNA border sequences essential for plant transformation, called the binary vector (Schell and Van Montagu 1977; Joos et al. 1983). When these two plasmids were within the same *Agrobacterium* cell, the products of *vir* genes on the disarmed Ti or Ri plasmid could act in *trans* on the T-region of the binary vector and make it transfer to plant cells. Subsequently, *Agrobacterium* strains harboring disarmed Ti or Ri plasmids, such *A. tumefaciens* GV3101 (Van Larebeke et al. 1974; Koncz and Schell 1986) and *A. rhizogenes* A4 (White and Nester 1980), were developed. In this case, one just need to insert the gene of interest into an expression binary vector containing a selectable marker gene cassette (Koncz and Schell 1986).

A. rhizogenes is a soil-borne bacterium that causes hairy root disease at the infection sites. Such hairy roots are capable of grow rapidly in vitro without addition of exogenous plant growth regulators (Mugnier 1988). It makes hairy root culture a widely used tool to produce plant secondary metabolites (Rao and Ravishankar 2002; Sharma et al. 2013), such as tanshinones and phenolic acids in *S. miltiorrhiza* (Zhi and Alfermann 1993; Chen et al. 1997, 1999; Ge and Wu 2005; Xiao et al. 2009). Since *A. rhizogenes* can also transfer the T-DNA of binary vectors carrying foreign genes, it offers great potential to investigate gene functions, enhance secondary metabolites, or produce recombinant proteins (Georgiev et al. 2012). A general scheme for *A. rhizogenes*-mediated hairy root transformation of *S. miltiorrhiza* is shown in Fig. 13.1a (Hao et al. 2013; Xu and Song 2017). Sterile young leaves are excised from *S. miltiorrhiza* plants grown in vitro and dipped into *Agrobacterium* suspension harboring binary vector for explant infection. After blotted dry on sterile filter paper and cocultivated on solid MS medium for 2–3 days, they are transferred to solid MS medium containing selective antibiotic and cefotaxime for inducing hairy root formation. Hairy roots emerged from the wound sites are cut off and transferred on solid MS medium with 0.1 mg/L indole-3-acetic acid (IAA) for 2–3 days to stimulate lateral root production. Then, they are subcultured on fresh solid MS medium with selective antibiotic and gradual decrease of cefotaxime to remove *Agrobacterium* cells. Vigorous transgenic root lines with no bacterial contamination are used to mass culture in liquid 6,7-V medium containing selective antibiotic before harvest. All the cultures are maintained in dark at 25 °C.

A. rhizogenes-mediated hairy root transformation is widespread to study metabolic pathways in *S. miltiorrhiza* due to its simplicity and a short period of time. Nevertheless, it presents several limitations. For instance, not all the hairy roots are co-transformed with the foreign genes, which makes it complicated to select the transgenic hairy roots (Crane et al. 2006). *A. rhizogenes rol* gene on Ri plasmids may also have effects on plant host secondary metabolism (Chandra 2012). In some cases, it is required to analyze other plant organs, such as leaves, flowers, or stems, but 'composite' plants regenerated from hairy roots typically show an altered morphology due to expression of *rol* genes (Tepfer 1990). One method solving this problem is to obtain normal plants by segregation in the subsequent generations when T-DNAs of Ri plasmid and vector insert into different chromosomes (Hatamoto et al. 1990; Boulter et al. 1990; Webb et al. 1994), while it is time-consuming. An alternative strategy for obtaining the whole transgenic plants is to use *A. tumefaciens*-mediated transformation approach, since oncogenes have been removed from the Ti plasmid (Hood et al. 1993). Figure 13.1b shows a simple and high-frequency *A. tumefaciens*-mediated transformation system of *S. miltiorrhiza* (Yan and Wang 2007; Liu et al. 2015). Leaf disks are infected with *A. tumefaciens* harboring binary vector and cocultivated for 2–3 days on solid MS medium supplemented with 1 mg/L 6-Benzylaminopurine (BAP) and 0.1 or 0.01 mg/L α-Naphthalene acetic acid (NAA) in the dark at 25 °C. After that, explants are transferred to the same medium supplemented with selective antibiotic and cefotaxime for bud regeneration in the light at 25 °C. Then, regenerated buds are cut off and transferred to fresh MS or 1/2 MS medium with the selective antibiotic for rooting. Finally, the rooted plantlets are acclimated and successfully transplanted to soil for further analysis.

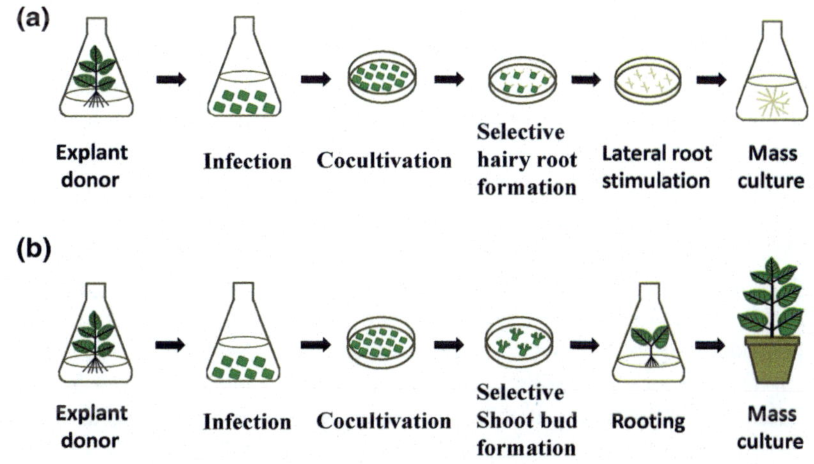

Fig. 13.1 An outline for *Agrobacterium*-mediated transformation of *S. miltiorrhiza*. **a** *A. rhizogenes*-mediated hairy root transformation; **b** *A. tumefaciens*-mediated 'composite' plant transformation

Taken together, two *Agrobacterium*-mediated *S. miltiorrhiza* transformation systems have been successfully developed. They are powerful tools for gene functional studies and metabolic engineering of *S. miltiorrhiza*.

13.3 Application of Genetic Transformation in *S. miltiorrhiza*

Establishment of *S. miltiorrhiza* transformation system will be helpful to enhance secondary metabolite production by genetic engineering strategies. So far, the specific tanshinone and phenolic acid biosynthesis pathways in *S. miltiorrhiza* are not fully understood. However, many efforts have been made to improve their production by metabolic engineering (Tables 13.1, 13.2, 13.3 and 13.4). Overexpression of one or two tanshinone biosynthetic pathway genes, such as *SmHMGR*, *SmDXS* and *SmDXR*, *SmHDR*, *SmIPI*, *SmGGPPS*, *SmCPS*, *SmKSL* and *CYP76AH1*, could significantly improve tanshinones, while knockdown of them has opposite effects (Dai et al. 2011; Kai et al. 2011; Hao et al. 2013; Cheng et al. 2014; Shi et al. 2014; Cui et al. 2015; Zhang et al. 2015c; Guo et al. 2016; Ma et al. 2016; Shi et al. 2016a; Zhou et al. 2016a). Similarly, transgenic lines overexpressing the upstream genes of phenolic acid biosynthesis, such as *SmPAL1*, Sm*C4H*, Sm*TAT*, and *SmHPPR*, could obviously increase the accumulation of phenolic acids, including rosmarinic acid, salvia acid B, and lithospermic acid B (Song and Wang 2011; Xiao et al. 2011; Wang et al. 2017a). Since the biosynthesis of lignin, flavonoid, plastoquinone, and phenolic acid shares the same upstream pathway, it provides an alternative approach to enhance phenolic acid accumulation by changing the metabolic flux from the competitive branch. For example, downregulation of *SmCCR*, *SmCHS,* and *SmHPPD* responsible for lignin, flavonoid, and plastoquinone biosynthesis, respectively, could strongly induce phenolic acid accumulation (Xiao et al. 2011; Wang et al. 2012b; Zhang et al. 2015b). Transcription factors (TFs) offer great advantages in metabolic engineering due to their regulation of multiple genes at the same time (Gantet and Memelink 2002; Lu et al. 2016). It has been reported that overexpression of transcription factor genes, such as *SmbHLH148*, *AtDREB1A*, Sm*MYC2a,* and Sm*MYC2b*, results in simultaneous enhancement of tanshinones and phenolic acids in *S. miltiorrhiza* (Wei et al. 2016b; Zhou et al. 2016b, Xing et al. 2018a). However, some transcription factors, such as *ZmC1*, *SmMYB36*, *SmERF6,* and *ERF1L1*, act as activator on tanshinones and repressor on phenolic acids, whereas some others, such as *SmERF115* and *AtEDT1*, play opposite functions (Zhao et al. 2015; Ding et al. 2017; Bai et al. 2018; Huang et al. 2019; Sun et al. 2018; Liu et al. 2017b). It implies the complex regulatory network of secondary metabolite biosynthesis. Further work is required to elucidate the pathway genes and their regulatory mechanisms. Other type genes, such as carbon metabolic genes *SmSnRK2.3* and *SmSnRK2.6* (Jia et al. 2017), hormone-regulated gene *Sm2OGD5* (Xu and Song 2017), and epidermal hair development gene *SmTTG1* (Li et al. 2018), are also used to study their roles in the biosynthesis of tanshinones or phenolic acids.

Jasmonates (JAs) are plant-specific signaling molecules that are involved in secondary metabolites biosynthesis (Wasternack 2007). It has been shown that methyl jasmonate (MeJA) treatment can stimulate tanshinones and phenolic acids accumulation in *S. miltiorrhiza* (Xiao et al. 2009; Zhao et al. 2010). Hence, the combination of elicitor treatment and transgenic technology could produce a synergistic effect on stimulating secondary metabolites production. One application is the further enhancement of tanshinones in *SmGGPPS* transgenic hairy root line G50 after MeJA or salicylic acid (SA) treatment. The results showed that tanshinone content was significantly enhanced (Hao et al. 2015). Moreover, metabolic engineering based on jasmonate biosynthesis or signaling pathway genes, such as *SmAOC*, *SmJMT*, *SmJAZ,* and *LePS*, could also be an effective strategy for enhancing tanshinone and phenolic acid production in *S. miltiorrhiza* (Gu et al. 2012; Ge et al. 2015; Chen et al. 2016;

Table 13.1 Metabolic engineering of tanshinone biosynthesis in *S. miltiorrhiza*

Gene name	*Agrobacterium*		Selection	Overexpression/Knockdown	Results	References
	Species	Strain				
SmHMGR2	*A. rhizogenes*	ACCC10060	Hygromycin	Overexpression	Enhanced tanshinone I, dihydrotanshinone I, cryptotanshinone and tanshinone IIA and squalene	Dai et al. (2011)
SmHMGR	*A. rhizogenes*	C58C1	Hygromycin	Overexpression	Enhanced cryptotanshinone	Kai et al. (2011)
SmHMGR/SmGGPPS	*A. rhizogenes*	C58C1	Hygromycin	Overexpression	Enhanced cryptotanshinone, tanshinone I, and tanshinone IIA	
SmGGPPS	*A. rhizogenes*	C58C1	Hygromycin	Overexpression	Enhanced cryptotanshinone and tanshinone I	
SmDXS	*A. rhizogenes*	C58C1	Hygromycin	Overexpression	Enhanced cryptotanshinone, tanshinone I, and tanshinone IIA	
SmGGPPS	*A. rhizogenes*	C58C1	Hygromycin	Overexpression	Higher levels of tanshinone	Shi et al. (2016a)
SmDXSII	*A. rhizogenes*	C58C1	Hygromycin	Overexpression	Higher levels of tanshinone	
SmGGPPS/SmDXSII	*A. rhizogenes*	C58C1	Hygromycin	Overexpression	Much higher levels of tanshinone compared with single *SmGGPPS* or *SmDXSII* transformed lines	
SmHMGR	*A. rhizogenes*	C58C1	Hygromycin	Overexpression	Enhanced tanshinone	Shi et al. (2014)
SmDXR	*A. rhizogenes*	C58C1	Hygromycin	Overexpression	Enhanced tanshinone	
SmHMGR/SmDXR	*A. rhizogenes*	C58C1	Hygromycin	Overexpression	Enhanced tanshinone	
SmHDR1	*A. rhizogenes*	ACCC10060	Hygromycin	Overexpression	Enhanced cryptotanshinone, tanshinone I, and tanshinone IIA	Hao et al. (2013)
SmDXS1	*A. rhizogenes*	C58C1	Hygromycin	Overexpression	Increased tanshinone contents	Zhou et al. (2016a)
				Knockdown	Slightly decreased tanshinone contents	
SmDXS2	*A. rhizogenes*	C58C1	Hygromycin	Overexpression	Increased tanshinone contents	
				Knockdown	Obviously decreased tanshinone contents	
SmIPI1	*A. rhizogenes*	ACCC10060	Kanamycin	Knockdown	Reduced dihydrotanshinone I, cryptotanshinone, tanshinone I, and tanshinone IIA	Zhang et al. (2015c)
SmCPS	*A. rhizogenes*	ACCC10060	Kanamycin	Knockdown	Reduced dihydrotanshinone I, cryptotanshinone, and tanshinone IIA	Cheng et al. (2014)

(continued)

Table 13.1 (continued)

Gene name	*Agrobacterium*		Selection	Overexpression/Knockdown	Results	References
	Species	Strain				
SmCPS1	*A. tumefaciens*	EHA105	Kanamycin	Knockdown	Reduced cryptotanshinone, tanshinone IIA, tanshinone I, and 21 potential intermediates	Cui et al. (2015)
CYP76AK1	*A. rhizogenes*	C58C1	Kanamycin	Knockdown	Reduced cryptotanshinone, tanshinone IIA, tanshinone I,and the direct CYP76AK1 products 11,20-dihydroxy ferruginol, and 11,20-dihydroxy sugiol	Guo et al. (2016)
CYP76AH3	*A. rhizogenes*	C58C1	Kanamycin	Knockdown	Reduced the direct CYP76AH3 products 11-hydroxy ferruginol and 11-hydroxy sugiol, and downstream metabolites 10-hydroxymethyl tetrahydromiltirone, 11,20-dihydroxy ferruginol and 11,20-dihydroxy sugiol	
CYP76AH1	*A. rhizogenes*	C58C1	Kanamycin	Knockdown	Reduced tanshinone (tanshinone IIA, cryptotanshinone, dihydrotanshinone, and tanshinone I), the direct CYP76AH1 product ferruginol, and downstream metabolites (sugiol, 11-hydroxy sugiol, 11-hydroxy ferruginol, 11,20-dihydroxy ferruginol, 11,20-dihydroxy sugiol, 10-hydroxymethyl tetrahydromiltirone, and miltirone)	Ma et al. (2016)

NA means not available

Table 13.2 Metabolic engineering of phenolic acid biosynthesis in *S. miltiorrhiza*

Gene classification	Gene name	*Agrobacterium*		Selection	Overexpression/Knockdown	Results	References
		Species	Strain				
Phenolic acid biosynthetic pathway genes	*SmPAL1*	*A. tumefaciens*	EHA105	Kanamycin	Knockdown	Exhibited several phenotypes (stunted growth, delayed root formation, altered leaves); reduced lignin and phenolic acids (rosmarinic acid and salvianolic acid B)	Song and Wang (2011)
	C4H	*A. rhizogenes*	C58C1	Hygromycin	Overexpression	Grew slowly with slender branches; increased t-cinnamic acid, 4-coumaric acid, L-tyrosine, 4-hydroxyphenylpyruvic acid, and the end-product phenolic acids (rosmarinic acid and lithospermic acid B)	Xiao et al. (2011)
	TAT	*A. rhizogenes*	C58C1	Hygromycin	Overexpression	Grew fast and vigorously with thick branches; accumulated phenolic acids only at similar levels to those of wt	
	HPPR	*A. rhizogenes*	C58C1	Hygromycin	Overexpression	Grew slowly with slender branches; presented much higher rosmarinic acid and lithospermic acid B	
	TAT/HPPR	*A. rhizogenes*	C58C1	Hygromycin	Overexpression	Grew fast and vigorously with thick branches; presented the greatest abundance of rosmarinic acid and lithospermic acid B, and the lowest accumulation of homogentisic acid	
	SmHPPR	*A. rhizogenes*	C58C1	Kanamycin	Overexpression	Enhanced phenolic acids (salvianolic acid B and lithospermic acid B)	Wang et al. (2017a)

(continued)

Table 13.2 (continued)

Gene classification	Gene name	*Agrobacterium*		Selection	Overexpression/Knockdown	Results	References
		Species	Strain				
Plastoquinone biosynthetic pathway genes	*HPPD*	*A. rhizogenes*	C58C1	Hygromycin	Knockdown	Grew slowly with slender branches; increased its substrate 4-hydroxyphenylpyruvic acid, its product homogentisic acid, and the target products rosmarinic acid and lithospermic acid B	Xiao et al. (2011)
Flavonoid biosynthetic pathway genes	*SmCHS1-3*	*A. rhizogenes*	ATCC15834	Kanamycin	Knockdown	Decreased fresh weight; enhanced contents of phenolic acids (*t*-cinnamic acid, 4-coumaric acid, RA, SAB, salvianolic acid A); decreased total flavonoids	Zhang et al. (2015b)
Lignin biosynthetic pathway genes	*SmCCR1*	*A. tumefaciens*	EHA105	NA	Knockdown	Exhibited dwarfing phenotypes; decreased syringyl and guaiacyl lignin monomers; enhanced phenolic acids (danshensu, rosmarinic acid, and salvianolic acid B)	Wang et al. (2012b)
	SmCCoAOMT	*A. tumefaciens*	EHA105	Kanamycin	Knockdown	Reduced phenolic acids (rosmarinic acid and salvianolic acid B) and total lignin	Wang et al. (2017b)

NA means not available

Table 13.3 Application of transcription factor genes in metabolic engineering of *S. miltiorrhiza*

Gene classification	Gene name	*Agrobacterium*		Selection	Overexpression/ Knockdown	Results	References
		Species	Strain				
bHLH	*SmbHLH148*	*A. rhizogenes*	ATCC15834	Hygromycin	Overexpression	Enhanced tanshinone (tanshinone I, cryptotanshinone, dihydrotanshinone I) and phenolic acids (caffeic acid, rosmarinic acid, salvianolic acid B)	Xing et al. (2018a)
	SmbHLH10	*A. rhizogenes*	ATCC15834	Hygromycin	Overexpression	Enhanced tanshinone (tanshinone I, dihydrotanshinone I, tanshinone IIA, cryptotanshinone)	Xing et al. (2018b)
	SmbHLH51	*A. tumefaciens*	NA	Kanamycin	Overexpression	Enhanced phenolic acid (rosmarinic acid and salvianolic acid B); enhanced total phenolics and total flavonoids; no change of anthocyanin	Wu et al. (2018)
					Knockdown	Reduced phenolic acid (rosmarinic acid and salvianolic acid B), total phenolics, total flavonoids, and anthocyanins	
	ZmR	*A. rhizogenes*	LBA9402	Hygromycin	Overexpression	Thicker hairy roots; slower propagation; nearly equal levels of rosmarinic acid and salvianolic acid B; produced nearly equal amounts of anthocyanins; had almost no impact on tanshinones	Zhao et al. (2015)
MYB/ bHLH	*ZmC1/R*	*A. rhizogenes*	LBA9402	Hygromycin	Overexpression	Longer but with fewer lateral hairy roots; significantly lower biomass; be prone to dedifferentiation to form a callus; produced less rosmarinic acid and salvianolic acid B, but more anthocyanins; produced higher levels of tanshinone (15,16-dihydrotanshinone I, 1,2-didehydromiltirone, methyltanshinonate, cryptotanshinone, tanshinone I, 1,2-dihydrotanshinone I and tanshinone IIA)	
	AmROS1/ AmDEL	*A. tumefaciens*	EHA105	Hygromycin	Overexpression	Enhanced rosmarinic acids, salvianolic acid B, total phenolics, total flavonoids, and anthocyanins; stronger radical scavenging capacity	Wang et al. (2013)

(continued)

Table 13.3 (continued)

Gene classification	Gene name	*Agrobacterium*		Selection	Overexpression/ Knockdown	Results	References
		Species	Strain				
MYB	*ZmC1*	*A. rhizogenes*	LBA9402	Hygromycin	Overexpression	Longer but with fewer lateral hairy roots; significantly lower biomass; be prone to dedifferentiation to form a callus; produced less rosmarinic acid and salvianolic acid B, but more anthocyanins; produced higher levels of tanshinone (15,16-dihydrotanshinone I, 1,2-didehydromiltirone, methyltanshinonate, cryptotanshinone, tanshinone I, 1,2-dihydrotanshinone I, and tanshinone IIA)	Zhao et al. (2015)
	AtPAP1	*A. tumefaciens*	EHA105	Hygromycin	Overexpression	Enhanced phenolic acids (rosmarinic acid and salvianolic acid B), total phenolics, total flavonoids, anthocyanin and lignin	Zhang et al. (2010)
	AtPAP1/SmCCR/ SmCOMT	*A. tumefaciens*	EHA105	Hygromycin	Overexpression (AtPAP1) Knockdown (SmCCR and SmCOMT)	Enhanced phenolic acids (rosmarinic acid and salvianolic acid B); reduced lignin concentration	Zhang et al. (2014)
	*SmPAP*1	*A. tumefaciens*	LBA4404	Hygromycin	Overexpression	Enhanced rosmarinic acid, salvianolic acid B, total phenolics, and total flavonoids	Hao et al. (2016)
	SmMYB36	*A. rhizogenes*	ATCC15834	Kanamycin	Overexpression	Enhanced tanshinone (dihydrotanshinone I, cryptotanshinone, tanshinone I, and tanshinone II A); reduced phenolic acids (rosmarinic acid and salvianolic acid B), total phenolics, and total flavonoids	Ding et al. (2017)
	SmMYB9b	*A. rhizogenes*	LBA9402	Hygromycin	Overexpression	Enhanced tanshinone (15,16-dihydrotanshinone I, 1,2-didehydromiltirone, methyltanshinonate, cryptotanshinone, tanshinone I, 1,2-dihydrotanshinone I, and tanshinone IIA); reduced biomass	Zhang et al. (2017)
	SmMYB111	*A. tumefaciens*	NA	NA	Overexpression	Enhanced phenolic acid (rosmarinic acid and salvianolic acid B)	Li et al. (2018)

(continued)

Table 13.3 (continued)

Gene classification	Gene name	*Agrobacterium*		Selection	Overexpression/ Knockdown	Results	References
		Species	Strain				
					Knockdown	Reduced phenolic acid (rosmarinic acid and salvianolic acid B)	
	SmMYB39	*A. tumefaciens*	EHA105	Kanamycin	Overexpression	Reduced phenolic acids (4-coumaric acid, rosmarinic acid, salvianolic acid B, salvianolic acid A, and total phenolics)	Zhang et al. (2013b)
					Knockdown	Enhanced phenolic acids (4-coumaric acid, rosmarinic acid, salvianolic acid B, salvianolic acid A, and total phenolics)	
MYC	*SmMYC2a*	*A. rhizogenes*	C58C1	Hygromycin	Knockdown	Reduced tanshinone (tanshinone I, dihydrotanshinone I, tanshinone IIA, cryptotanshinone), phenolic acids (salvianolic acid A and salvianolic acid B), and intermediate products (L-phenylalanine, homogentisic acid, rosmarinic acid)	Zhou et al. (2016b)
	*Sm*MYC2b	*A. rhizogenes*	C58C1	Hygromycin	Knockdown	Reduced tanshinone (tanshinone I, dihydrotanshinone I, tanshinone IIA, cryptotanshinone), phenolic acids (salvianolic acid A and salvianolic acid B), and intermediate products (L-phenylalanine, homogentisic acid, rosmarinic acid)	
	SmMYC2	*A. tumefaciens*	GV3101	Glufosinate-ammonium	Overexpression	Enhanced phenolic acids (rosmarinic acid and salvianolic acid B), total phenolics, and total flavonoids	Yang et al. (2017)
WRKY	*SmWRKY1*	*A. rhizogenes*	C58C1	Kanamycin	Overexpression		

(continued)

Table 13.3 (continued)

Gene classification	Gene name	*Agrobacterium*		Selection	Overexpression/ Knockdown	Results	References
		Species	Strain				
						Increased tanshinones (cryptotanshinone, dihydrotanshinone I, tanshinone I, tanshinone IIA) by positively regulate *SmDXR*	Cao et al. (2018)
	SmWRKY2	*A. rhizogenes*	C58C1	Kanamycin	Overexpression	Enhanced tanshinone (tanshinone I, dihydrotanshinone I, tanshinone IIA, cryptotanshinone)	Deng et al. (2019)
					Knockdown	Reduced tanshinone IIA	
ERF	*SmERF115*	*A. rhizogenes*	C58C1	Hygromycin	Overexpression	Enhanced salvianolic acid B, but reduced tanshinone (dihydrotanshinone I, cryptotanshinone, tanshinone I, and tanshinone II A)	Sun et al. (2018)
					Knockdown	Reduced salvianolic acid B, but enhanced tanshinone (dihydrotanshinone I and cryptotanshinone)	
	SmERF6	*A. rhizogenes*	ATCC15834	Hygromycin	Overexpression	Enhanced tanshinone (tanshinone I, dihydrotanshinone I, tanshinone IIA, cryptotanshinone); reduced flavonoid and phenolic acid; decreased the fresh weight of the transgenic hairy roots	Bai et al. (2018)
					Knockdown	Reduced tanshinone (tanshinone I, dihydrotanshinone I, tanshinone IIA, cryptotanshinone); no change on flavonoid, phenolic acid, and the fresh weight of the transgenic hairy roots	
	SmERF1L1	*A. rhizogenes*	C58C1	NA	Overexpression	Enhanced dihydrotanshinone I and cryptotanshinone), but no change of tanshinone I and tanshinone IIA; reduced phenolic acid (rosmarinic acid and salvianolic acid A)	Huang et al. (2019)

(continued)

Table 13.3 (continued)

Gene classification	Gene name	*Agrobacterium*		Selection	Overexpression/ Knockdown	Results	References
		Species	Strain				
AREB	*SmAREB1*	*A. rhizogenes*	ATCC15834	NA	Overexpression	Enhanced phenolic acids (rosmarinic acid and salvianolic acid B)	Jia et al. (2017)
DREB	*AtDREB1A*	*A. tumefaciens*	LBA4404	Phosphinothricin	Overexpression	Enhanced drought tolerance; enhanced phenolic acids (rosmarinic acid and salvianolic acid B) and tanshinones (tanshinone I, dihydrotanshinone I, tanshinone IIA, cryptotanshinone)	Wei et al. (2016b)
	AtDREB1B	*A. tumefaciens*	LBA4404	Phosphinothricin	Overexpression	Enhanced drought tolerance; no growth inhibition	Wei et al. (2016a)
	AtDREB1C	*A. tumefaciens*	LBA4404	Basta	Overexpression	p35S::AtDREB1C transgenic lines showed significant growth retardation; pRD29A::AtDREB1C lines showed normal growth; showed increased drought resistance that related to their reduced MDA content, increased antioxidant enzyme (superoxide dismutase, peroxidase, and catalase) activity, higher chlorophyll content, and photosynthetic capacity	Tao et al. (2017)
HD-ZIP	*AtEDT1*	*A. tumefaciens*	EHA105	Phosphinothricin	Overexpression	Promoted root elongation and drought tolerance; enhanced phenolic acids (rosmarinic acid, lithospermic acid, and salvianolic acid B); reduced tanshinone	Liu et al. (2017b)

NA means not available

Table 13.4 Application of other genes in metabolic engineering of *S. miltiorrhiza*

Gene classification	Gene name	*Agrobacterium*		Selection	Overexpression/Knockdown	Results	References
		Species	Strain				
Jasmonate biosynthetic/signaling genes	*SmAOC*	*A. rhizogenes*	C58C1	Hygromycin	Overexpression	Enhanced tanshinone IIA, phenolic acids (rosmarinic acid and lithospermic acid B	Gu et al. (2012)
	SmJMT	*A. tumefaciens*	EHA105	Glufosinate-ammonium	Overexpression	Enhanced phenolic acid (rosmarinic acid and salvianolic acid B); enhanced total phenolics and total flavonoids	Wang et al. (2018a)
	SmJAZ8	*A. tumefaciens*	EHA105	Kanamycin	Overexpression	Weak effects on root growth; likely to be JA-hyposensitive; enhanced rosmarinic acid, salvianolic acid B, total phenolics and flavonoids; enhanced antioxidant capacity	Ge et al. (2015)
	SmJAZ3	*A. rhizogenes*	C58C1	NA	Overexpression	Decreased tanshinone contents	Shi et al. (2016b)
					Knockdown	Enhanced tanshinone production	
	SmJAZ9	*A. rhizogenes*	C58C1	NA	Overexpression	Slightly decreased tanshinone contents	
					Knockdown	Enhanced tanshinone production	
	LePS	*A. tumefaciens*	EHA105	Hygromycin	Overexpression	Enhanced tanshinone (tanshinone I, tanshinone IIA, and cryptotanshinone), jasmonate, and pest resistance	Chen et al. (2016)
Carbon metabolic genes	*SmSnRK2.3*	*A. rhizogenes*	ATCC15834	NA	Overexpression	No change of phenolic acids	Jia et al. (2017)
	SmSnRK2.6	*A. rhizogenes*	ATCC15834	NA	Overexpression	Enhanced phenolic acids (rosmarinic acid and salvianolic acid B)	
	SmGols2	*A. tumefaciens*	EHA105	Kanamycin	Knockdown	Significantly higher malondialdehyde contents under NaCl or cold stress; more rapid rate of water loss	Wang et al. (2012a)

(continued)

Table 13.4 (continued)

Gene classification	Gene name	*Agrobacterium*		Selection	Overexpression/Knockdown	Results	References
		Species	Strain				
Hormone-regulated gene	*SmGASA4*	*A. tumefaciens*	LBA4404	Glufosinate	Overexpression	Promote growth condition (larger taproots, more and longer lateral roots); enhanced salt, drought, and paclobutrazol (PBZ) tolerance; promote responses to GA3	Wang et al. (2018b)
	Sm2OGD5	*A. rhizogenes*	ACCC10060	NA	Knockdown	Reduced miltirone, cryptotanshinone, and tanshinone IIA	Xu and Song (2017)
Other genes	*SmTTG1*	*A. tumefaciens*	NA	NA	Overexpression	More trichomes; no change of phenolic acids (rosmarinic acid and salvianolic acid B)	Li et al. (2018)
					Knockdown	Lower anthocyanin; reduced phenolic acids (rosmarinic acid and salvianolic acid B)	
	Human fgf-1	*A. tumefaciens*	LBA4404	Kanamycin	Overexpression	Promote cell proliferation; speed up the growth of new blood vessels,accelerate the burn wound healing process	Tan et al. (2014)

NA means not available

Shi et al. 2016b; Wang et al. 2018a). Combination of metabolic engineering and elicitor treatments may be a more promising strategy to produce secondary metabolites in the future.

In addition, several genes, such as *SmGols2*, *SmGASA4*, and *AtDREB1A/B/C*, have been transformed into *S. miltiorrhiza* to promote stress resistance (Wang et al. 2012a, 2018b; Wei et al. 2016b; Tao et al. 2017). Transgenic *S. miltiorrhiza* has also been used to produce recombinant protein human FGF-1 (Tan et al. 2014). Application of genes used in genetic engineering of *S. miltiorrhiza* is summarized in Table 13.4.

13.4 Recent Advances in Gene Mutagenesis of *S. miltiorrhiza*

Activation tagging mutagenesis (ATM) uses T-DNA or a transposable element harboring CaMV35S enhancers to generate random insertional mutagenesis, leading to transcriptional activation of nearby genes (Hayashi et al. 1992; Suzuki et al. 2001). Such gain-of-function mutation system can be employed to isolate mutants with increased secondary metabolites. Six ATM transgenic *S. miltiorrhiza* callus lines and one ATM plant line (SH41) with enhanced tanshinone levels have been successfully produced (Lee et al. 2008; Ho et al. 2013). However, a large ATM population of transgenic lines has not been established for *S. miltiorrhiza*.

Compared with ATM, CRISPR/Cas is a more useful tool to edit target genes causing site-specific mutagenesis. It has been paid considerable attention due to its simple manipulation and high efficiency in gene mutation and transcriptional regulation (Liu et al. 2017a). For medicinal plants, it can be used for biosynthesis of effective components or reduce toxicity of toxic constituents. Knockout of *SmCPS1* in *S. miltiorrhiza* using CRISPR/Cas9 technology resulted in completely missing of tanshinones in homozygous mutants and reduce of them in chimeric mutants (Li et al. 2017). The contents of phenolic acids, including RA and LAB, were significantly reduced in the successfully edited hairy root lines, while the RA precursor 3,4-dihydroxyphenyllactic acid was highly accumulated (Zhou et al. 2018). CRISPR/Cas system is a new path for the analysis of gene function in *S. miltiorrhiza*. It presents a promising potential in metabolic engineering of bioactive compounds in *S. miltiorrhiza* in the near future.

Acknowledgements This work was supported by the National Natural Science Foundation of China (81872964, 81773836) and the CAMS Innovation Fund for Medical Sciences (CIFMS) (2016-I2M-3-016).

References

Bai Z, Li W, Jia Y, Yue Z, Jiao J, Huang W, Xia P, Liang Z (2018) The ethylene response factor SmERF6 co-regulates the transcription of *SmCPS1* and *SmKSL1* and is involved in tanshinone biosynthesis in *Salvia miltiorrhiza* hairy roots. Planta 248:243–255

Barker RF, Idler KB, Thompson DV, Kemp JD (1983) Nucleotide sequence of the T-DNA region from the *Agrobacterium tumefaciens* octopine Ti plasmid pTi15955. Plant Mol Biol 2(6):335–350

Boulter ME, Croy E, Simpson P, Shields R, Shirsat AH (1990) Transformation of *Brassica napus* L. (oilseed rape) using *Agrobacterium tumefaciens* and *Agrobacterium rhizogenes*—a comparison. Plant Sci 70(1): 91–99

Byrne MC, Koplow J, David C, Tempé J, Chilton MD (1983) Structure of T-DNA in roots transformed by *Agrobacterium rhizogenes*. J Mol Appl Genet 2(2): 201–209

Cao W, Wang Y, Shi M, Hao X, Zhao W, Wang Y, Ren J, Kai G (2018) Transcription factor SmWRKY1 positively promotes the biosynthesis of tanshinones in *Salvia miltiorrhiza*. Front Plant Sci 9:554

Chandra S (2012) Natural plant genetic engineer *Agrobacterium rhizogenes*: role of T-DNA in plant secondary metabolism. Biotechnol Lett 34(3):407–415

Chen C, Zhang Y, Qiakefu K, Zhang X, Han L, Hua W, Yan Y, Wang Z (2016) Overexpression of tomato Prosystemin (LePS) enhances pest resistance and the production of tanshinones in *Salvia miltiorrhiza* Bunge. J Agric Food Chem 64(41):7760–7769

Chen H, Chen F, Zhang Y, Song J (1999) Production of lithospermic acid B and rosmarinic acid in hairy root cultures of *Salvia miltiorrhiza*. J Ind Microbiol Biotechnol 22(3):133–138

Chen H, Yuan J, Chen F, Zhang Y, Song J (1997) Tanshinone production in Ti-transformed *Salvia miltiorrhiza* cell suspension cultures. J Biotechnol 58(3): 147–156

Cheng Q, Su P, Hu Y, He Y, Gao W, Huang L (2014) RNA interference-mediated repression of *SmCPS* (*copalyldiphosphate synthase*) expression in hairy roots of *Salvia miltiorrhiza* causes a decrease of tanshinones and sheds light on the functional role of SmCPS. Biotechnol Lett 36(2):363–369

Crane C, Wright E, Dixon RA, Wang Z (2006) Transgenic *Medicago truncatula* plants obtained from *Agrobacterium tumefaciens*-transformed roots and *Agrobacterium rhizogenes*-transformed hairy roots. Planta 223(6):1344–1354

Cui G, Duan L, Jin B, Qian J, Xue Z, Shen G, Snyder JH, Song J, Chen S, Huang L, Peters RJ, Qi X (2015) Functional divergence of diterpene syntheses in the medicinal plant *Salvia miltiorrhiza*. Plant Physiol 169 (3):1607–1618

Dai S, Zheng P, Marmey P, Zhang S, Tian W, Chen S, Beachy RN, Fauquet C (2001) Comparative analysis of transgenic rice plants obtained by *Agrobacterium*-mediated transformation and particle bombardment. Mol Breed 7:25–33

Dai Z, Cui G, Zhou S, Zhang X, Huang L (2011) Cloning and characterization of a novel 3-hydroxy-3-methylglutaryl coenzyme A reductase gene from *Salvia miltiorrhiza* involved in diterpenoid tanshinone accumulation. J Plant Physiol 168(2):148–157

de Frammond AJ, Barton KA, Chilton M (1983) Mini-Ti: a new vector strategy for plant genetic engineering. Nat Biotechnol 5:262–269

Deng C, Hao X, Shi M, Fu R, Wang Y, Zhang Y, Zhou W, Feng Y, Makunga NP, Kai G (2019) Tanshinone production could be increased by the expression of *SmWRKY2* in *Salvia miltiorrhiza* hairy roots. Plant Sci 284:1–8

Ding K, Pei T, Bai Z, Jia Y, Ma P, Liang Z (2017) SmMYB36, a novel R2R3-MYB transcription factor, enhances tanshinone accumulation and decreases phenolic acid content in *Salvia miltiorrhiza* hairy roots. Sci Rep 7(1):5104

Gantet P, Memelink J (2002) Transcription factors: tools to engineer the production of pharmacologically active plant metabolites. Trends Pharmacol Sci 23:563–569

Gao W, Sun H, Xiao H, Cui G, Hillwig ML, Jackson A, Wang X, Shen Y, Zhao N, Zhang L, Wang X, Peters RJ, Huang L (2014) Combining metabolomics and transcriptomics to characterize tanshinone biosynthesis in *Salvia miltiorrhiza*. BMC Genom 15:73

Ge Q, Zhang Y, Hua W, Wu Y, Jin X, Song S, Wang Z (2015) Combination of transcriptomic and metabolomic analyses reveals a JAZ repressor in the jasmonate signaling pathway of *Salvia miltiorrhiza*. Sci Rep 5:14048

Ge X, Wu J (2005) Tanshinone production and isoprenoid pathways in *Salvia miltiorrhiza* hairy roots induced by Ag^{+} and yeast elicitor. Plant Sci 168(2):487–491

Gelvin SB (2003) *Agrobacterium*-mediated plant transformation: the biology behind the "gene-jockeying" tool. Microbiol Mol Biol Rev 67(1):16–37

Georgiev MI, Agostini E, Ludwig-Müller J, Xu J (2012) Genetically transformed roots: from plant disease to biotechnological resource. Trends Biotechnol 30 (10):528–537

Gu X, Chen J, Xiao Y, Di P, Xuan H, Zhou X, Zhang L, Chen W (2012) Overexpression of allene oxide cyclase promoted tanshinone/phenolic acid production in *Salvia miltiorrhiza*. Plant Cell Rep 31(12):2247–2259

Guo J, Ma X, Cai Y, Ma Y, Zhan Z, Zhou YJ, Liu W, Guan M, Yang J, Cui G, Kang L, Yang L, Shen Y, Tang J, Lin H, Ma X, Jin B, Liu Z, Peters RJ, Zhao ZK, Huang L (2016) Cytochrome P450 promiscuity leads to a bifurcating biosynthetic pathway for tanshinones. New Phytol 210(2):525–534

Guo J, Zhou YJ, Hillwig ML, Shen Y, Yang L, Wang Y, Zhang X, Liu W, Peters RJ, Chen X, Zhao ZK, Huang L (2013) CYP76AH1 catalyzes turnover of miltiradiene in tanshinones biosynthesis and enables heterologous production of ferruginol in yeasts. Proc Natl Acad Sci 110(29):12108–12113

Hao G, Jiang X, Feng L, Tao R, Li Y, Huang L (2016) Cloning, molecular characterization and functional analysis of a putative R2R3-MYB transcription factor of the phenolic acid biosynthetic pathway in *S. miltiorrhiza* Bge. f. *alba*. Plant Cell Tiss Org Cult 124(1):151–168

Hao G, Shi R, Tao R, Fang Q, Jiang X, Ji H, Feng L, Huang L (2013) Cloning, molecular characterization and functional analysis of 1-hydroxy-2-methyl-2-(E)-butenyl-4-diphosphate reductase (HDR) gene for diterpenoid tanshinone biosynthesis in *Salvia miltiorrhiza* Bge. f. *alba*. Plant Physiol Biochem 70:21–32

Hao X, Shi M, Cui L, Xu C, Zhang Y, Kai G (2015) Effects of methyl jasmonate and salicylic acid on tanshinone production and biosynthetic gene expression in transgenic *Salvia miltiorrhiza* hairy roots. Biotechnol Appl Biochem 62(1):24–31

Hatamoto H, Boulter ME, Shirsat AH, Croy EJ, Ellis JR (1990) Recovery of morphologically normal transgenic tobacco from hairy roots co-transformed with *Agrobacterium rhizogenes* and a binary vector plasmid. Plant Cell Rep 9:88–92

Hayashi H, Czaja I, Lubenow H, Schell J, Walden R (1992) Activation of a plant gene by T-DNA tagging: auxin-independent growth in vitro. Science 258:1350–1353

Ho HS, Vishwakarma RK, Chen ECF, Tsay HS (2013) Activation tagging in *Salvia miltiorrhiza* can cause increased leaf size and accumulation of tanshinone I and IIA in its roots. Bot Stud 54(1):37

Hoekema A, Hirsh PR, Hooykaas PJJ, Schilperoort RA (1983) A binary plant vector strategy based on separation of *vir-* and T-region of the *Agrobacterium tumefaciens* Ti-plasmid. Nature 303:179–180

Hood EE, Gelvin SB, Melchers LS, Hoekema A (1993) New *Agrobacterium* helper plasmids for gene transfer to plants. Transgenic Res 2(4):208–218

Huang Q, Sun M, Yuan T, Wang Y, Shi M, Lu S, Tang B, Pan J, Wang Y, Kai G (2019) The AP2/ERF transcription factor SmERF1L1 regulates the biosynthesis of tanshinones and phenolic acids in *Salvia miltiorrhiza*. Food Chem 274:368–375

Jia Y, Bai Z, Pei T, Ding K, Liang Z, Gong Y (2017) The protein kinase SmSnRK2.6 positively regulates phenolic acid biosynthesis in *Salvia miltiorrhiza* by interacting with SmAREB1. Front Plant Sci 8:1384

Joos H, Timmerman B, Montagu MV, Schell J (1983) Genetic analysis of transfer and stabilization of *Agrobacterium* DNA in plant cells. EMBO J 2 (12):2151–2160

Kai G, Xu H, Zhou C, Liao P, Xiao J, Luo X, You L, Zhang L (2011) Metabolic engineering tanshinone biosynthetic pathway in *Salvia miltiorrhiza* hairy root cultures. Metab Eng 13(3):319–327

Kohli A, Gahakwa D, Vain P, Laurie DA, Christou P (1999) Transgene expression in rice engineered through particle bombardment: molecular factors controlling stable expression and transgene silencing. Planta 208:88–97

Koncz C, Schell J (1986) The promoter of TL-DNA gene 5 controls the tissue-specific expression of chimaeric genes carried by a novel type of *Agrobacterium* binary vector. Mol Gen Genet 204:383–396

Lee CY, Agrawal DC, Wang CS, Yu SM, Chen JJ, Tsay HS (2008) T-DNA activation tagging as a tool to isolate *Salvia miltiorrhiza* transgenic lines for higher yields of tanshinones. Planta Med 74(07):780–786

Lemmers M, De Beuckeleer M, Holsters M, Zambryski P, Depicker A, Hernalsteens JP, Van Montagu M, Schell J (1980) Internal organization, boundaries and integration of Ti-plasmid DNA in nopaline crown gall tumours. J Mol Biol 144(3):353–376

Li B, Cui G, Shen G, Zhan Z, Huang L, Chen J, Qi X (2017) Targeted mutagenesis in the medicinal plant *Salvia miltiorrhiza*. Sci Rep 7:43320

Li S, Wu Y, Kuang J, Wang H, Du T, Huang Y, Zhang Y, Cao X, Wang Z (2018) SmMYB111 is a key factor to phenolic acid biosynthesis and interacts with both SmTTG1 and SmbHLH51 in *Salvia miltiorrhiza*. J Agric Food Chem 66(30):8069–8078

Liu X, Wu S, Xu J, Sui C, Wei J (2017a) Application of CRISPR/Cas9 in plant biology. Acta Pharm Sin B 7 (3):292–302

Liu Y, Sun G, Zhong Z, Ji L, Zhang Y, Zhou J, Zheng X, Deng K (2017b) Overexpression of *AtEDT1* promotes root elongation and affects medicinal secondary metabolite biosynthesis in roots of transgenic *Salvia miltiorrhiza*. Protoplasma 254(4):1617–1625

Liu Y, Yang S, Cheng Y, Liu D, Zhang Y, Deng K, Zheng X (2015) Production of herbicide-resistant medicinal plant *Salvia miltiorrhiza* transformed with the *bar* gene. Appl Biochem Biotechnol 177(7):1456–1465

Lu X, Tang KX, Li P (2016) Plant metabolic engineering strategies for the production of pharmaceutical terpenoids. Front Plant Sci 7:1647

Luo H, Zhu Y, Song J, Xu L, Sun C, Zhang X, Xu Y, He L, Sun W, Xu H, Wang B, Li X, Li C, Liu J, Chen S (2014) Transcriptional data mining of *Salvia miltiorrhiza* in response to methyl jasmonate to examine the mechanism of bioactive compound biosynthesis and regulation. Physiol Plant 152(2):241–255

Ma Y, Ma X, Meng F, Zhan Z, Guo J, Huang L (2016) RNA interference targeting *CYP76AH1* in hairy roots of *Salvia miltiorrhiza* reveals its key role in the biosynthetic pathway of tanshinones. Biochem Bioph Res Commun 477(2):155–160

Mei XD, Cao YF, Che YY, Li J, Shang ZP, Zhao WJ, Qiao YJ, Zhang JY (2019) Danshen: a phytochemical and pharmacological overview. Chin J Nat Med 17 (1):59–80

Mugnier J (1988) Establishment of new axenic hairy root lines by inoculation with *Agrobacterium rhizogenes*. Plant Cell Rep 7:9–12

Ngo TM, Tran PT, Hoang LS, Lee JH, Min BS, Kim JA (2019) Diterpenoids isolated from the root of *Salvia miltiorrhiza* and their anti-inflammatory activity. Nat Prod Res. https://doi.org/10.1080/14786419.2019.1596098

Rao SR, Ravishankar GA (2002) Plant cell cultures: chemical factories of secondary metabolites. Biotechnol Adv 20(2):101–153

Schell J, Van Montagu M (1977) The Ti-plasmid of *Agrobacterium tumefaciens*, a natural vector for the introduction of *nif* genes in plants? In: Hollaender A, Burris RH, Day PR, Hardy RW, Helinski DR, Lamborg MR, Owens L, Valentine RC (eds) Genetic engineering for nitrogen fixation. Basic Life Sci 9 (2):159

Sharma P, Padh H, Shrivastava N (2013) Hairy root cultures: a suitable biological system for studying secondary metabolic pathways in plants. Eng Life Sci 13(1):62–75

Shi M, Luo X, Ju G, Li L, Huang S, Zhang T, Wang H, Kai G (2016a) Enhanced diterpene tanshinone accumulation and bioactivity of transgenic *Salvia miltiorrhiza* hairy roots by pathway engineering. J Agric Food Chem 64(12):2523–2530

Shi M, Luo X, Ju G, Yu X, Hao X, Huang Q, Xiao J, Cui L, Kai G (2014) Increased accumulation of the cardio-cerebrovascular disease treatment drug tanshinone in *Salvia miltiorrhiza* hairy roots by the enzymes 3-hydroxy-3-methylglutaryl CoA reductase and 1-deoxy-D-xylulose 5-phosphate reductoisomerase. Funct Integr Genomic 14(3):603–615

Shi M, Zhou W, Zhang J, Huang S, Wang H, Kai G (2016b) Methyl jasmonate induction of tanshinone biosynthesis in *Salvia miltiorrhiza* hairy roots is mediated by jasmonate zim-domain repressor proteins. Sci Rep 6:20919

Shimomura K, Kitazawa T, Okamura N, Yagi A (1991) Tanshinone production in adventitious roots and regenerates of *Salvia miltiorrhiza*. J Nat Prod 54:1583

Song J, Wang Z (2011) RNAi-mediated suppression of the phenylalanine ammonia-lyase gene in *Salvia miltiorrhiza* causes abnormal phenotypes and a reduction in rosmarinic acid biosynthesis. J Plant Res 124:183

Song J, Zhang Y, Qi J, Lu G (1997) Selection of a high tanshinone–producing crown gall strain and production of tanshinone in the strain. Chin J Biotechnol 13 (3):207–210

Su C, Ming Q, Rahman K, Han T, Qin L (2015) *Salvia miltiorrhiza*: Traditional medicinal uses, chemistry, and pharmacology. Chin J Nat Med 13(3):163–182

Sun M, Shi M, Wang Y, Huang Q, Yuan T, Wang Q, Wang C, Zhou W, Kai G (2018) The biosynthesis of phenolic acids is positively regulated by the JA-responsive transcription factor ERF115 in *Salvia miltiorrhiza*. J Exp Bot 70(1):243–254

Suzuki K, Hattori Y, Uraji M, Ohta N, Iwata K, Murata K, Kato A, Yoshida K (2000) Complete nucleotide sequence of a plant tumor-inducing Ti plasmid. Gene 242:331–336

Suzuki Y, Uemura S, Saito Y, Murofushi N, Schmitz G, Theres K, Yamaguchi I (2001) A novel transposon tagging element for obtaining gain-of-function mutants based on a self-stabilizing Ac derivative. Plant Mol Biol 45:123–131

Tan Y, Wang KY, Wang N, Li G, Liu D (2014) Ectopic expression of human acidic fibroblast growth factor 1 in the medicinal plant, *Salvia miltiorrhiza*, accelerates the healing of burn wounds. BMC Biotechnol 14:74

Tao W, Deng K, Zhang Q, Gao Y, Liu Y, Yang M, Zhang L, Zheng X, Wang C, Liu Z, Chen C, Zhang Y (2017) Modulating *AtDREB1C* expression improves drought tolerance in *Salvia miltiorrhiza*. Front Plant Sci 8:52

Tepfer D (1990) Genetic transformation using *Agrobacterium rhizogenes*. Physiol Plant 79:140–146

Tsutomu N, Hitoshi M, Masao N, Hideko H, Kaisuke Y (1983) Production of cryptotanshinone and ferruginol in cultured cells of *Salvia miltiorrhiza*. Phytochem 22 (3):721–722

Van Larebeke N, Engler G, Holsters M, Van den Elsacker S, Zaenen J, Schilperoort RA, Schell J (1974) Large plasmid in *Agrobacterium tumefaciens* essential for crown gall-inducing ability. Nature 252:169–170

Wang B, Niu J, Li B, Huang Y, Han L, Liu Y, Zhou W, Hu S, Li L, Wang D, Wang S, Cao X, Wang Z (2018a) Molecular characterization and overexpression of *SmJMT* increases the production of phenolic acids in *Salvia miltiorrhiza*. Int J Mol Sci 19(12):3788

Wang D, Song Y, Chen Y, Yao W, Li Z, Liu W, Yue S, Wang Z (2013) Metabolic pools of phenolic acids in *Salvia miltiorrhiza* are enhanced by co-expression of *Antirrhinum majus Delila* and *Rosea1* transcription factors. Biochem Eng J 74(Complete):115–120

Wang D, Yao W, Song Y, Liu W, Wang Z (2012a) Molecular characterization and expression of three galactinol synthase genes that confer stress tolerance in *Salvia miltiorrhiza*. J Plant Physiol 169(18):1838–1848

Wang G, Chen J, Yi B, Tan H, Zhang L, Chen W (2017a) *HPPR* encodes the hydroxyphenylpyruvate reductase required for the biosynthesis of hydrophilic phenolic acids in *Salvia miltiorrhiza*. Chin J Nat Med 15 (12):0917–0927

Wang H, Wei T, Wang X, Zhang L, Yang M, Chen L, Song W, Wang C, Chen C (2018b) Transcriptome analyses from mutant *Salvia miltiorrhiza* reveals important roles for SmGASA4 during plant development. Int J Mol Sci 19(7):2088

Wang Z, Cui L, Chen C, Liu X, Yan Y, Wang Z (2012b) Downregulation of cinnamoyl CoA reductase affects lignin and phenolic acids biosynthesis in *Salvia miltiorrhiza* Bunge. Plant Mol Biol Rep 30:1229

Wang Z, Ge Q, Chen C, Jin X, Cao X, Wang Z (2017b) Function analysis of caffeoyl-COA O-methyltransferase for biosynthesis of lignin and phenolic acid in *Salvia miltiorrhiza*. Appl Biochem Biotechnol 181(2):562–572

Wasternack C (2007) Jasmonates: an update on biosynthesis, signal transduction and action in plant stress response, growth and development. Ann Bot 100:681–697

Webb KJ, Robbins MP, Mizen S (1994) Expression of *GUS* in primary transformants and segregation patterns of GUS, TL- and TR-DNA in the T1 generation of hairy root transformants of *Lotus corniculatus*. Transgenic Res 3:232–240

Wei T, Deng K, Gao Y, Liu Y, Yang M, Zhang L, Zheng X, Wang C, Song W, Chen C, Zhang Y (2016a) *Arabidopsis DREB1B* in transgenic *Salvia miltiorrhiza* increased tolerance to drought stress without stunting growth. Plant Physiol Biochem 104:17–28

Wei T, Deng K, Liu D, Gao Y, Liu Y, Yang M, Zhang L, Zheng X, Wang C, Song W, Chen C, Zhang Y (2016b) Ectopic expression of DREB transcription factor, *AtDREB1A*, confers tolerance to drought in transgenic *Salvia miltiorrhiza*. Plant Cell Physiol 57 (8):1593–1609

Wei WJ, Zhou PP, Lin CJ, Wang WF, Li Y, Gao K (2017) Diterpenoids from *Salvia miltiorrhiza* and their immune-modulating activity. J Agric Food Chem 65 (29):5985–5993

White FF, Nester EW (1980) Relationship of plasmids responsible for hairy root and crown gall tumorigenicity. J Bacteriol 144:710–720

Wu C, Karioti A, Rohr D, Bilia AR, Efferth T (2016) Production of rosmarinic acid and salvianolic acid B from callus culture of *Salvia miltiorrhiza* with cytotoxicity towards acute lymphoblastic leukemia cells. Food Chem 201:292–297

Wu Y, Zhang Y, Li L, Guo X, Wang B, Cao X, Wang Z (2018) AtPAP1 interacts with and activates SmbHLH51, a positive regulator to phenolic acids biosynthesis in *Salvia miltiorrhiza*. Front Plant Sci 9:1687

Xiao Y, Gao S, Di P, Chen J, Chen W, Zhang L (2009) Methyl jasmonate dramatically enhances the accumulation of phenolic acids in *Salvia miltiorrhiza* hairy root cultures. Physiol Plant 137(1):1–9

Xiao Y, Gao S, Di P, Chen J, Chen W, Zhang L (2010) Lithospermic acid B is more responsive to silver ions (Ag^+) than rosmarinic acid in *Salvia miltiorrhiza* hairy root cultures. Biosci Rep 30(1):33–40

Xiao Y, Zhang L, Gao S, Saechao S, Di P, Chen J, Chen W (2011) The *c4 h, tat, hppr* and *hppd* genes prompted engineering of rosmarinic acid biosynthetic pathway in *Salvia miltiorrhiza* hairy root cultures. PLoS One 6(12):e29713

Xing B, Liang L, Liu L, Hou Z, Yang D, Yan K, Zhang X, Liang Z (2018a) Overexpression of *SmbHLH148* induced biosynthesis of tanshinones as well as phenolic acids in *Salvia miltiorrhiza* hairy roots. Plant Cell Rep 37(12):1681–1692

Xing B, Yang D, Yu H, Zhang B, Yan K, Zhang X, Han R, Liang Z (2018b) Overexpression of *SmbHLH10* enhances tanshinones biosynthesis in *Salvia miltiorrhiza* hairy roots. Plant Sci 276:229–238

Xu H, Song J, Luo H, Zhang Y, Li Q, Zhu Y, Xu J, Li Y, Song C, Wang B, Sun W, Shen G, Zhang X, Qian J, Ji A, Xu Z, Luo X, He L, Li C, Sun C, Yan H, Cui G, Li X, Li X, Wei J, Liu J, Wang Y, Hayward A, Nelson D, Ning Z, Peters RJ, Qi X, Chen S (2016) Analysis of the genome sequence of the medicinal plant *Salvia miltiorrhiza*. Mol Plant 9:949–952

Xu Z, Peters RJ, Weirather J, Luo H, Liao B, Zhang X, Zhu Y, Ji A, Zhang B, Hu S, Au K, Song J, Chen S (2015) Full-length transcriptome sequences and splice variants obtained by a combination of sequencing platforms applied to different root tissues of *Salvia miltiorrhiza* and tanshinone biosynthesis. Plant J 82 (6):951–961

Xu Z, Song J (2017) The 2-oxoglutarate-dependent dioxygenase superfamily participates in tanshinone production in *Salvia miltiorrhiza*. J Exp Bot 68 (9):2299–2308

Yan Y, Wang Z (2007) Genetic transformation of the medicinal plant *Salvia miltiorrhiza* by *Agrobacterium tumefaciens*-mediated method. Plant Cell Tiss Org Cult 88(2):175–184

Yang N, Zhou W, Su J, Wang X, Li L, Wang L, Cao X, Wang Z (2017) Overexpression of *SmMYC2* increases the production of phenolic acids in *Salvia miltiorrhiza*. Front Plant Sci 8:1804

Zambryski P, Holsters M, Kruger K, Depicker A, Schell J, Van Montagu M, Goodman HM (1980) Tumor DNA structure in plant cells transformed by *A. tumefaciens*. Science 209:1385–1391

Zhang C, Yan Q, Cheuk WK, Wu J (2004) Enhancement of tanshinone production in *Salvia miltiorrhiza* hairy root culture by Ag^+ elicitation and nutrient feeding. Planta Med 70(02):147–151

Zhang D, Liu X, Xie D, Chen R, Tao X, Zou J, Dai J (2013a) Two new diterpenoids from cell cultures of *Salvia miltiorrhiza*. Chem Pharm Bull 61(5):576–580

Zhang G, Tian Y, Zhang J, Shu L, Yang S, Wang W, Sheng J, Dong Y, Chen W (2015a) Hybrid *de novo* genome assembly of the Chinese herbal plant danshen (*Salvia miltiorrhiza* Bunge). GigaScience 4:62

Zhang J, Zhou L, Zheng X, Zhang J, Yang L, Tan R, Zhao S (2017) Overexpression of *SmMYB9b* enhances tanshinone concentration in *Salvia miltiorrhiza* hairy roots. Plant Cell Rep 36(8):1297–1309

Zhang S, Li H, Liang X, Yan Y, Xia P, Jia Y, Liang Z (2015b) Enhanced production of phenolic acids in *Salvia miltiorrhiza* hairy root cultures by combing the RNAi-mediated silencing of chalcone synthase gene with salicylic acid treatment. Biochem Eng J 103:185–192

Zhang S, Ma P, Yang D, Li W, Liang Z, Liu Y, Liu F (2013b) Cloning and characterization of a putative R2R3 MYB transcriptional repressor of the rosmarinic acid biosynthetic pathway from *Salvia miltiorrhiza*. PLoS ONE 8(9):e73259

Zhang X, Guan H, Dai Z, Guo J, Shen Y, Cui G, Gao W, Huang L (2015c) Functional analysis of the isopentenyl diphosphate isomerase of *Salvia miltiorrhiza* via color complementation and RNA interference. Molecules 20(11):20206–20218

Zhang Y, Yan Y, Wang Z (2010) The *Arabidopsis* PAP1 transcription factor plays an important role in the enrichment of phenolic acids in *Salvia miltiorrhiza*. J Agr Food Chem 58(23):12168–12175

Zhang Y, Yan Y, Wu Y, Hua W, Chen C, Ge Q, Wang Z (2014) Pathway engineering for phenolic acid accumulations in *Salvia miltiorrhiza* by combinational genetic manipulation. Metab Eng 21:71–80

Zhao J, Zhou L, Wu J (2010) Effects of biotic and abiotic elicitors on cell growth and tanshinone accumulation in *Salvia miltiorrhiza* cell cultures. Appl Microbiol Biotechnol 87(1):137–144

Zhao S, Zhang J, Tan R, Yang L, Zheng X (2015) Enhancing diterpenoid concentration in *Salvia miltiorrhiza* hairy roots through pathway engineering with maize C1 transcription factor. J Exp Bot 66(22):7211

Zhi B, Alfermann AW (1993) Diterpenoid production in hairy root cultures of *Salvia miltiorrhiza*. Phytochemistry 32(3):699–703

Zhou L, Zuo Z, Chow MS (2005) Danshen: an overview of its chemistry, pharmacology, pharmacokinetics, and clinical use. J Clin Pharmacol 45(12):1345–1359

Zhou W, Huang F, Li S, Wang Y, Zhou C, Shi M, Wang J, Chen Y, Wang Y, Wang H, Kai G (2016a) Molecular cloning and characterization of two 1-deoxy-d-xylulose-5-phosphate synthase genes involved in tanshinone biosynthesis in *Salvia miltiorrhiza*. Mol Breed 36(9):124

Zhou Y, Sun W, Chen J, Tan H, Xiao Y, Li Q, Ji Q, Gao S, Chen L, Chen S, Zhang L, Chen W (2016b) SmMYC2a and SmMYC2b played similar but irreplaceable roles in regulating the biosynthesis of tanshinones and phenolic acids in *Salvia miltiorrhiza*. Sci Rep 6:22852

Zhou Z, Tan H, Li Q, Chen J, Gao S, Wang Y, Chen W, Zhang L (2018) CRISPR/Cas9-mediated efficient targeted mutagenesis of *RAS* in *Salvia miltiorrhiza*. Phytochemistry 148:63–70

Printed by Printforce, the Netherlands